THIRD EDITION

Introduction to Classical and Modern
Optics

JURGEN R. MEYER-ARENDT, M.D.

Pacific University

Prentice Hall, Englewood Cliffs, New Jersey 07632

Library of Congress Cataloging-in-Publication Data

MEYER-ARENDT, JURGEN R.
  Introduction to classical and modern optics/Jurgen R.
Meyer-Arendt.—3rd ed.
      p.   cm.
  Includes bibliographies and index.
  ISBN 0-13-499039-0
  1. Optics. I. Title.
QC355.2.M49 1989   88-31431
535—dc19   CIP

Editorial/production supervision
  and interior design: *Kathleen M. Lafferty*
Cover design: *Ben Santora*
Manufacturing buyer: *Paula Massenaro*

© 1989, 1984, 1972 by Jurgen R. Meyer-Arendt
Published by Prentice-Hall, Inc.
A Division of Simon & Schuster
Englewood Cliffs, New Jersey 07632

Printed in the United States of America

10 9 8 7 6 5 4 3 2 1

ISBN 0-13-499039-0

Prentice-Hall International (UK) Limited, *London*
Prentice-Hall of Australia Pty. Limited, *Sydney*
Prentice-Hall Canada Inc., *Toronto*
Prentice-Hall Hispanoamericana, S.A., *Mexico*
Prentice-Hall of India Private Limited, *New Delhi*
Prentice-Hall of Japan, Inc., *Tokyo*
Simon & Schuster Asia Pte. Ltd., *Singapore*
Editora Prentice-Hall do Brasil, Ltda., *Rio de Janeiro*

Contents

Preface   xi

Introduction   1

PART 1   GEOMETRICAL IMAGE FORMATION

1.1 The Propagation of Light   4

Characteristics of Light   *4*
Shadows   *7*
Path Length   *9*
Suggestions for Further Reading   *11*
Problems   *11*

1.2 Reflection and Refraction   13

Fermat's Principle   *14*
Prisms   *17*
Dispersion   *21*
Suggestions for Further Reading   *25*
Problems   *25*

1.3 Thin Lenses   27

Single Refracting Surface   *27*
Thin Lenses   *35*
Magnification   *41*
Suggestions for Further Reading   *44*
Problems   *45*

1.4 Thick Lenses and Lens Combinations   47

Thick Lenses   *47*
Angular Magnification   *56*
Suggestions for Further Reading   *64*
Problems   *65*

1.5 Mirrors   67

Spherical Mirrors   *67*
Plane Mirrors   *73*
Aspheric Mirrors   *74*
Suggestions for Further Reading   *76*
Problems   *77*

1.6 Stops and Pupils   79

Stops   *81*
Pupils   *82*
Suggestion for Further Reading   *89*
Problems   *89*

PART 2   OPTICAL DESIGN

2.1 Ray Tracing   91

Trigonometric Ray Tracing   *92*
Skew Rays   *99*
Suggestions for Further Reading   *106*
Problems   *106*

2.2 Aberrations   108

Spherical Aberration   *109*
Coma   *115*
Oblique Astigmatism   *117*
Curvature of Field   *120*
Distortion   *121*

Chromatic Aberration   *122*
Suggestions for Further Reading   *127*
Problems   *127*

2.3 Gradient-Index, Fiber, and Integrated Optics   129

Theory of Refractive Gradients   *130*
Gradient-Index Lenses   *132*
Experimental Gradient-Index Optics   *134*
Fiber Optics   *137*
Integrated Optics   *140*
Suggestions for Further Reading   *142*
Problems   *142*

2.4 Computer-Aided Lens Design   144

Approaching the Problem   *145*
Programming in BASIC   *146*
Tracing a Ray through a Plane Surface   *149*
Tracing a Ray through a Spherical Surface   *152*
Tracing a Ray through a System   *154*
Conclusions   *156*
Suggestions for Further Reading   *156*
Problems   *156*

2.5 Optical Systems   158

Telescopes   *158*
The Microscope   *167*
The Eye   *172*
Camera Lenses   *174*
Reflecting Systems   *179*
Suggestions for Further Reading   *183*
Problems   *184*

2.6 Systems Evaluation   186

Contrast   *186*
Transfer Functions   *188*
The Experimental Determination of Transfer Functions   *192*
Patent Considerations   *194*
Suggestions for Further Reading   *195*
Problems   *195*

PART 3   WAVE OPTICS

3.1 Light as a Wave Phenomenon   197

Simple Harmonic Motion   *197*
Moving Waves   *200*
The General Wave Equation   *202*
Suggestions for Further Reading   *203*
Problems   *203*

3.2 Interference   205

Young's Double-Slit Experiment   *205*
Superposition of Waves   *207*
The Michelson Interferometer   *213*
Application and Evaluation   *216*
Suggestions for Further Reading   *218*
Problems   *218*

3.3 Thin Films   221

Plane-Parallel Plates   *221*
The Fabry–Perot Interferometer   *223*
Newton's Rings   *228*
Interference Filters   *230*
Antireflection Coatings   *232*
Suggestions for Further Reading   *234*
Problems   *235*

3.4 Coherence   237

Spatial Coherence   *237*
Temporal Coherence   *239*
Partial Coherence   *240*
Applications   *242*
Suggestions for Further Reading   *245*
Problems   *245*

3.5 Diffraction   247

Fraunhofer Diffraction   *249*
Fresnel Diffraction   *255*
Suggestions for Further Reading   *268*
Problems   *268*

3.6 Diffraction Gratings   270

The Grating Equation   *270*
Three-Dimensional Gratings   *278*
Moiré Fringes   *280*
Suggestions for Further Reading   *282*
Problems   *282*

PART 4   LIGHT AND MATTER

4.1 Light as an Electromagnetic Phenomenon   285

Maxwell's Equations   *285*
Fresnel's Equations   *289*
Suggestions for Further Reading   *292*
Problems   *293*

4.2 Light Scattering   295

Rayleigh Scattering   *295*
Mie Scattering   *300*
Light Scattering and Meteorology   *301*
Suggestions for Further Reading   *302*
Problems   *302*

4.3 Polarization of Light   304

Types of Polarized Light   *304*
Production of Polarized Light   *306*
The Analysis of Light of Unknown Polarization   *312*
Optical Activity   *316*
Electrooptics and Magnetooptics   *317*
Suggestions for Further Reading   *319*
Problems   *319*

4.4 Fourier Transform Spectroscopy   321

Fringe Contrast Variations   *321*
Michelson Interferometer Spectroscopy   *324*
Advantages of Fourier Transform Spectroscopy   *330*
Suggestions for Further Reading   *331*
Problems   *331*

4.5 Optical Data Processing   332

Abbe's Theory of Image Formation   *332*
Two-Dimensional Transforms   *335*
Optical versus Electronic Data Processing   *342*
Suggestions for Further Reading   *343*
Problems   *343*

4.6 Holography   346

Producing the Hologram   *348*
Reconstruction   *349*
Applications of Holography   *351*
Suggestions for Further Reading   *355*
Problems   *355*

PART 5   QUANTUM OPTICS

5.1 Light Sources and Detectors   356

Light Sources   *356*
Detectors   *361*
Practical Quantum Detectors   *364*
Suggestions for Further Reading   *365*
Problems   *365*

5.2 Radiometry/Photometry   367

Terms and Units   *367*
Radiant-Luminous Conversion   *370*
Radiometers and Photometers   *378*
Suggestions for Further Reading   *379*
Problems   *379*

5.3 Atomic Spectra   381

Line Emission Spectra   *381*
Atomic Transitions   *386*
Suggestions for Further Reading   *388*
Problems   *389*

5.4 Absorption   390

Absorption Spectra   *390*
Experimental Methods   *396*

Suggestions for Further Reading   *398*
Problems   *399*

5.5 Lasers   400

Stimulated Emission of Radiation   *400*
Practical Realization   *404*
Types of Lasers   *409*
Applications   *414*
Laser Safety   *420*
Suggestions for Further Reading   *421*
Problems   *422*

PART 6   EPILOGUE

6.1 Relativistic Optics   424

Transformations   *424*
Optics and the Special Theory of Relativity   *429*
Optics and the General Theory of Relativity   *441*
Suggestions for Further Reading   *442*
Problems   *442*

Answers to Odd-Numbered Problems   445

Index   449

Preface

THE PURPOSE AND EMPHASIS in this third edition remain the same: to provide a concise, and still readable, *Introduction to Classical and Modern Optics,* written for advanced undergraduates and for a course spanning two semesters or the equivalent. I have made the text as self-contained as practical, set out the motivation for each step, and avoided shortcuts of the it-can-be-shown-that type.

Compared with the earlier editions, I have again made major changes. I continue using the rational *Cartesian sign convention,* long familiar from ray tracing and from ophthalmic optics but until recently somewhat slighted in physics. This convention is essential for the concept of *vergence* and mandatory in any *computer-aided lens design.* This dual connection identifies the two groups of readers to which this *Introduction to Optics* is addressed in particular: those interested in the *scientific and engineering applications of optics* and those preparing for the *ophthalmic professions.*

Virtually every chapter on *geometrical optics* opens with the same triplet combination of lenses. Each time, the light progresses a little further, showing the logic in the sequence of topics: from thin lenses to a combination of lenses, to stops and pupils, to ray tracing, to aberrations, and to computer-aided lens design (CALD).

Many chapters have been rewritten, most extensively the chapters on the propagation of light, ray tracing, gradient-index and fiber optics, systems evaluation, interference, coherence, optical data processing, atomic spectra, and lasers. Completely new are the chapters on CALD and on Light Sources and Detectors.

As before, ample space has been allocated to classical topics such as thin lenses, optical systems, and polarization. But modern subjects such as holography, and of course lasers and laser safety, as well as Fourier transform spectroscopy and radiometry and photometry, are also presented in reasonable detail. Relativistic optics, likely to play an important part in celestial navigation, is fun to read, aside from its utilitarian aspect.

As a prerequisite, all that is needed is a good background in general physics and a working knowledge of how to use a calculator. A concurrent course in calculus, and some knowledge of a computer language, are desirable though not essential.

Numerous worked-out *examples,* from penumbras to relativistic reflection, are interspersed throughout the text. *Problems,* ranging from easy to difficult, are found at the end of each chapter, following the *Suggestions for Further Reading.* Most of the examples and problems are new. All have been use-tested and modified where needed. (Answers to the odd-numbered problems are found in the back of the book.) None of the problems is intended as "busy work"; all are realistic and apply to practical situations; some, in fact, were drawn from consulting work I do from time to time. *Lecture demonstrations* are referred to on occasion. A great many *historical footnotes* have been included to reveal the human side of optics' great masters, adding color to the description of their accomplishments. (For the statistical-minded, there are 30 chapters, 344 illustrations, 89 examples, 7 computer programs, 121 historical footnotes, 549 problems, and 154 Suggestions for Further Reading.)

I have enjoyed writing this new edition. It took hundreds of hours of lecturing. It also took the patience of a great many students, questioning, challenging, attentive, and at times not so attentive, for me to discover—often by trial and error—how to get a concept across. How do I best present the idea of principal planes, the vector form of Snell's law, the use of Cornu's spiral, the theory of stimulated emission? How do I present the fantastically wide field of optics at a reasonable level and within a reasonable length of time? There is no final answer to that, just steps of successive improvement.

ACKNOWLEDGMENTS

It is a pleasure to acknowledge the assistance I have received from others. Many of my colleagues have helped, in ways large and small, with the preparation and revision of this new edition. Much appreciation also goes to my students. But comments, both laudatory and critical, have come also from readers I have never met. In particular, I wish to thank Thomas B. Greenslade, Jr., David A. Cantley, and Peter A. Barnes (geometrical optics), Silverio P. Almeida and Michael E. Mickelson (physical and modern optics), Laurel Gregory, and Patricia R. Wakeling *(Applied Optics).* I also express my gratitude to the editorial and production staff of Prentice Hall, especially to Holly Hodder, acquisitions editor, and to Kathleen Lafferty, production editor; because of them my manuscript became the book it now is.

Jurgen R. Meyer-Arendt

Introduction

I BEGIN THIS INTRODUCTION TO OPTICS by presenting right away a complex practical problem. Look at Figure 0-1. This is a combination of lenses, a lens *system,* containing a positive lens in front (on the left), a negative lens behind it, and another positive lens in the back (on the right). These lenses have certain surface characteristics, they have certain thicknesses, and they are certain distances apart. Probably, they are made out of different types of glass. By choosing these parameters correctly, we can obtain a system of superior performance. In fact, the system shown is the basis for some of the best, and best known, photographic camera lenses.

How does the system work? Why are the lenses of the type shown preferable to other lenses? Why is this system superior to other systems? These are questions that we are not yet ready to answer. We will use this system as an example and a guide to introduce many of the concepts of optics.

More specifically, we will use this model to introduce *geometrical optics.* Geometrical

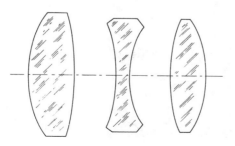

Figure 0-1 Combination of three lenses, used for introducing various aspects of geometrical optics.

optics is that part of optics where the wave nature of light can be neglected. *Physical optics* is more inclusive: it is the branch of optics concerned with the wave properties of light, the superposition of waves *(interference),* and the deviation of light from its rectilinear propagation by means other than geometrical optics *(diffraction).* Physical optics also includes the concept of *transformation,* which is fundamental to optical data processing, pattern recognition, and holography. In *quantum optics* we discuss light sources and detectors, spectra, absorption, and the ubiquitous laser. *Relativistic optics* points to the future.

Some phenomena in optics are easy to see. Others are very subtle. Look at a street lamp through the fabric of an open umbrella. You will see light *fans,* extending in various directions. These are due to diffraction. Or look at a fairly bright star (without the umbrella). You may see light fans or "points" that seem to emanate from the star. In reality, there are no such points. They are merely due to occasional straight segments in the otherwise round pupil of the eye caused by a hardening of the smallest arteries, arteriolosclerosis.

And so, in the art of all ages, stars have traditionally been represented as objects with a multitude of points. However, a star with an odd number of points, as shown in Figure 0-2, is wave-optically impossible because the light fans must, by necessity, always occur in pairs.

Figure 0-2 Artist's conception of a star. Stained-glass window in St. John's Church, Herford, Germany, fourteenth century.

Optics is a field of science that is particularly lucid, logical, challenging, and beautiful. Most of our appreciation of the outside world—nature, art—comes to us through light. We will now proceed to discuss its many aspects.

A BIBLIOGRAPHY ON OPTICS

E. HECHT, *Optics,* 2nd edition (Reading, MA: Addison-Wesley Publishing Company, Inc., 1987).

F. A. JENKINS and H. E. WHITE, *Fundamentals of Optics,* 4th edition (New York: McGraw-Hill Book Company, 1976).

M. YOUNG, *Optics and Lasers,* 3rd edition (New York: Springer-Verlag New York, Inc., 1986).

W. H. A. FINCHAM and M. H. FREEMAN, *Optics,* 9th edition (Woburn, MA: Butterworth Publishers, Inc., 1980).

If you want to work on some research project in optics, you should consult, in addition,

M. BORN and E. WOLF, *Principles of Optics,* 6th edition (Elmsford, NY: Pergamon Press, Inc., 1980).

R. S. LONGHURST, *Geometrical and Physical Optics,* 3rd edition (New York: Longman, Inc., 1974).

M. V. KLEIN and TH. E. FURTAK, *Optics,* 2nd edition (New York: John Wiley & Sons, Inc., 1986).

H. HAFERKORN, *Optik,* 2nd edition (Berlin: VEB Deutscher Verlag der Wissenschaften, 1984).

You should also read, at least, the following journals:

Applied Optics
Journal of the Optical Society of America
Scientific American

1.1

The Propagation of Light

THE FIELD OF OPTICS IS OFTEN DIVIDED into Geometrical Optics and Physical Optics, as if there were a dichotomy, a division of optics into separate entities. But, take the example of image formation, a topic that surely pervades much of optics. Both geometrical optics and physical optics discuss image formation, albeit at different levels of sophistication. The more elementary approach is by geometrical optics. Much more comprehensive is to discuss it in terms of diffraction and optical transformation.

Let us begin with the *propagation* of light. Look at sunlight that, on a misty morning, breaks through the dense foliage of a tree. The light, made visible by the moisture in the air, travels along straight lines called *rays*. Rays follow the law of *rectilinear propagation*. But, why are the rays in Figure 1.1-1 diverging? That is merely an illusion; in reality, the rays are parallel, just as railroad tracks are parallel extending to the horizon; they only *seem* to be converging.

CHARACTERISTICS OF LIGHT

Under certain conditions, light behaves like a sequence of *waves*. If we set a rope in oscillatory motion, waves will propagate along the rope. Actually, the rope moves only up and down. It does not move forward. What moves forward is the *wave configuration*. Wave configurations can move along in one, two, or three dimensions. In a rope, the wave moves along in one dimension. Surface ripples on a pond expand in two dimensions. Sound and

Figure 1.1-1 Sunlight passing through the foliage of a tree.

light propagate through space in three dimensions. Terms we need to know in this context are the following:

Amplitude, **A**, is the *height* of the wave above the average. Amplitudes vary between a maximum of $+A$ and a maximum of $-A$ (Figure 1.1-2).

Wavelength, λ, is a measure of *length.* It is the length of a full wave, from peak to peak. Wavelengths are measured in the same units as length in general. The basic unit of length in the Système International d'Unités, the International System of Units, SI, is the meter, m. But since the wavelength of light is rather small, fractional units of a meter are needed. These are

1 millimeter, mm, 10^{-3} m.

1 mi′crometer, μm, 10^{-6} m, 1/1000 of a millimeter. The prefix μ alone must not be used. (Do not confuse with micro′meter, accent on the *o*, which is an instrument for measuring small distances.)

1 nanometer, nm, 10^{-9}. This is the *preferred unit of wavelength of light* (in the visible part of the spectrum).

1 Ångström, Å, 10^{-10} m.*

*Named for Anders Jonas Ångström (1814–1874), Swedish spectroscopist and professor of physics and astronomy at the University of Uppsala. Ångström found that a cold gas absorbs the same wavelengths of light that it emits when hot, discovered hydrogen in the spectrum of the sun, carefully determined the wavelengths of all other lines he could see, and published the results in *Recherches sur le spectra solaire* (Uppsala, 1869).

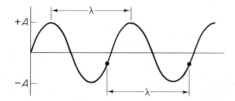

Figure 1.1-2 Amplitude, **A,** and wavelength, λ, of a sinusoidal wave.

Frequency, ν, is the number of oscillations, or waves, or cycles, *per second.* The unit of frequency is the hertz,* Hz. For example, if an object makes 12 complete (back and forth) oscillations per minute, its frequency is

$$\nu = \frac{12 \text{ osc}}{1 \text{ min}} = \frac{12 \text{ osc}}{60 \text{ s}} = 0.2 \text{ Hz}$$

The *period, T,* of a wave is a measure of *time.* It is the time it takes the object to make one oscillation. Period and frequency, therefore, are reciprocal to one another:

$$T = \frac{1}{\nu} \qquad\qquad [1.1\text{-}1]$$

Velocity, v, in general, is the ratio of distance to time:

$$v = \frac{s}{t}$$

If the distance is 1 λ, and the time is 1 *T*, then

$$v = \frac{\lambda}{T}$$

But $1/T = \nu$ and hence the velocity of propagation of a wave is the product of wavelength and frequency,

$$\boxed{v = \lambda\nu} \qquad\qquad [1.1\text{-}2]$$

This is an important equation; it holds for any wave.

The *velocity of light* in free space is the same at all wavelengths and the same as that of any other radiation throughout the electromagnetic spectrum: *The velocity of light is a natural constant.* Indeed, the special theory of relativity (Chapter 6.1) *requires* the speed of light to be constant. That has far-reaching consequences: it has led to international agreement to make the velocity of light a *defined,* rather than a measured, quantity. The value adopted is

$$\boxed{c = 299\ 792\ 458 \text{ m s}^{-1}} \qquad\qquad [1.1\text{-}3]$$

---

*Named after Heinrich Rudolf Hertz (see page 286 for biographical footnote). When in 1924 the German Physical Society proposed the name "hertz" for the unit cycles per second, Walter Hermann Nernst (1864–1941), German physical chemist, objected, saying: "I do not see the need for introducing a new name; by the same reasoning one could as well call one liter per second a 'falstaff.'"

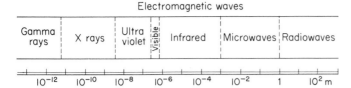

Figure 1.1-3 Electromagnetic spectrum. The spectrum extends from gamma and X rays (*left*) to radiowaves (*right*). Note how narrow the visible part is.

Conversely, the meter has become a *derived* quantity; it is defined as "the length of the path traveled by light in a vacuum during a time interval of 1/299 792 458 of a second."* This means that there is no need anymore to rely on a laboriously calibrated meter bar; the standard meter can be reproduced using the speed of light, and any future refinement would change, not the velocity of light, but the length of the meter.

The electromagnetic spectrum.   Light is only part of the electromagnetic spectrum. The wavelength of visible light extends from about 380 nm for violet-blue to about 750 nm for deep red. "Pure" blue has a peak wavelength of about 475 nm, green of about 520 nm, yellow 575 nm, and red 630 nm. Below 380 nm there is the *ultraviolet,* UV, above 750 nm the *infrared,* IR (Figure 1.1-3).

X rays have wavelengths from as short as 10^{-12} m to about 50 nm. Ultraviolet of less than 200 nm is absorbed by air and therefore is called *vacuum ultraviolet. Microwaves* extend from 1 mm to 30 cm and *radio waves* from there up to perhaps 30 km wavelength.

All these limits are didactic and for tabulation only. They do not represent division lines. The properties of one category merge with those of the next, and the methods of production and detection overlap.

SHADOWS

One of the consequences of the rectilinear propagation of light is the formation of *shadows.* With a single point source, an opaque object will cast a uniformly dark shadow. Outside the shadow the screen is fully illuminated (Figure 1.1-4, top). With two point sources, there is a central dark shadow or *umbra,* and next to it two partial shadows or *penumbras,* each of the penumbras receiving additional light from the source not casting the shadow (center). With an extended source, the penumbra assumes the shape of a ring, surrounding the umbra (bottom).

Example

Light from an extended source, 10 cm in diameter, falls on an opaque object, 12 cm in diameter. If on a screen 6 m from the source the full shadow (umbra) is 16 cm in diameter, find:
(a) The distance from the source to the object.
(b) The width of the half-shadow (penumbra).
(c) The *total* diameter of the half-shadow.

---

*Definition adopted by the 17th General Conference on Weights and Measures (Conférence Générale des Poids et Mesures, CGPM), Paris, France, 1983.

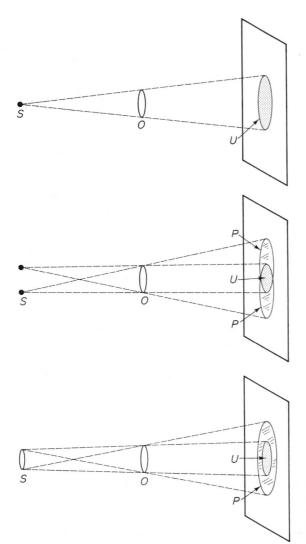

Figure 1.1-4   Shadows as formed by a single point source (*top*), by two point sources (*center*), and by an extended source (*bottom*). *S*, light source; *O*, obstacle; *U*, shadow; *P*, penumbra.

Solution.   Draw two parallel lines from the edges of the source to the screen (the dashed lines in Figure 1.1-5).

(a) From the construction it follows that

$$\frac{(B - S)/2}{L} = \frac{(U - S)/2}{L'}$$

and, inserting the actual figures,

$$\frac{(12 - 10)/2}{L} = \frac{(16 - 10)/2}{600}$$

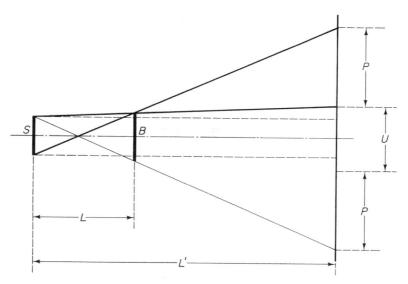

Figure 1.1-5

Cross-multiplication and solving for L gives

$$(12 - 10)(600) = (16 - 10)L$$

$$L = \frac{1200}{6} = \boxed{200 \text{ cm}}$$

which is the distance from the source to the object.

(b) The width of the half-shadow is found from

$$\frac{S}{L} = \frac{P}{L' - L}$$

$$P = \frac{S(L' - L)}{L} = \frac{(10)(600 - 200)}{200} = \boxed{20 \text{ cm}}$$

(c) The total diameter of the half-shadow is

$$P + U + P = 20 + 16 + 20 = \boxed{56 \text{ cm}}$$

PATH LENGTH

Consider two points, A and B, that are a certain distance apart from one another. The distance as such has nothing to do with the medium that may be present between A and B. For example, if the path is in air and the distance between A and B is L, then, if the path were in water, the distance would still be L.

But now let *light* travel from A to B. In air it takes the light a certain length of time to go from A to B. But if the path is in water, it takes the light *longer*. In fact, it takes the light n

times as long (where n is the refractive index of the water). Therefore, another "length" (in place of L) is needed to account for the delay. This length is the *optical path length, S,* the product of length (distance) and refractive index:

$$S = Ln \qquad\qquad [1.1\text{-}4]$$

Refractive index, in turn, is defined as the ratio of the velocity of light in a given medium to the velocity of light in free space, $n \equiv v/c$. (The symbol "$\equiv$" means "defined as.")

But then look at the situation from the opposite point of view (Figure 1.1-6). An observer may view an object A through a medium of index n. In reality the object is in plane A, but to the observer the object *appears* to be in plane A'. Both distances, the path length in air, Ln_0, and the path length in matter, $L'n$, are equal:

$$Ln_0 = L'n$$

and therefore, since the index of air $n_0 \approx 1$,

$$L' = \frac{L}{n} \qquad\qquad [1.1\text{-}5]$$

where L' is called the *reduced distance.*

Since refractive index is a dimensionless quantity, all distances, actual distance, optical path length, and reduced distance are measured in the same units as length in general, preferably in meters or millimeters.

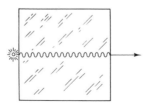

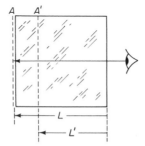

Figure 1.1-6 Whereas light passing through a block of glass encounters a *longer* path (*top*), to an observer it appears as if A has moved to A' and hence is *closer* (*bottom*).

Example

If two points, A and B, are 10 cm apart, their *actual distance* is $L = 10$ cm. But if the two points and the path between them are immersed in a liquid of refractive index 1.6:
(a) What is the *optical path length?*
(b) How much closer will point A appear to an observer at B?

Solution.   (a) Whereas the actual distance from A to B is 10 cm, with a medium of index 1.6 present, it will take the light 1.6 times as long to go from A to B, and thus

$$S = (10)(1.6) = \boxed{16 \text{ cm}}$$

(b) To an observer at B, point A will appear at a distance of

$$L' = \frac{10}{1.6} = 6.25 \text{ cm}$$

and, hence, it will appear

$$10 - 6.25 = \boxed{3.75 \text{ cm}}$$

closer to B.

Sometimes it may seem difficult to decide when to use optical path length and when reduced distance. From the "point of view" of the light, use optical path length. From the point of view of the observer, use reduced distance.

SUGGESTIONS FOR FURTHER READING

P. Giacomo, "The New Definition of the Meter," *Am. J. Phys.* **52** (1984), 607–13.

M. Young, "Pinhole Imagery," *Am. J. Phys.* **40** (1972), 715–20.

PROBLEMS

1.1-1.   A 4-m-long rope is set in oscillatory motion so that $2\frac{1}{2}$ waves are present on the rope at any one time. If the waves travel at a velocity of 10 m/s, what is the frequency of oscillation?

1.1-2.   A 50-cm-long string is oscillating in such a way that $3\frac{1}{3}$ waves are present on the string at any one time. If the waves travel at a velocity of 9 m/s, what is the frequency of oscillation?

1.1-3.   Yellow sodium light has an average wavelength of 589.3 nm.
    (a) How many waves are there in 1 mm?
    (b) What is the frequency of the light?

1.1-4.   Determine the frequency and the period of light of 625-nm wavelength.

1.1-5.   A small hole is made in the shutter of a dark room, and a screen is placed at a distance of 1.5 m from the shutter. If a tree outside is 30 m away from the shutter and if the tree casts on the screen an image 20 cm high, how tall is the tree?

1.1-6.   If in a *pinhole camera* (Figure 1.1-7) the distance between an object, left, and its image, right, is 1.5 m, how long must the camera be in order to make the image one-fifth the size of the object?

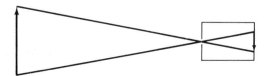

Figure 1.1-7

1.1-7. A pinhole camera produces a 5-cm-high image of a telephone pole. Moving the camera 5 m farther away from the pole reduces the image size to 4 cm. To make the image 5 cm high again, the camera has to be made 4 cm longer. How tall is the telephone pole?

1.1-8. A light source is 10 cm in diameter and located 3 m to the left of a screen. A solid round object, 15 cm in diameter, is placed halfway between the light source and the screen.
(a) What is the diameter of the (central) shadow?
(b) What is the width of the penumbra?

1.1-9. An extended light source 5 cm in diameter is located 3 m from an opaque disk $11\frac{2}{3}$ cm in diameter. If, on a screen some distance away, the total outside diameter of the penumbra is 45 cm, what is the diameter of the umbra?

1.1-10. A solid disk 20 cm in diameter is suspended from the center of a circular light source mounted flush with the ceiling of a room. When 80 cm above the floor, the disk casts a (central) shadow 12.5 cm in diameter. If the penumbra has a (total) diameter of 47.5 cm:
(a) How high is the ceiling?
(b) What is the diameter of the source?

1.1-11. A sheet of newsprint is viewed through a plane-parallel plate of glass of $n = 1.5$. When looking through the plate, the image appears to be 5 mm closer to the observer than without it. How thick is the plate?

1.1-12. Two beams of light pass through a distance of 8.5 cm. In the path of one beam is placed a 1.5-cm-thick plate of a transparent material. If the optical path *difference* is 7.2 mm, what is the refractive index of the material?

Reflection
and Refraction

WHEN LIGHT IS INCIDENT ON A SURFACE between two media, part of the light is *reflected*, and part is *refracted* (Figure 1.2-1). Reflection means that the light is returned to the first medium from which it came. Refraction means that on entering the second medium the light follows a direction different from its direction in the first medium.

The point where the light intersects the surface is the *point of incidence*. A line con-

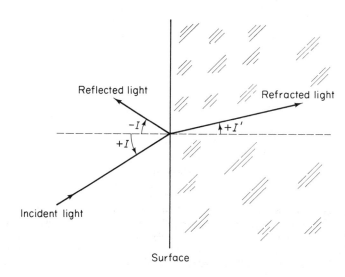

Reflected light

Refracted light

$-I$

$+I'$

$+I$

Incident light

Surface

Figure 1.2-1 Reflection and refraction of light at the surface between two media (air to the left of the surface, glass to the right).

structed at this point, perpendicular (*normal*) to the surface, is the *surface normal.* The angle subtended by the surface normal and the incident ray is the *angle of incidence,* $+I$ in Figure 1.2-1. The angle subtended by the surface normal and the reflected ray is the *angle of reflection,* $-I$, and the angle subtended by the normal and the refracted ray is the *angle of refraction,* $+I'$. These angles are measured *from* the surface normal, *toward* the ray.

FERMAT'S PRINCIPLE

According to Fermat's principle, *light takes the path of least time.** This is an important theorem from which, in fact, all laws of reflection and refraction can be derived.

Consider first the case of *reflection*. Light coming from a point A is reflected at a mirror M-M toward a point B (Figure 1.2-2). The distance from A to M-M is called a, and the distance from M-M to B is called b. The angles $+I$ and $-I$, as before, are the angles of incidence and reflection. From the construction and from Pythagoras' theorem it follows that the total path length L, from A to M-M to B, is

$$L = \sqrt{a^2 + x^2} + \sqrt{b^2 + (d - x)^2} \qquad [1.2\text{-}1]$$

For Fermat's principle to hold, the derivative of L with respect to x must be zero:

$$\frac{dL}{dx} = \frac{1}{2} \frac{1}{\sqrt{a^2 + x^2}} 2x + \frac{1}{2} \frac{1}{\sqrt{b^2 + (d - x)^2}} 2(d - x)(-1) = 0$$

which is equivalent to

$$\frac{x}{\sqrt{a^2 + x^2}} = \frac{d - x}{\sqrt{b^2 + (d - x)^2}} \qquad [1.2\text{-}2]$$

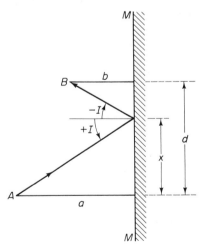

Figure 1.2-2 Fermat's principle for reflection.

---

*This theorem was first formulated, in 1657, by Pierre de Fermat (1601–1665), French jurist and mathematician, counsel to the Toulouse parliament. Fermat used to write his thoughts about analytic geometry, calculus, probability, and number theory into the margins of other books, pursuing his avocation mostly for his own enjoyment. Five years after his death his notes were published by his son. Fermat justified his principle of least time on the grounds that nature is "economical." Indeed, the time spent by light traveling from one point to another is most often a minimum.

The two fractions are the sines of the angles of incidence and reflection, respectively. The angles, therefore, are *numerically equal* (although their signs are opposite),

$$+I = |-I| \qquad [1.2\text{-}3]$$

In addition, the incident ray, the surface normal, and the reflected ray all lie in the same plane, the *plane of incidence.*

Next we turn to *Fermat's principle for refraction* (Figure 1.2-3). Now we have a surface, *S-S*, that separates two media but again the light goes from *A* to *B*. If the refractive index on either side of the surface were the same, no matter what its magnitude, the path from *A* to *B* would be a straight line. But if the two media have different indices, the path is no longer straight. Instead, the light is *refracted.*

Consider the velocities of the light, v and v', on both sides of the surface. Since $v = L/t$, the time it takes the light to travel from *A* to the point of incidence and from there to *B* is

$$t = \frac{\sqrt{a^2 + x^2}}{v} + \frac{\sqrt{b^2 + (d - x)^2}}{v'} \qquad [1.2\text{-}4]$$

Again we differentiate,

$$\frac{dt}{dx} = \frac{x}{v\sqrt{a^2 + x^2}} - \frac{d - x}{v'\sqrt{b^2 + (d - x)^2}} = 0$$

from which

$$\frac{\sin I}{v} - \frac{\sin I'}{v'} = 0$$

and

$$\frac{\sin I}{\sin I'} = \frac{v}{v'} \qquad [1.2\text{-}5]$$

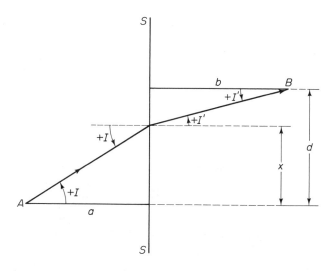

Figure 1.2-3 Fermat's principle for refraction.

If we now substitute $v = c/n$ and $v' = c/n'$, respectively, in Equation [1.2-5], we obtain

$$\frac{\sin I}{\sin I'} = \frac{c/n}{c/n'} = \frac{n'}{n}$$

and therefore

$$\boxed{n \sin I = n' \sin I'} \qquad [1.2\text{-}6]$$

which is *Snell's law of refraction.* *

Snell's law tells us that when light passes from a rarer medium to a denser medium (so that $n' > n$), $\sin I'$ must by necessity be less than $\sin I$ and the light be *bent toward the surface normal.* When the light goes from dense to rare, it is bent *away* from the surface normal. And, as before, the incident ray, the surface normal, and the refracted ray all lie in the same plane, the *plane of incidence.* Table 1.2-1 lists some representative data.

TABLE 1.2-1 INDICES OF REFRACTION
(FOR HELIUM d LIGHT, 587.6 nm)

Vacuum	1.000000
Air	1.0003
Water	$\frac{4}{3}$
Spectacle crown, C-1	1.5230
Carbon disulfide	1.63
Extra dense flint, EDF-3	1.7200
Methylene iodide	1.74
Diamond	2.42

Critical angle.   Assume that the light is incident on a transparent medium at an angle of incidence $I = 0$. The light will go through without deviation. As I is made gradually larger, the angle of refraction, I', becomes larger too, although not as rapidly as I. As I reaches 90°, its sine becomes unity and Snell's law reduces to

$$n = n' \sin I' \qquad [1.2\text{-}7]$$

where I' is now called the *critical angle of refraction* (Figure 1.2-4, left).

But then let the light come from the side of the denser medium, which means the angle of incidence is *below* the surface and the light is bent *away* from the normal. As I is made larger, it will reach a value such that I' becomes 90° and

$$n \sin I = n' \qquad [1.2\text{-}8]$$

---

*Named after Willebrord Snel van Royen (1591–1626), Dutch astronomer and mathematician. At age 21, Snell succeeded his father as professor of mathematics at the University of Leiden. At 26, he determined the size of the earth from measurements of its curvature between Alkmaar and Bergen-op-Zoom in The Netherlands. Snell's original statement, based on empirical observation, was that the ratio of the cosecants of the angles of incidence and refraction is constant. The law in its present form is due to René Descartes (Renatus Cartesius) (1596–1650), French philosopher and mathematician, who published it in his *La Dioptrique,* 1637.

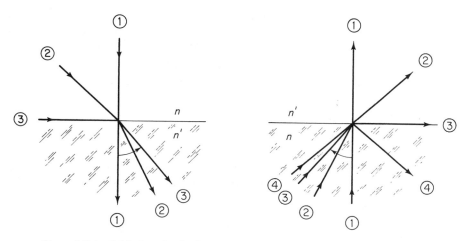

Figure 1.2-4   Critical angle of refraction, with the light passing from rare to dense (*left*), and angle of total internal reflection, with the light passing from dense to rare (*right*).

where *I* is now the *minimum angle of total internal reflection* (Figure 1.2-4, right). At this angle, and at angles larger than that, the light is returned to the first medium. This time, the index of the first medium, *n*, is higher than that of the second, *n′*; hence, the critical angle (of refraction) and the minimum angle (of total internal reflection) are numerically equal; they both lie in the denser medium.

PRISMS

Reflecting prisms.   We distinguish two major groups of prisms, *reflecting prisms* and *refracting prisms*. Reflecting prisms, unless they have a reflective coating, make use of total internal reflection. In Figure 1.2-5 we show several examples. In a *right-angle*

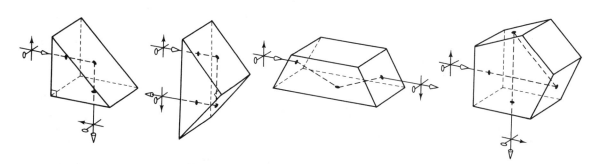

Figure 1.2-5   Some reflecting prisms, based on total internal reflection are (*from left to right*) right-angle prism, roof prism, Dove prism, and penta prism.

prism the light is reflected at the hypotenuse. If the same prism is oriented so that the light is passing (twice) through the hypotenuse, we have a *roof* or *Porro prism*. Still the same prism, with or without its apex cut off, is called a *Dove prism*. As the Dove prism is rotated about the line of sight, the image rotates through twice the angle. A *pentagonal prism* deflects the light through a constant 90°, without changing the "handedness" of the image.

Refracting prisms.   A refracting prism has two plane surfaces that subtend a certain angle A, the *apex angle*. The face opposite the apex is called the *base*. The total angle by which the light changes direction is the *angle of deviation, D*.

When the light passes through the prism *symmetrically* (with equal angles of incidence and emergence), then from the construction in Figure 1.2-6 it follows that

$$I_1 = I_2'$$

$$I_1' = I_2 = \frac{A}{2}$$

$$D_1 = D_2 = \frac{D}{2} = I_2' - I_2 = I_2' - \frac{A}{2}$$

so that the angle of emergence is equal to one-half the apex angle plus one-half the angle of deviation,

$$I_2' = \frac{A}{2} + \frac{D}{2}$$

Now we apply Snell's law, $n \sin I = n' \sin I'$, to the second (right-hand) surface:

$$n_{\text{prism}} \sin I_2 = n_0 \sin I_2'$$

where n_0 is the index outside the prism. Solving for $\sin I_2'$ gives

$$\sin I_2' = \left(\frac{n_{\text{prism}}}{n_0}\right) \sin I_2$$

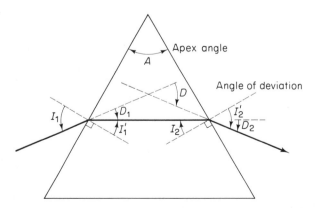

Figure 1.2-6   Deriving the prism equation.

and thus

$$\sin\left(\frac{A + D}{2}\right) = \left(\frac{n_{prism}}{n_0}\right)\sin\left(\frac{A}{2}\right)$$

$$\boxed{\frac{n_{prism}}{n_0} = \frac{\sin\frac{1}{2}(A + D)}{\sin\frac{1}{2}A}} \qquad [1.2\text{-}9]$$

which is the *prism equation*. It determines the *angle of minimum deviation*. For light not passing through symmetrically, the deviation is larger.

Example

Look at the moon, best on a cold evening and with the sky slightly overcast. If you see a *halo*, it is due to the refraction of light in ice crystals floating in the upper atmosphere. The crystals have the hexagonal shape shown in Figure 1.2-7 but, in essence, they are 60° prisms with the corners cut off. If we assume that the refractive index of ice is 1.31, somewhat less than the index of water (why?), what is the angle of minimum deviation?

Solution.   From the prism equation, setting $n = 1.31$ and $A = 60°$, we find that

$$\frac{1.31}{1.00} = \frac{\sin\frac{1}{2}(60° + D)}{\sin\frac{1}{2}(60°)}$$

We solve $\sin\frac{1}{2}(60°) = \sin 30° = 0.5$ and multiply with the ratio on the left:

$$0.655 = \sin\tfrac{1}{2}(60° + D)$$

Taking the arcsine on both sides gives

$$40.92° = \tfrac{1}{2}(60° + D) = 30° + \frac{D}{2}$$

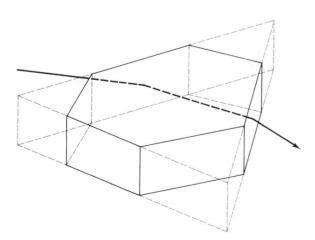

Figure 1.2-7   Hexagonal ice crystal (*solid lines*) and the triangular 60° prism equivalent to it (*dashed lines*).

and thus

$$10.92° = \frac{D}{2}$$

$$D = \boxed{21.84°}$$

That means that the halo subtends with the axis of observation an angle of approximately 22°, the *22° halo*. Because of additional *dispersion* (to be discussed shortly), the halo is faintly red inside, blue outside.

Thin prisms.   Ophthalmic or *thin prisms* have apex angles and angles of deviation much smaller than those of other prisms. The sines of these angles may be set equal to the angles themselves, measured in radians, and if the prism is in air ($n_0 \approx 1$), Equation [1.2-9] reduces to

$$D = A(n_{\text{prism}} - 1) \qquad\qquad [1.2\text{-}10]$$

As the prism is rotated about the line of sight, the displacement changes with the angle of rotation, as illustrated in Figure 1.2-8. We note that light passing through the prism is bent toward the base, but an object viewed through the prism appears to be deviated toward the apex.

Figure 1.2-8   Displacement of line as seen through a thin prism.

Example

A prism is 1 mm thick at the apex and 6 mm at the base. The distance between apex and base is 50 mm and the refractive index 1.61.
(a) Determine the apex angle.
(b) Determine the angle of deviation and the power of the prism.
(c) What is the power of the prism if it is immersed in water ($n = \frac{4}{3}$)?

Solution.   (a) First, imagine the thickness of the prism at its apex is reduced to zero, taking 1 mm off on one face; that makes the base 5 mm thick. Then we determine the apex angle,

$$A = (2)\arctan\frac{2.5}{50} = \boxed{5.7°}$$

(b) The angle of deviation, in units of *centrad*, $^\triangledown$, defined as $\frac{1}{100}$ of a radian, is found from Equation [1.2-10],

$$D = 5.7°(1.61 - 1) = \boxed{3.5°} = 0.06 \text{ rad} = \boxed{6^\triangledown}$$

(c) Because the prism is "thin," we may simplify Equation [1.2-9] and write

$$\frac{n_{\text{prism}}}{n_0} = \frac{A + D}{A} = 1 + \frac{D}{A}$$

Thus

$$D = A\left(\frac{n_{\text{prism}}}{n_0} - 1\right) = 5.7°\left(\frac{1.61}{\frac{4}{3}} - 1\right) = 0.02 \text{ rad} = \boxed{2^{\triangledown}}$$

DISPERSION

When light containing more than one wavelength passes through a refracting prism, the prism will separate the light according to wavelengths; such separation is called *chromatic dispersion*. Expressed in another way, the refractive index of matter for light varies as a function of wavelength. In fact, there is no material known whose refractive index does not change throughout the spectrum. As a rule, the index *decreases* as the wavelength *increases*.

With the light at normal incidence, blue light (for which the index is higher and the velocity less) will merely lag behind red light (Figure 1.2-9, left). With the light passing obliquely through a plane-parallel plate, the colors become separated but the rays remain parallel (center). But with a prism, the rays spread apart (right). The angle subtended by the two colors is called the *angular dispersion*.

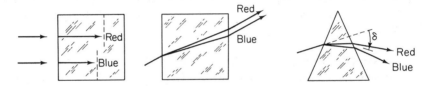

Figure 1.2-9   Dispersion in a block of glass at normal incidence (*left*) and at oblique incidence (*center*); dispersion caused by a prism (*right*).

Example

A hollow 60° prism is filled with carbon disulfide, whose index of refraction for blue light is 1.652, for red light 1.618. What is the angular dispersion?

Solution.   Using Equation [1.2-9], we determine first the deviation for blue:

$$\frac{1.652}{1.000} = \frac{\sin \frac{1}{2}(60° + D)}{\sin \frac{1}{2}(60°)}$$

$$0.826 = \sin \tfrac{1}{2}(60° + D)$$

$$111.38° = 60° + D$$

$$D = 51.38°$$

For red,

$$0.809 = \sin \tfrac{1}{2}(60° + D')$$

$$108° = 60° + D'$$

$$D' = 48.0°$$

The difference between the two angles is the angular dispersion between blue and red,

$$51.38° - 48.00° = \boxed{3.38°}$$

But refraction and dispersion bear no simple relationship to one another: some glasses have a high index of refraction and little dispersion, others have just the opposite (Figure 1.2-10). Clearly, to characterize a glass we need more than one refractive index. In practice, three indices are chosen, one for each of three colors, the blue F line produced by hydrogen, wavelength 486.1 nm, defining the refractive index n_F; the yellow d line of helium, 587.6 nm, defining n_d; and the red C line of hydrogen, 656.3 nm, defining n_C. (Note that today we use the single helium d line, rather than the sodium D doublet.)

The difference $n_F - n_C$ is the *mean dispersion,* and the ratio

$$\frac{n_d - 1}{n_F - n_C} = \nu \qquad [1.2\text{-}11]$$

is *Abbe's number,* also called ν *value* or *V-number.* A small Abbe's number, in other words, means high dispersion.

Glasses of low dispersion (where the difference $n_F - n_C$ is small and the V-number high, above 55) are customarily called *crowns.* Glasses of high dispersion (with ν below 50) are called *flints.* Some typical examples are shown in Table 1.2-2, and a *glass chart* in Figure 1.2-11.

Optical glass.   Crown glasses, as well as flint glasses, can be further subdivided according to their chemical composition. The designation BK7, for example, refers to borosilicate crown, the type of optical glass produced in larger quantities than any other glass. The term SK refers to barium-oxide glass; it has a high index and little dispersion. Flint

Figure 1.2-10 (*Left*) Prism of high refraction and low dispersion (note how much the light is deviated but how close the two colors are). (*Right*) Prism of low refraction and high dispersion (little deviation but the two colors are spread apart much farther).

TABLE 1.2-2 REFRACTIVE INDICES OF CROWN AND FLINT AT DIFFERENT WAVELENGTHS

Fraunhofer line	Color	Wavelength (nm)	Spectacle crown, C-1	Extra dense flint, EDF-3
F	Blue	486.1	1.5293	1.7378
d	Yellow	587.6	1.5230	1.7200
C	Red	656.3	1.5204	1.7130
			V-number	
			58.8	29.0

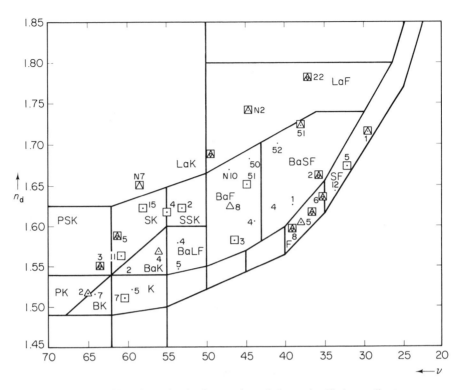

Figure 1.2-11   Glass chart, showing frequently used glasses classified according to n_d and ν.

glasses contain lead oxide, light flint relatively small amounts, heavy flint, SF, up to 70%. Some new types of glasses have low dispersion and extraordinarily little thermal expansion (important for lasers and satellite-borne telescopes), or high refraction and light weight (good for eyeglasses). Plastics have refractive indices generally lower than those of glasses. Figure 1.2-11 shows various types and designations of optical glasses.

The *glass number* is a six-digit code identifying the glass. The first three digits refer to the refractive index $n_d - 1$, and the last three digits refer to Abbe's number multiplied by 10. A glass with the number 523588, for example, has a refractive index $n_d = 1.523$ and a dispersion of 58.8.

I had mentioned before that all transparent media have refractive indices that are higher at shorter wavelengths and lower at longer wavelengths. A plot of n versus λ, therefore, follows the dashed curve in Figure 1.2-12. Such curves can be represented by any of several formulas. Probably the most precise is Herzberger's equation,*

$$n = A + B\lambda^2 + \frac{C}{\lambda^2 - \lambda_0^2} + \frac{D}{(\lambda^2 - \lambda_0^2)^2} \qquad [1.2\text{-}12]$$

---

*M. Herzberger, "Colour correction in optical systems and a new dispersion formula," *Opt. Acta (London)* **6** (1959), 197–215.

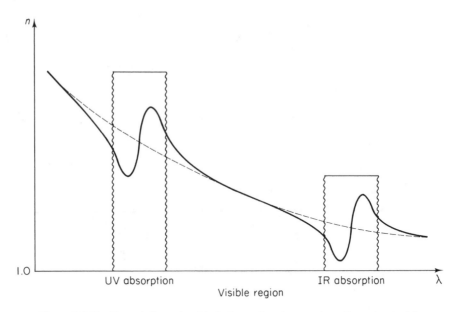

Figure 1.2-12   Normal dispersion (*dashed curve*) and anomalous dispersion (*solid curve*). Note absorption bands.

where *n* is the refractive index, λ the wavelength, and *A* through *D* are empirically determined constants. The wavelength λ_0 is set at 168 nm, which determines the UV absorption. The near IR is covered by the $B\lambda^2$ term.

This formula, and others, holds for what in earlier times has been called *normal dispersion*. Normal dispersion is seen in colorless, transparent matter like water or glass, away from any absorption. But if such absorption is present, the dispersion curve follows the solid line in Figure 1.2-12 and is called *anomalous* (meaning "abnormal"). Today the terms "normal dispersion" and "anomalous dispersion" are of historic interest only. If we include the UV and the IR, most any substance will show some absorption—and hence "anomalous" dispersion—somewhere within the spectrum.

Applications of dispersing prisms.   A prism spectro*scope* is used for viewing a spectrum. A spectro*meter* is used for measuring. A spectro*graph* is built for photography (Figure 1.2-13). In a spectro*photometer* a photodetector takes the place of the photographic film. A spectro*gram* is the result of spectography or spectrophotometry.

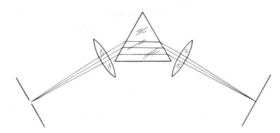

Figure 1.2-13   Diagram of a prism spectrograph. The two lenses (collimating lens left, focusing lens right) allow parallel light to pass through the prism, oriented for minimum deviation.

An *achromatic prism* deviates the light but gives no dispersion. This is possible by using two prisms that have equal dispersions but unequal apex angles, one made of crown and the other of flint, and combining them so that the base of one prism rests against the apex of the other. A *direct-vision prism,* in contrast, gives dispersion but no deviation (for one wavelength). A *monochromator* is an instrument for selecting light of different wavelengths.

SUGGESTIONS FOR FURTHER READING

L. LEVI, *Applied Optics, A Guide to Optical System Design,* Vol. 1, Chap. 8.3 and Appendix 8.2, "Prisms," pp. 354–89 (New York: John Wiley & Sons, Inc., 1968).

T. S. IZUMITANI, *Optical Glass* (New York: American Institute of Physics, 1986).

W. L. WOLFE, "Properties of Optical Materials," in W. G. Driscoll and W. Vaughan, editors, *Handbook of Optics,* Sec. 7, pp. 1–157 (New York: McGraw-Hill Book Company, 1978).

R. GREENLER, *Rainbows, Halos, and Glories* (New York: Cambridge University Press, 1980).

PROBLEMS

1.2-1. Light is entering a plate of glass at an angle of incidence of 20°. If the angle of refraction is 12°, what is the index of refraction of the glass?

1.2-2. A layer of oil of unknown refractive index is floating on top of a layer of carbon disulfide. If the angle of incidence at the air–oil boundary is 60°, what is the angle of refraction in the carbon disulfide?

1.2-3. An open cylindrical container, resting on its circular base, is 20 cm in diameter and 11.2 cm deep. An observer is looking into the empty container from such a direction that he or she can just see the opposite bottom corner (Figure 1.2-14). When the container is filled with a liquid and the observer keeps looking from the same direction, he or she can see the *center* of the bottom. What is the refractive index of the liquid?

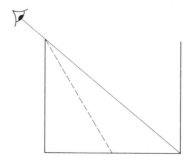

Figure 1.2-14

1.2-4. How far from true vertical does the setting sun appear to an observer under water ($n = \frac{4}{3}$)?

1.2-5. What is the minimum angle of total internal reflection for light passing from glass of $n = 1.5238$ to water ($n = \frac{4}{3}$)?

1.2-6. What is the least refractive index that a right-angle prism must have to reflect light through an angle of 90°?

1.2-7. A prism of 72° apex angle and 1.66 refractive index is immersed in water ($n = \frac{4}{3}$). Find the angle of minimum deviation.

1.2-8. A hollow (and empty) 60° prism is immersed in a liquid of index 1.74. What is the angle of minimum deviation?

1.2-9. A 30° prism made out of BK7 glass ($n = 1.515$) is oriented first so that the light is incident at right angles. The prism is then turned to the position of minimum deviation. By how much do the two angles of deviation differ?

1.2-10. A prism, made of glass of $n = 1.62$, shows an angle of minimum deviation of 48.2°. What is the apex angle?

1.2-11. What is the apex angle of a prism made of glass of index 1.5 which produces a minimum deviation of 7 centrads?

1.2-12. An object 1.5 m away is viewed through a $3^{\triangledown}$ prism. By how much, and in which direction, will the object appear to be displaced?

1.2-13. If white light is incident at an angle of 30° on a slab of glass that for blue light has a refractive index of 1.7 and for red light of 1.6, what is the angular dispersion between blue and red inside the glass?

1.2-14. A 6° crown prism is combined with a flint prism so as to be achromatic for blue and yellow. What should be the apex angle of the flint prism?

1.3

Thin Lenses

Now WE COME TO THE FIRST LENS of an optical system. As an example, look at the front lens of the system shown in Figure 1.3-1. Note that a lens has two surfaces that enclose a medium of a refractive index different from the index outside the lens. Hence, before we discuss lenses as such, we ask what happens to the light as it passes through a single *surface*.

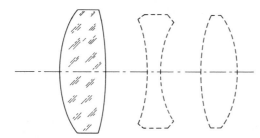

Figure 1.3-1 Front lens (solid outline) of a three-lens system.

SINGLE REFRACTING SURFACE

Vergence. As the light is incident on the first surface, it has a certain *vergence. Vergence is a term whose significance to optics can hardly be overemphasized.* Light that spreads out is called *divergent.* If the light comes from a source on the left, the light's wavefronts are curved concave to the left (Figure 1.3-2). Light that comes together is called *convergent;* if the light converges to the right, its wavefronts are concave to the right.

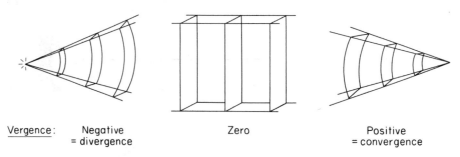

Vergence: Negative   Zero   Positive
             = divergence   = convergence

Figure 1.3-2 Vergence of light coming from a point source (*left*) and proceeding toward a point image (*right*).

Ordinarily, the *curvature* of a surface, C, is the reciprocal of the length of the radius (of curvature), R, of that surface:

$$C = \frac{1}{R} \qquad\qquad [1.3\text{-}1]$$

But the light may travel in a medium other than air. A medium of index n, for example, will delay the light by a factor n and the vergence, $\mathbf{V}$, becomes

$$\mathbf{V} = \frac{n}{R} \qquad\qquad [1.3\text{-}2]$$

Now the light is reaching the surface. Any refracting surface has a certain *refractive power*. Refractive power is the reciprocal of the focal length of the surface, in meters. If the power is positive and *higher* than the (negative) vergence, some excess (positive) vergence remains and the light after refraction becomes *con*vergent, as in Figure 1.3-3, top. Such light will form a *real image*. A real image can be received, and seen, on a screen.

If the object is moved closer to the surface, the radius of curvature of the wavefront entering the surface becomes shorter, the (negative) vergence of the light becomes more negative, and may just be counterbalanced by the (positive) power of the surface. The resultant vergence then is zero, which means that the light emerges parallel and the image is projected out to infinity (center).

If the object is brought even closer, the vergence of the light becomes still more negative, the surface does not have enough power to compensate for it, and the emergent light *remains divergent* (bottom). Now the rays will *not* come together on the right-hand side. Instead, the light seems to come from a point, S'', to the left of the surface. Such a point is part of a *virtual image*.*

Next, we consider the *focal points* of a surface. If *parallel light* is incident on a positive surface, as shown in Figure 1.3-4, top left, the light comes together at the *second focal point*,

---

*A good example of a virtual image is what you see in a mirror. The object and the observer are both in front of the mirror, but the image is behind it. No light is actually going to where the image *appears to be*; the image is *virtual*. In a way, the terms "real" and "virtual" are misleading. The word "real" seems to imply that "virtual" has something "unreal" to it. That is not so; a virtual image can be seen, and be photographed, just as well as a real image; the only difference is that it cannot be received on a screen.

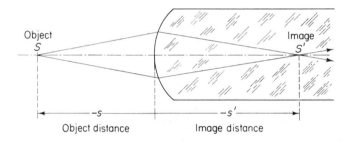

Object
S

Image
S'

|← −s →|← −s' →|

Object distance   Image distance

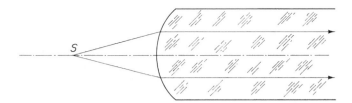

S

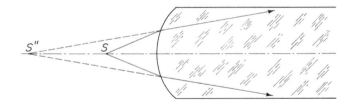

S''   S

Figure 1.3-3 When an object, *S,* is gradually brought closer to a positive refracting surface, the image is first real, *S'* (*top*); projected out to infinity (*center*); and then virtual, *S''* (*bottom*).

F_2. If the light is *made parallel* by the surface, the light comes from the *first focal point, F_1* (top right). The same convention applies to a negative surface: parallel light that is incident relates to F_2 (bottom left), and parallel light that emerges relates to F_1 (bottom right).

The point where the optic axis intersects the surface is called the *vertex* (see page 35). The distance from the vertex to the object is the *object distance, s*. The distance from the vertex to the image is the *image distance, s'*. Note that these distances are measured *from* the vertex, *toward* the object or the image, respectively. But the image may be either real or virtual, and therefore may either lie to the right or the left of the surface; the *image space,* in other words, extends from infinity on one side to infinity on the other. The same holds for the *object space;* both spaces completely overlap. Whether a given point is in the object space or the image space depends on whether it is part of a ray *before* or *after* refraction.

Objects, and images, are composed of a great many tiny "*pic*ture *cells,*" called *pixels*. Object pixels and image pixels that correspond (belong) to each other are called *conjugate*. Distances and other parameters that correspond to each other are called conjugate also.

Sign convention.   There are two major sign conventions in optics. In the *empirical system,* the distance of a real object is taken as positive. Intuitively, that seems to be a good choice. However, there are compelling reasons, from the definition of vergence to com-

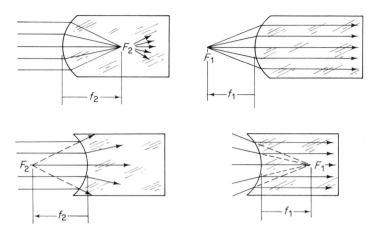

Figure 1.3-4   Focal points, F, and focal lengths, f.

puter-aided lens design, why it is better to consider that distance *negative* and to use the *rational, Cartesian sign convention,* * as we doing it here. In the Cartesian sign convention:

1. All figures and ray diagrams are drawn with the light traveling from left to right.

2. Distances measured to the left, in a direction opposite to the direction of the light, are considered *negative*. Distances measured to the right are considered *positive*. Radii of curvature are measured *from* the surface, *toward* the center of curvature. Radii that extend to the left are negative; radii that extend to the right are positive. Divergent light, therefore, has negative vergence; convergent light has positive vergence. Parallel light has zero vergence. Likewise, the *distance of a real object is measured to the left and is negative.* The distance of a *real image* is measured to the right and is positive. The distance of a virtual image is negative.

3. The *refractive power* of a surface, or a lens, that makes light more convergent, or less divergent, is positive. The *second focal length* of such a surface is positive also. The power and the (second) focal length of a diverging surface are negative.

4. *Heights* above the optic axis are positive; distances below the axis are negative.

5. *Angles* measured *counterclockwise* from the optic axis, or from a surface normal, are *positive*. Angles measured clockwise are negative.

Gauss' formula and the surface power equation.   With this sign convention in hand, we consider the single surface shown in Figure 1.3-5. The radius of curvature of the surface extends to the right, and therefore is positive. A ray originates at an axial

---

*Named after René Descartes (1596–1650), French mathematician and philosopher. Educated at a Jesuit college, Descartes graduated with a law degree, then joined the army to work on problems of ballistics and navigation. Shortly after the discovery of Snell's law, and most likely independent of it, he found the same law and published it, the first to do so. Next he showed how every point in space can be represented by a system of three numbers, the *Cartesian system of coordinates* (from the latinized version of the name he used in his writings, Renatus Cartesius). Descartes often started out with a central principle and moved outward to find the phenomena of nature which follow it, an approach opposite to that promulgated by Newton.

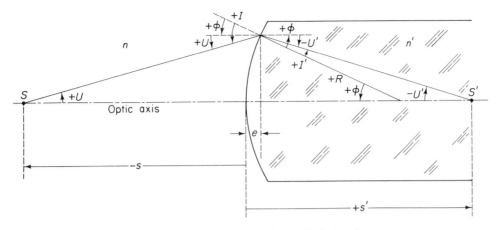

Figure 1.3-5 Refraction at a single spherical surface.

point object, S. At S, the ray subtends with the axis an angle $+U$. At the image, S', the ray subtends an angle $-U'$. From the construction it follows that

$$(+U) + (+\phi) = +I$$

which is the same as

$$I = U + \phi$$

and that

$$+\phi = (-U') + (+I')$$

which is the same as

$$I' = U' + \phi$$

If the ray is close to the optic axis, the angles involved are small and their sines and tangents are nearly equal to the angles themselves, measured in radians. Such rays are called *paraxial rays.** Snell's law,

$$n \sin I = n' \sin I'$$

then becomes

$$nI = n'I'$$

Substituting for I and I' the terms just derived (and considering that for small angles $\sin U \approx u$), we obtain

$$nu + n\phi = n'u' + n'\phi$$

---

*From the Greek $\pi\alpha\rho\grave{\alpha}$ = nearby, closely adjacent; paraxial = close to the optic axis.

For paraxial rays, furthermore, the distance e, compared to s and s', is small and can be neglected; thus

$$+u = \frac{+h}{+s}, \qquad -u' = \frac{+h}{-s'}, \qquad \text{and} \qquad +\phi = \frac{+h}{-R} \qquad [1.3\text{-}3]$$

Substituting these terms in the preceding equation gives

$$\frac{nh}{s} - \frac{nh}{R} = \frac{n'h}{s'} - \frac{n'h}{R}$$

Canceling h and rearranging then leads to

$$\boxed{\frac{n}{s} + \frac{n' - n}{R} = \frac{n'}{s'}} \qquad [1.3\text{-}4]$$

which is *Gauss' formula for refraction at a single surface,** written in a form that is easy to remember: Just think of the light propagating from object (s) to surface (R) to image (s'). Although I have derived Gauss' formula for a convex surface, and for a real object and image, the formula holds as well for all other conditions.

If the object is moved to infinity, $-s$ becomes $-\infty$, the first term in Equation [1.3-4] drops out, and the image moves to the second focal point, $s' \rightarrow f_2$; thus

$$\frac{n' - n}{R} = \frac{n'}{f_2} \qquad [1.3\text{-}5]$$

The right-hand term in this equation is actually the inverse of the reduced focal length which, by definition, is the refractive power, P, of the surface. Hence,

$$\boxed{P = \frac{n' - n}{R}} \qquad [1.3\text{-}6]$$

which is the *surface power equation.*

Recall from Equation [1.3-2] that vergence is $\mathbf{V} = n/R$. Therefore, the first term in Gauss' formula is simply the vergence of the light as it enters the surface, the *entrance vergence*, $\mathbf{V}$. The last term is the vergence of the light as it leaves the surface, the *exit vergence*, $\mathbf{V}'$. Substituting for the second term the surface power equation, we conclude that

$$\boxed{\mathbf{V} + P = \mathbf{V}'} \qquad [1.3\text{-}7]$$

---

*Named after Karl Friedrich Gauss (1777–1855), German mathematician and astronomer. A child prodigy even before entering elementary school, Gauss discovered the prime number theorem and the method of least squares (used in statistics). As an astronomer he found a way to calculate the orbits of asteroids and designed the eyepiece named after him. Other accomplishments that bear his name are the derivation of the surface power equation, the thin-lens equation, the law of electrostatics, and the divergence theorem, the latter two leading up to Maxwell's equations. His book, *Dioptrische Untersuchungen,* was published in 1843. Gauss was the first to use the paraxial-ray approximation; this is known as *first-order, paraxial,* or *Gaussian optics.*

The units of vergence and power are the same, *reciprocal meter,* m^{-1}, commonly called *diopter.**

Example 1

A point light source is placed 50 cm to the left of a single surface of $+6.00$ diopters power, ground on a rod of glass of index 1.6 (Figure 1.3-6). What is the image distance (inside the rod)?

Solution. The *entrance vergence,* as the light enters the surface, is

$$\mathbf{V} = \frac{1}{-0.5 \text{ m}} = -2 \text{ m}^{-1}$$

The *exit vergence,* from Equation [1.3-7], is

$$\mathbf{V}' = \mathbf{V} + P = (-2) + (+6) = +4 \text{ m}^{-1}$$

Note the reversal of the signs of the two vergences. If the exit vergence is positive, as here, the image is inverted and real.

The light now proceeds in the medium of higher index, $n' = 1.6$. The image distance, therefore, is

$$s' = \frac{n'}{\mathbf{V}'} = \frac{1.6}{+4} = \boxed{+40 \text{ cm}}$$

If the surface had only $+1.50$ diopters power, instead of $+6.00$, how would the image change? In that case

$$\mathbf{V}' = (-2) + (+1.5) = -0.5 \text{ m}^{-1}$$

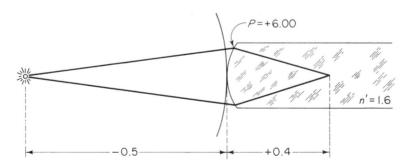

Figure 1.3-6 Image formation by a positive surface. Distances in meters.

---

*The unit of diopter was introduced in 1872 by Ferdinand Monoyer (1836–1912), French ophthalmologist. The son of an army physician, Monoyer in 1870 succeeded his stepfather as chairman of the Department of Ophthalmology at Strassburg, in 1872 became professor of ophthalmology in Nancy, and in 1876 professor of medical physics in Lyon. It was in Lyon where he did most of his optics work, introducing the metric system into ophthalmic measurements. The unit of diopter occurs first on page 111 of his contribution "Sur l'introduction du système métrique dans le numérotage des verres de lunette, et sur le choix d'une unité de réfraction," *Ann. Ocul.* **68** (1872), 105–117.

which means that the light remains divergent, and the image is virtual and *to the left* of the surface (as in Figure 1.3-3, bottom). The image distance in that case would be

$$s' = \frac{1.6}{-0.5} = -3.2 \text{ m}$$

This is an important case. While the medium to the left of the surface is air ($n \approx 1.0$), the *index related to the image space is 1.6.* In other words, with any image, real or virtual, *always use the image space index.* Do not use the index of the medium where the image merely *seems* to be.

Example 2

A *reduced eye* is a much simplified model of the human eye; it has only *one* refracting surface, assumed to have a radius of curvature of 5.7 mm (Figure 1.3-7). The refractive index to the left of the surface is 1.00, the index to the right 1.34. Find the focal length(s) and power(s) of the surface.

Solution. The focal length inside the eye, to the right of the surface, is found from Equation [1.3-5]:

$$f_2 = \frac{n'R}{n' - n} = \frac{(1.34)(0.0057)}{1.34 - 1.00} = \boxed{+22.5 \text{ mm}}$$

In order to find f_1, to the left of the surface, we let the image distance in Gauss' formula, Equation [1.3-4], become infinite, $s' \rightarrow \infty$. Then by definition $s \rightarrow f_1$, and therefore

$$f_1 = \frac{nR}{n - n'} = \frac{(1)(0.0057)}{1.00 - 1.34} = \boxed{-16.8 \text{ mm}}$$

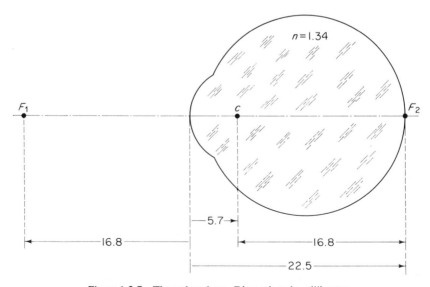

Figure 1.3-7 The reduced eye. Dimensions in millimeters.

In short, we have *two focal lengths*. Their ratio, neglecting the sign, is equal to the ratio of the refractive indices,

$$\frac{22.5}{16.8} = \frac{1.34}{1.00}$$

But we have only *one power*,

$$P = \frac{n' - n}{R} = \frac{1.34 - 1.00}{0.0057} = \boxed{+60 \text{ diopters}}$$

THIN LENSES

We distinguish two general classes of lenses, depending on whether they make parallel light convergent or divergent (Figure 1.3-8). A converging, positive lens is thicker in the center than at the periphery (provided that its refractive index is higher than the index outside the lens); a diverging, negative lens is thinner.

A lens has two centers of curvature and two radii of curvature, one for each surface. The line connecting the two centers is the *optic axis of the lens*. The points where the axis intersects the two surfaces are the *front vertex* and the *back vertex,* respectively. If parallel light is incident on the lens, it forms the *second* focal point, no matter whether the lens is converging or diverging. The only difference is that for a converging lens F_2 lies to the right of the lens, and for a diverging lens it lies to the left, the same as for a single surface shown earlier in Figure 1.3-4. If the light is made parallel by the lens, converging or diverging, it comes from, or is related to, the *first* focal point. As the object is brought closer to the lens, or moved away from it, the image distance changes accordingly (Figure 1.3-9).

The power of a converging lens is positive, that of a diverging lens negative. But whereas the second focal length of a converging lens is positive, the first, following the Cartesian sign convention, is negative. With a diverging lens, the reverse is true. The two focal lengths of a lens, therefore, are numerically equal but of opposite sign. Customarily, whenever reference is made to "*the* focal length" of a lens, that always means the second focal length.

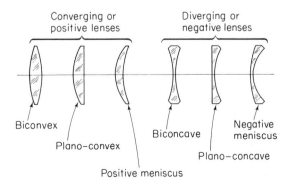

Figure 1.3-8 Types of lenses.

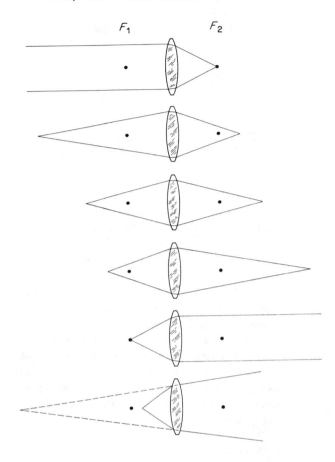

Figure 1.3-9 As an object at infinity (*top*) is brought closer to a converging lens, the image moves away from it. In the last example (*bottom*), the image is *virtual* and to the left of the lens.

Lens equations: the lens-makers formula.   If, compared with the lengths of the radii of curvature of the two surfaces, the thickness of the lens can be neglected, the lens is considered *thin*. In that case, the refractive power of the lens is simply the algebraic sum of the powers of the two surfaces:

$$P_{\text{lens}} = P_1 + P_2 \qquad\qquad [1.3\text{-}8]$$

We replace both P_1 and P_2 by the surface power equation,

$$P_1 = \frac{n_{\text{lens}} - n_0}{R_1} \qquad \text{and} \qquad P_2 = \frac{n_0 - n_{\text{lens}}}{R_2} = \frac{n_{\text{lens}} - n_0}{-R_2}$$

and substitute these terms in Equation [1.3-8]. That gives

$$\boxed{P_{\text{lens}} = (n_{\text{lens}} - n_0)\left(\frac{1}{R_1} - \frac{1}{R_2}\right)} \qquad\qquad [1.3\text{-}9]$$

which is the *lens-makers formula*.

Example

A biconvex lens of 50-mm focal length is made of glass of $n = 1.52$. If the second radius of curvature is twice as long as the first, what are the two radii?

Solution.   First we convert the focal length into power:

$$P = \frac{n}{f} = \frac{1.00}{0.05 \text{ m}} = 20 \text{ m}^{-1}$$

The second radius, R_2, is twice as long as R_1. But the lens is biconvex, hence the signs are opposite:

$$R_2 = -2R_1$$

Then from the lens-makers formula:

$$20 = (1.52 - 1.00)\left(\frac{1}{R_1} - \frac{1}{R_2} \right)$$

$$= (0.52)\left(\frac{1}{R_1} - \frac{1}{-2R_1} \right)$$

$$= 0.78 \frac{1}{R_1}$$

$$R_1 = \frac{0.78}{20} = 0.039 = \boxed{+39 \text{ mm}}$$

$$R_2 = -2R_1 = -2(39) = \boxed{-78 \text{ mm}}$$

Newton's lens equation.   The lens-makers formula tells us what the lens-maker needs to know to grind a lens of a given power: the refractive index of the glass and the two radii of curvature. Now we derive two "lens-users equations"; they connect the focal length of the lens with the object and image distances.

From Figure 1.3-10 we see that the object size, y, and the image size, y', are connected through similar triangles. To the left of the lens we have

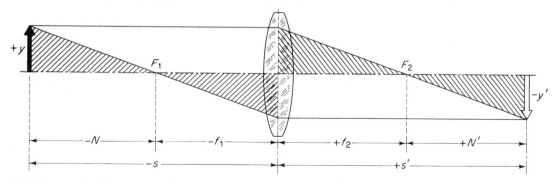

Figure 1.3-10   Deriving Newton's lens equation.

$$\frac{y}{y'} = \frac{N}{f_1}$$

and to the right of the lens

$$\frac{y}{y'} = \frac{f_2}{N'}$$

Combining the two equations by eliminating y/y' gives

$$\frac{N}{f_1} = \frac{f_2}{N'}$$

and thus

$$\boxed{NN' = f_1 f_2}$$ [1.3-10]

which is *Newton's lens equation*. Note that N and N' are measured from the focal points of the lens, rather than from the lens as such; they are *extrafocal* distances. Consequently, Newton's equation can be used with both "thin" lenses and "thick" lenses.

Gauss' thin-lens equation.   This time we begin with Gauss' formula for a single surface,

$$\frac{n}{s} + \frac{n' - n}{R} = \frac{n'}{s'}$$

Then we substitute for the center term the surface power equation,

$$\frac{n' - n}{R} = P$$

But P is the reciprocal of the focal length and thus if the lens is in air,

$$\boxed{\frac{1}{s} + \frac{1}{f_2} = \frac{1}{s'}}$$ [1.3-11]

which is *Gauss' thin-lens equation*, where, as before, we proceed from object (s) to lens (f) to image (s'). Solving Equation [1.3-11] for the individual terms gives for the

Object distance: $s = \dfrac{s'f}{f - s'}$ [1.3-12a]

Image distance: $s' = \dfrac{sf}{s + f}$ [1.3-12b]

Focal length: $f = \dfrac{ss'}{s - s'}$ [1.3-12c]

Example 1

When focusing at a target 24.5 cm from the object-side focal plane of a camera lens, the film in the camera must be moved 5 mm away from the position used for objects at infinity. What is the focal length of the lens?

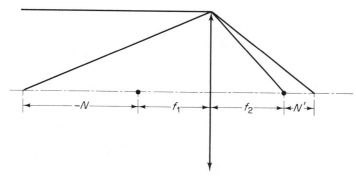

Figure 1.3-11

Solution. For the second of the two positions, and from Figure 1.3-11 and *Newton's equation,*

$$NN' = f_1 f_2$$

Since $-f_1 = f_2,$

$$-NN' = f_2^2$$

and thus,

$$f_2 = \sqrt{-NN'} = \sqrt{-(-245\ \text{mm})(5\ \text{mm})} = \sqrt{1225} = \boxed{+35\ \text{mm}}$$

Example 2

When an object is placed 9 cm in front of a converging lens, its image is three times as far away from the lens as when the object is at infinity. What is the focal length of the lens?

Solution. While for the object at infinity,

$$s' = f$$

for the close-by object,

$$s'' = 3f$$

Then from *Gauss' thin-lens equation* solved for $f,$

$$f = \frac{ss''}{s - s''} = \frac{(s)(3f)}{(s) - (3f)} = \frac{(-9)(3f)}{(-9) - (3f)}$$

$$f(-9 - 3f) = (-9)(3f)$$

$$-9f - 3f^2 = -27f$$

$$3f = 18$$

$$f = \frac{18}{3} = \boxed{+6\ \text{cm}}$$

Graphical construction.   In recent years the art of graphical ray tracing has lost some of its earlier prominence. This is due to the growing sophistication of computer-aided numerical ray tracing which we discuss later, in Chapter 2.4.

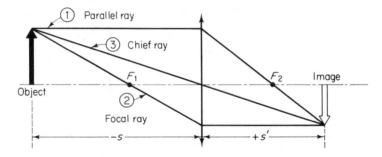

Figure 1.3-12 Designation of rays used for image construction. The object is *outside* the focal length of a converging lens.

Assume that the object is located *outside* the focal length. To find the image, we need at least two rays, out of three shown in Figure 1.3-12.

1. The *parallel ray* is first parallel to the axis and then, after refraction, passes through F_2.
2. The *focal ray* passes through F_1 and then is rendered parallel to the axis.
3. The *chief ray* goes through the center of the lens, without deviation and (with a *thin* lens) also without displacement.

The resulting image is real and inverted.

Next assume that the object lies *inside* the focal length. In principle we follow the same procedure. However, now rays 1–2–3, after refraction, are still divergent and do not intersect on the right. Instead, they seem to come from a point on the left (Figure 1.3-13). This point is part of a *virtual image;* that image is upright and magnified.

A virtual *object* is usually a virtual image formed by a preceding lens. Finally, with a *diverging* lens, the image, no matter where the object is placed, is always virtual, and reduced in size.

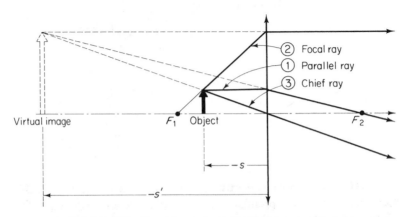

Figure 1.3-13 The object is *inside* the focal length of the lens.

MAGNIFICATION

The term "magnification" means many things. Whatever its type, though, magnification always refers to a *comparison*, comparing the size of the image to the size of the object. We discuss three types of magnification: *transverse magnification, axial magnification,* and *angular magnification.*

Transverse magnification.   Transverse, or *lateral*, magnification, M_T, is defined as the ratio of image size, y', to object size, y:

$$\boxed{M_T = \frac{y'}{y}}$$ [1.3-13]

From the similar triangles in Figure 1.3-14 we note that

$$\frac{+y}{-s} = \frac{-y'}{+s'}$$

and therefore, transverse magnification can also be defined as the ratio of image *distance*, s', to object *distance*, s:

$$M_T = \frac{s'}{s}$$ [1.3-14]

But this holds only if the refractive indices on both sides are the same. With a *single surface*, by necessity, they are different. Consider such a single surface and draw a ray from the point of the arrow to the vertex (Figure 1.3-15). The ray will be refracted at the surface according to Snell's law, which for paraxial rays simplifies to

$$nI = n'I'$$ [1.3-15]

From the construction it follows that the angles of incidence and refraction are

$$+I = \frac{-y}{-s} \qquad \text{and} \qquad +I' = \frac{+y'}{+s'}$$

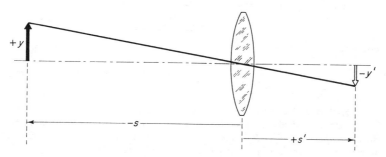

Figure 1.3-14   Transverse magnification.

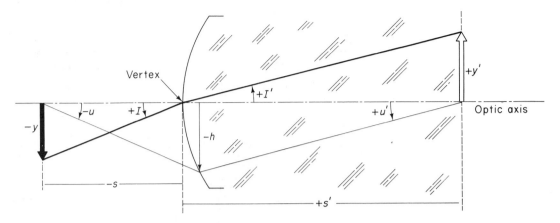

Figure 1.3-15   Transverse magnification with a single surface.

Substituting in Equation [1.3-15] gives

$$n \frac{y}{s} = n' \frac{y'}{s'}$$ [1.3-16]

But the ratio n/s is the entrance vergence of the light, **V**, and the ratio n'/s' is the exit vergence, **V'**; thus,

$$\mathbf{V}y = \mathbf{V}'y'$$

Rearranging and substituting in Equation [1.3-13] then shows that transverse magnification can also be defined as

$$\boxed{M_T = \frac{\mathbf{V}}{\mathbf{V}'}}$$ [1.3-17]

From Figure 1.3–15 we also see that

$$-u = \frac{-h}{-s} \quad \text{and} \quad +u' = \frac{-h}{+s'}$$

Solving for s and s', substituting in Equation [1.3-16], and eliminating h results in

$$\boxed{nyu = n'y'u'}$$ [1.3-18]

which is the *Smith–Helmholtz relationship,* also known as *Lagrange's theorem.** The prod-

---

*Robert Smith (1689–1768), British physicist, Plumian professor of physics at Trinity College. Smith seems to have been the first to find this relationship. He wrote two books, *A Compleat System of Opticks in Four Books, viz. A Popular, a Mathematical, a Mechanical, and a Philosophical Treatise* (Cambridge, 1738), a textbook widely used in the eighteenth century, and *Harmonics, or the Philosophy of Musical Sounds* (Cambridge, 1749).

The modern form, Equation [1.3-18], is due to Hermann Ludwig Ferdinand von Helmholtz (1821–1894),

uct *nyu* is called *Lagrange's invariant*. Its practical significance will come up later when we discuss zoom lenses.

If the object is farther away from a lens than twice its focal length, the image is reduced in size. If the object is closer, the image is magnified. If both object distance and image distance are equal, they are both equal to 2*f* and object size and image size are equal too. Finally, following Equation [1.3-13], if an image is inverted, its height, and therefore M_T, are negative. If the image is upright, its height, and M_T, are positive.

Axial magnification.   So far we have always assumed that object and image are both two-dimensional and that they lie in planes normal to the optic axis. But many objects also have *depth* (they are three-dimensional), and thus we should consider also the *axial* (longitudinal) *magnification*. Axial magnification, M_X, is defined as the ratio of a short length in the image, measured *along the axis*, to the conjugate length in the object:

$$M_X = \frac{\Delta s'}{\Delta s}$$
[1.3-19]

To see how axial magnification relates to transverse magnification, we take Gauss' lens equation, solve it for *f*, and using the notation in Figure 1.3-16, apply it first to the *head* of the arrow and then to its *foot:*

$$f = \frac{s_1 s_1'}{s_1 - s_1'} = \frac{s_2 s_2'}{s_2 - s_2'}$$

$$s_1 s_1'(s_2 - s_2') = s_2 s_2'(s_1 - s_1')$$

Multiplying and rearranging gives

$$s_1 s_2 s_2' - s_1 s_1' s_2 = s_1' s_2 s_2' - s_1 s_1' s_2'$$

$$(s_2' - s_1')(s_1 s_2) = (s_2 - s_1)(s_1' s_2')$$

and, since $s_2 - s_1 = \Delta s$ and $s_1' - s_2' = \Delta s'$,

$$\frac{s_2' - s_1'}{s_2 - s_1} = M_X = \frac{s_1' s_2'}{s_1 s_2} = M_{T1} M_{T2}$$

$$M_X = M_T^2$$
[1.3-20]

German physician and physicist. While still in medical school, von Helmholtz investigated heat production in animals and found the law of conservation of energy. After working for six years as an army surgeon, he became an instructor of physiology in Königsberg, professor of anatomy and physiology in Bonn, then professor of physics in Berlin. Von Helmholtz wrote two books, covering similar ground as Smith before him, *Handbuch der physiologischen Optik* (Leipzig: L. Voss, 1867) and *Die Lehre von den Tonempfindungen als physiologische Grundlage für die Theorie der Musik* (Braunschweig: F. Vieweg & Sohn, 1863). He invented the ophthalmoscope and the keratometer; extended Young's theory of color vision; contributed to diffraction theory, thermodynamics, fluid dynamics, and meteorology; and showed that the character of a musical tone depends on harmonics present in addition to the fundamental.

Joseph Louis Lagrange (1736–1813), French mathematician and astronomer. Lagrange is known for his work on the calculus of variations, which he summarized in his book *Analytical Mechanics,* Paris, 1788, worked on perturbation theory (investigating three celestial bodies, the sun, the earth, and the moon), introduced to optics the concept of conjugate elements. In 1793, after the French Revolution, Lagrange headed a commission to draw up a new system of weights and measures and, together with Pierre Simon Laplace and Adrien Marie Legendre, devised the metric system, now in use in most countries the world over.

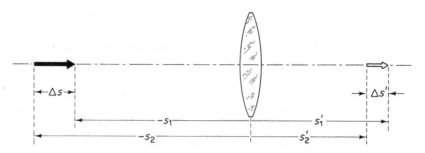

Figure 1.3-16 Deriving axial magnification.

which shows that axial magnification is the *square* of transverse magnification. Note also that axial magnification is always positive; if the object extends from left to right, the image likewise extends from left to right.

Example

The object is a transparent cube, each edge 4 mm in length, placed 60 cm in front of a lens of +20 cm focal length. Compare the transverse and axial magnifications.

Solution. From Gauss' equation, solved for s', we find for the rear surface of the cube (the face closer to the lens) that

$$s' = \frac{sf}{s+f} = \frac{(-60)(20)}{-60+20} = +30 \text{ cm}$$

For the front surface (the face farther away from the lens),

$$s'' = \frac{(-60.4)(20)}{-60.4+20} = +29.9 \text{ cm}$$

The *transverse* magnification for the rear surface is

$$M_T = \frac{+30}{-60} = \boxed{-0.5 \times}$$

But the axial magnification is

$$M_X = \frac{\Delta s'}{\Delta s} = \frac{30-29.9}{-60-(-60.4)} = \boxed{+0.25 \times}$$

which is the square of the transverse magnification.

The subject of *angular magnification* will be deferred to the next chapter because, instead of a single thin lens, it requires a *combination of lenses*.

SUGGESTIONS FOR FURTHER READING

W. H. A. Fincham and M. H. Freeman, *Optics,* 9th edition, pp. 61–109 (Woburn, MA: Butterworth Publishers, 1980).

D. D. Michaels, *Visual Optics and Refraction,* 2nd edition, pp. 30–53, 66–72 (St. Louis: The C. V. Mosby Company, 1980).

PROBLEMS

1.3-1. Light originates at a point light source. What is the vergence of the light at distances of:
(a) 12.5 mm?
(b) 20 cm?
(c) 1.6 m?

1.3-2. An object is placed 20 cm to the left of a solid rod ($n = 1.6$) with a convex surface facing the object. If an image is formed 40 cm to the right of the surface, inside the rod, what is the power of the surface?

1.3-3. An object is located 2 cm to the left of the convex end of a glass rod which has a radius of curvature of 1 cm. The index of refraction of the glass is $n = 1.5$. Find the image distance.

1.3-4. A glass sphere is used to form an image of a distant object. If the image is found to be located on the (back) surface of the sphere, what is the refractive index of the glass?

1.3-5. A thin negative meniscus has radii of curvature of $+50$ cm and $+25$ cm, respectively. If the refractive index of the glass is 1.5, and if air is to the left of the lens and oil ($n = 1.8$) to the right, what is the power of the meniscus?

1.3-6. A negative meniscus of index 1.5 and with radii of curvature of 50 cm and 25 cm is held horizontal and its concave side filled with a liquid of index 1.6. What is the power of the combination?

1.3-7. A -10.00-diopter biconcave lens has 12 cm radii of curvature for either surface. What is the refractive index of the glass?

1.3-8. A thin lens of refractive index 1.5 has a focal length of 20 cm. If the radius of curvature of the back surface is -8 cm, what is the radius of the front surface?

1.3-9. If a lens made of glass of $n = 1.5$ has a power of -10.00 diopters, what is the power of a lens of the same shape but made of plastic of $n = 1.4$?

1.3-10. A lens made of glass of $n = 1.5$ has, when in air, $+10.00$ diopters power. What is its power when immersed in a liquid of $n = 1.58$?

1.3-11. When an object is placed 75 mm in front of a converging lens, its image is three times as far away from the lens as when the object is at infinity. What is the focal length of the lens?

1.3-12. When an object is placed 15 cm in front of a thin lens, a virtual image is formed 5 cm away from the lens. What is the focal length of the lens?

1.3-13. An object is placed first at infinity and then 20 cm from the object-side focal plane of a converging lens. The two images thus formed are 8 mm apart from each other. What is the focal length of the lens?

1.3-14. An object is placed 30 cm from a lens having a focal length of $+10$ cm. Determine the image distance:
(a) By the Gaussian form of the lens equation.
(b) By the Newtonian form of the lens equation.

1.3-15. A parallel beam of light 16 mm in diameter is incident on a thin lens. Passing through the lens, the beam spreads out into a cone with a (total) apex angle of $90°$. What is the power of the lens?

1.3-16. Light is made to converge toward a certain point. If then a -6.00-diopter lens is placed 25 cm ahead of this point, where will the light converge, or appear to come from?

1.3-17. A ray of light originates at an axial point object located 25 cm to the left of a $+10.00$-diopter lens. If the ray before entering the lens subtends with the axis an angle of $5°$, what is the angle, neglecting the sign, when leaving the lens?

1.3-18. A $+5.00$-diopter lens forms a real image on a screen placed 90 cm away from the object. Find the two object distances possible. Solve by the vergence method.

1.3-19. An object is located 30 cm to the left of a lens of $+20$-cm focal length. A second lens (of negative focal length) is placed 10 cm to the right of the first lens, both lenses together forming an image 60 cm to the right of the second lens. What is the focal length of the second lens?

1.3-20. A lens forms a real image on a screen 67.5 cm away. If then another lens is added 42.5 cm to the right of the first lens, the screen must be moved 25 cm farther away for the image to be in focus. What is the power of the second lens?

1.3-21. An axial point object is viewed through a diverging lens located $1\frac{1}{2}$ times its (absolute) focal length from the object. Find the conjugate image by ray tracing.

1.3-22. Light from a point object, located 2 cm off the optic axis, falls on a thin lens of $+10.00$ diopters power. After refraction the light is still divergent, in such a way that the ray, which emerges from the object parallel to the optic axis, after refraction subtends an angle of $7°$ with the chief ray. What is the image distance?

1.3-23. An object is located 1.25 m in front of a screen. Determine the focal length of a lens that forms on the screen a real, inverted image magnified four times.

1.3-24. If an object is placed 50 cm in front of a -3.00-diopter lens, what is the transverse magnification of the image?

1.3-25. If the real image formed by a lens is twice as large as the object, and if the distance between object and image is 90 cm, what is the power of the lens?

1.3-26. If a lens placed 4 cm to the right of an object produces a virtual image twice as large as the object, what is the refractive power of the lens?

1.3-27. An object 12 mm high is located 80 cm to the left of a single surface of $+2.50$ diopters power ground on a glass cylinder of $n = 1.5$. Find:
(a) The image *distance* from Gauss' formula.
(b) The image *size* by Lagrange's theorem.

1.3-28. The image formed on a screen 60 cm from an object is 2 cm in size. If the lens forming the image is moved a certain distance, without moving either object or screen, another image comes into focus that is 8 cm in size.
(a) How large is the object?
(b) What is the power of the lens?

1.4

Thick Lenses and Lens Combinations

Now WE COME TO THE MORE REALISTIC aspects of lenses. Most often, a real lens is by necessity a *thick lens*. It is a combination of two surfaces separated by a certain thickness. Furthermore, many optical systems are *combinations of lenses*. An example is the system shown in Figure 1.4-1.

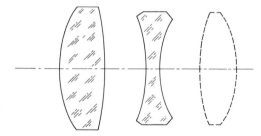

Figure 1.4-1 Introducing a combination of lenses.

THICK LENSES

Lenses are customarily divided into "thin lenses" and "thick lenses." But where does a thin lens end and a thick lens begin? That is entirely a matter of precision required for solving a given problem; the same lens may well be considered "thin" for a preliminary, and "thick" for a rigorous solution. Thin lenses, as we have seen in the preceding chapter, require no more than simple thin-lens equations. Thick lenses, of course, are more complex.

Assume, for example, that a bundle of light passes through a thick lens, or through a combination of thin or thick lenses. Then the problem arises from where to measure the focal length and the object distance and image distance. Probably we should not measure from the center of the lens because that might work only with lenses that are symmetrical, with equal powers on both sides. Instead, we measure these distances from certain hypothetical planes called *principal planes*. If we do this, we can continue using convenient thin-lens equations.

The principal planes of a lens, or a combination of lenses, are defined as the loci where we *assume* refraction to occur without reference as to where it actually does occur. Principal planes do not really exist within a lens; they are conceptual planes (actually they are slightly curved surfaces). In a symmetrical lens, the two principal planes, H_1 and H_2, straddle the center of the lens (such as in Figure 1.4-9). In a plano-convex lens, and even more so in a meniscus (as in Figure 1.4-4), the two **H** planes move to one side, and may in fact move outside the lens, toward the side of the steeper curvature. And, important to note, even a complex system with many lenses has only *two* principal planes. Their positions are invariant; they do not change with the object distance or image distance used in a given experiment.

As an example, consider three rays that come from an off-axis point object (on the left in Figure 1.4-2). The rays enter a "black box" containing the system. We do not know what is inside the black box but we find out that the system forms an image (on the right). The parallel ray, (1), as it emerges from the box, goes to the second focal point, F_2 (and continues). The focal ray, (2), goes through the first focal point, F_1, and emerges parallel to the axis. The point where the two rays intersect determines the position of the image. If we now continue the incident rays to the right, and backtrace the emergent rays to the left, we can locate the two principal planes, H_1 and H_2.

More specifically, the parallel ray is thought to be refracted at H_2 (and then passes through F_2). The focal ray is thought to be refracted at H_1 (and made parallel to the axis).

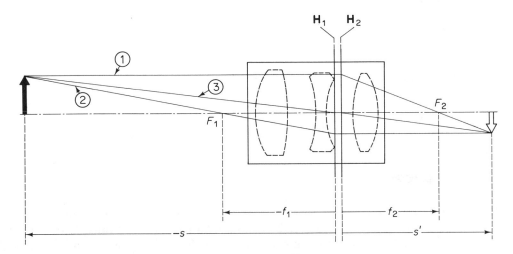

Figure 1.4-2 Principal planes, focal lengths, and object and image distances relative to a complex lens system.

The chief ray, (3), goes to the first *principal point* (the point where the axis intersects $\mathbf{H}_1$), is translated (shifted) along the axis to the second principal point, and continues from there parallel to its initial direction. In short: *The first focal length, f_1, and the object distance, s, are measured from $\mathbf{H}_1$, and the second focal length, f_2, and the image distance, s', are measured from $\mathbf{H}_2$.*

The concept of equivalent power.   How can we find the principal planes? This can be done in two ways: analytically (as I will show next) and experimentally, by the nodal slide method (as I will show later). In Figure 1.4-3 we have a system composed of two thick lenses. Each of these lenses has two principal planes, H_{11} and H_{12} for the left-hand lens, and H_{21} and H_{22} for the right-hand lens. The system as a whole also has two principal planes, $\mathbf{H}_1$ and $\mathbf{H}_2$, but only $\mathbf{H}_2$ is shown.* The second focal point *of the system* is $\mathbf{F}_2$.

We start with Gauss' formula for a single surface,

$$\frac{n}{s} + \frac{n' - n}{R} = \frac{n'}{s'}$$

We replace the second term by the surface power equation,

$$\frac{n' - n}{R} = P$$

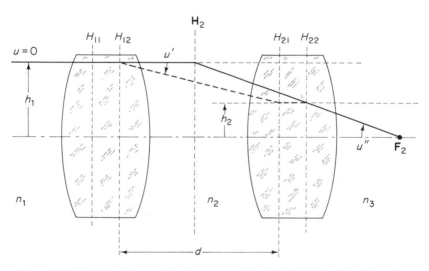

Figure 1.4-3   Deriving the equivalent-power equation. Note that H_{11}, H_{12}, H_{21}, and H_{22} are the principal planes of the individual lenses; $\mathbf{H}_2$ pertains to the *system*.

*Oddly enough, with a combination of *two* lenses, in contrast to a single thick lens, the sequence of the $\mathbf{H}$ planes is *reversed:* $\mathbf{H}_2$ is to the left of the center and $\mathbf{H}_1$ is to the right (as we also see in Figure 1.4-6).

We multiply both sides by h_1 and rearrange; that gives

$$\frac{n'h_1}{s'} - \frac{nh_1}{s} = Ph_1 \qquad [1.4\text{-}1]$$

By definition, and following Equations [1.3-3], the slope angles before and after refraction are

$$u = \frac{h_1}{s} \qquad \text{and} \qquad u' = \frac{h_1}{s'}$$

Substituting these terms in Equation [1.4-1] and setting $u = 0$ yields

$$n'u' = P_1 h_1 \qquad [1.4\text{-}2]$$

We repeat the process for the second lens:

$$n''u'' - n'u' = P_2 h_2 \qquad [1.4\text{-}3]$$

Adding Equations [1.4-2] and [1.4-3] gives

$$n''u'' = P_1 h_1 + P_2 h_2 \qquad [1.4\text{-}4]$$

Now consider both lenses together. The focal length, **f**, *of the system* is measured from $\mathbf{H}_2$. Since the height of the ray when incident on $\mathbf{H}_2$ is h_1,

$$n''u'' = \mathbf{P}h_1 \qquad [1.4\text{-}5]$$

where **P** is the power of the system. Substituting Equation [1.4-5] into [1.4-4] gives

$$\mathbf{P}h_1 = P_1 h_1 + P_2 h_2 \qquad [1.4\text{-}6]$$

From Figure 1.4-3 we also see that

$$u' = \frac{h_1 - h_2}{d}$$

Thus

$$h_2 - h_1 = -u'd$$

Substituting for u' Equation [1.4-2] gives

$$h_2 - h_1 = -\frac{P_1 h_1 d}{n'}$$

Solving this expression for h_2 and substituting in Equation [1.4-6] yields

$$\mathbf{P}h_1 = P_1 h_1 + P_2 h_1 - \frac{P_1 P_2 h_1 d}{n'}$$

and canceling h_1

$$\boxed{\mathbf{P}_{\text{equivalent}} = P_1 + P_2 - P_1 P_2 \frac{d}{n'}} \qquad [1.4\text{-}7]$$

which is *Gullstrand's equation for the equivalent power of the system.**

Converting the powers into focal length, $P \to 1/f$, and multiplying the converted second and third term by $f_1 f_2 / f_1 f_2$ leads to

$$\mathbf{f} = \frac{f_1 f_2}{f_1 + f_2 - d/n'} \qquad [1.4\text{-}8]$$

which is the *equivalent focal length of the system.* Note that n' in both Equations [1.4-7] and [1.4-8] is the refractive index of the medium *between* the lenses; it is *not* the refractive index of the glass.

If the two lenses are in air, $n' \approx 1$ and

$$\mathbf{P} = P_1 + P_2 - P_1 P_2 d \qquad [1.4\text{-}9]$$

and if the two lenses are in contact, $d = 0$ and

$$\mathbf{P} = P_1 + P_2 \qquad [1.4\text{-}10]$$

which is the *approximate power* of the combination. Table 1.4-1 summarizes these findings.

TABLE 1.4-1 EQUIVALENT (AND APPROXIMATE) POWERS AND FOCAL LENGTHS

Description	Power	Focal length
Two surfaces of power P_1 and P_2, separated by distance d, and enclosing medium of index n	$\mathbf{P} = P_1 + P_2 - P_1 P_2 \dfrac{d}{n}$	$\mathbf{f} = \dfrac{f_1 f_2}{f_1 + f_2 - d/n}$
Two *lenses*, of power P_1 and P_2, separated by distance d	$\mathbf{P} = P_1 + P_2 - P_1 P_2 d$	$\mathbf{f} = \dfrac{f_1 f_2}{f_1 + f_2 - d}$
Two elements, of power P_1 and P_2, in contact	$\mathbf{P} = P_1 + P_2$	$\mathbf{f} = \dfrac{f_1 f_2}{f_1 + f_2}$

Front vertex power and back vertex power. Equivalent power is measured *at* the respective principal plane. This statement makes more sense when we realize that the reciprocal of equivalent power is the equivalent focal length, and this focal length is measured *from* the respective principal plane. Unfortunately, we often do not know where the principal planes are, and when we do know, they often are inaccessible within the lens, or lens system. What we need, then, are clearly defined and easily accessible reference points

---

*Allvar Gullstrand (1862–1930), Swedish ophthalmologist and professor of physiological and physical optics at the University of Uppsala. Gullstrand is best known for the *schematic eye* and for his development of the slit lamp and the ophthalmoscope. He also contributed to the tracing of skew rays, caustic surfaces (in aberrations), accommodation, aspheric lenses for aphakics, and to the propagation of light through gradient-index media (such as the crystalline lens of the eye). In 1911 he received the Nobel Prize in physiology for his "work on the diffraction of light by lenses applied to the eye" and, a few years later, published a detailed exposition of geometric-optical image formation, replete with numerous equations, "Das allgemeine optische Abbildungssystem," *K. Sven. vetenskapsakad. handl* (4) **55** (1915-1916), 1–139.

from which to measure. These reference points are the *vertices* of the lens. In Figure 1.4-4, for example, v_1 is the *front vertex focal length* and v_2 is the *back vertex focal length*. The reciprocals of these focal lengths are the *front vertex power*, P_{FV}, and the *back vertex power*, P_{BV}.

Be sure to distinguish between back *surface* power and back *vertex* power. Back surface power is simply the power of the back surface. But back vertex power refers to the power of the whole lens (front surface plus back surface), *as measured at the back vertex*.

The two focal lengths v_1 and v_2 can be determined, very conveniently, by autocollimation. Use an illuminated slit or other target in a screen to the left of the lens and a plane mirror to the right. Move the lens back and forth and align the mirror until the image of the slit appears in focus on the screen, in the same plane as the slit. Then the screen is at F_1 and it is easy to measure v_1. Next turn the lens around so that its right-hand side faces the target, giving v_2.

With a lens of appreciable thickness, or with two lenses separated by an appreciable distance, we can no longer use the approximate power equation, $\mathbf{P} = P_1 + P_2$. At the front vertex, the first power term, P_1, would be measured correctly but, because of the thickness or separation, the second power term, P_2, would not be. Therefore, P_2 in Equation [1.4-10] has to change.

To find out how P_2 has to change, consider the vergence in two planes, 1 and 2 in Figure 1.4-5. In plane 1 the vergence is

$$\mathbf{V}_1 = \frac{n}{L}$$

In plane 2 it is

$$\mathbf{V}_2 = \frac{n}{L - d}$$

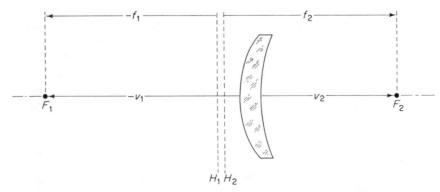

Figure 1.4-4 Conventional focal lengths, f_1 and f_2, and front vertex and back vertex focal lengths, v_1 and v_2, of a thick positive meniscus.

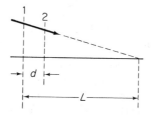

Figure 1.4-5   Change of vergence.

Solving for L in the first equation, substituting in the second equation, and multiplying both numerator and denominator by $\mathbf{V}_1/n$ gives

$$\mathbf{V}_2 = \frac{n}{(n/\mathbf{V}_1) - d} = \frac{n(\mathbf{V}_1/n)}{(n/\mathbf{V}_1 - d)(\mathbf{V}_1/n)}$$

from where

$$\boxed{\mathbf{V}_2 = \frac{\mathbf{V}_1}{1 - \mathbf{V}_1(d/n)}}$$

[1.4-11]

which is the *change-of-vergence equation.*

Now replace the vergences by powers and substitute for P_2 in the approximate-power equation. This gives for the front vertex power

$$\boxed{\mathbf{P}_{FV} = P_1 + \frac{P_2}{1 - P_2(d/n)}}$$

[1.4-12]

which is approximately equal to

$$\mathbf{P}_{FV} \approx P_1 + P_2 + P_2^2 \frac{d}{n}$$

[1.4-13]

Back vertex power or, as it is sometimes called, "effective power," is the most important of all. Virtually all ophthalmic procedures, from determining visual acuity to writing lens prescriptions to "verifying" a lens using a lensometer, refer to the back vertex power. Similar to the derivation of Equation [1.4-12] we now replace the P_1 term; this gives

$$\boxed{\mathbf{P}_{BV} = \frac{P_1}{1 - P_1(d/n)} + P_2}$$

[1.4-14]

or, approximately,

$$\mathbf{P}_{BV} \approx P_1 + P_2 + P_1^2 \frac{d}{n}$$

[1.4-15]

Example 1

What happens if two positive lenses of equal power are first in contact and then are gradually pulled apart from each other?

Solution.   With the lenses in contact, the power of the combination is highest, the focal length least, and the two **H** planes are close together inside the system (Figure 1.4-6a). Then, as we separate the lenses, the **H** planes separate too. For instance, when the distance between the lenses is equal to their focal length, the **H** planes coincide with the lenses (b). When the distance is larger than the focal length, the **H** planes have moved outside the lenses (c); and when the distance is equal to $2f$ and the system becomes *afocal* (zero power), $\mathbf{H}_1$ and $\mathbf{H}_2$ have moved out to infinity (d). Predictably, as we separate the lenses even farther, the power of the system becomes less than zero; that is, it becomes *negative*. Note that the ray that emerges on the right-

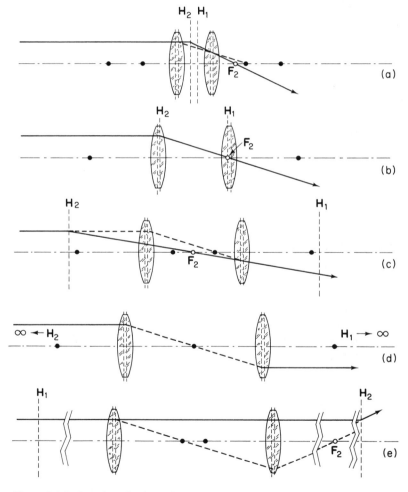

Figure 1.4-6   Location of principal planes of a system of two lenses as a function of distance between the lenses. Focal points of individual lenses are identified by dots, second focal points of system, $\mathbf{F}_2$ by hollow circles.

hand side has continued to turn counterclockwise and now, in (e), is pointing *upward,* apparently refracted at H_2, as it should.

The question of whether a system has positive or negative power is easy to answer: If the emergent ray, after refraction at H_2, points toward the axis and crosses it at F_2 [(a), (b), and (c)], the system is positive. If the ray turns away from the axis, without crossing it (e), the system is negative.

Assume, for example, that the two lenses have $+8.00$ diopters power each and that they are separated by a distance d. At which distance does the power of the combination turn from positive to negative?

Numerical solution.   We take the equivalent-power equation and set it equal to zero:

$$(+8.00) + (+8.00) - (+8.00)(+8.00)(d) = 0$$

Then

$$16 = 64d$$

and

$$d = \frac{16}{64} = 0.25m = \boxed{25 \text{ cm}}$$

Since $1/8.00 = 12.5$ cm, this is twice the focal length of either lens; at that distance the system is *afocal.*

Example 2

Two plano-convex lenses of $+5.00$ diopters power each, their plane sides facing each other, are separated by a distance of 8 cm.

(a) Determine the positions of the principal planes.

(b) Let the space between the lenses be filled with glass of $n = 1.534$, that is, consider a rather thick lens. Again find the principal planes.

Solution.   (a) First, using Equation [1.4-14], determine the *back vertex power:*

$$\mathbf{P}_{BV} = \frac{P_1}{1 - P_1(d/n)} + P_2 = \frac{+5}{1 - (+5)(0.08)} + (+5) = +13.33 \text{ diopters}$$

The reciprocal (multiplied by 100) is the *back vertex focal length:*

$$\mathbf{v}_2 = \left(\frac{1}{+13.33}\right)(100) = +7.5 \text{ cm}$$

Then, using Equation [1.4-7], determine the *equivalent power:*

$$\mathbf{P} = P_1 + P_2 - P_1 P_2 \frac{d}{n} = (+5) + (+5) - (+5)(+5)(0.08) = +8 \text{ diopters}$$

The reciprocal (multiplied by 100) is the *equivalent focal length:*

$$\mathbf{f} = \left(\frac{1}{+8}\right)(100) = +12.5 \text{ cm}$$

Since $\mathbf{v}_2$ is measured from the last surface, and $\mathbf{f}$ from H_2, the difference between the two,

$$12.5 - 7.5 = \boxed{5 \text{ cm}}$$

gives the location of $\mathbf{H}_2$: 5 cm to the left of the last lens, or 1 cm *to the left* of the midpoint between the two lenses.

(b) With the space between the lenses filled with glass, the back vertex power is

$$\mathbf{P}_{BV} = \frac{+5}{1 - (+5)(0.08/1.534)} + (+5) = +11.76 \text{ diopters}$$

the back vertex focal length,

$$\mathbf{v}_2 = +8.5 \text{ cm}$$

the equivalent power,

$$\mathbf{P} = (+5) + (+5) - (+5)(+5)\left(\frac{0.08}{1.534}\right) = +8.70 \text{ diopters}$$

and the equivalent focal length,

$$\mathbf{f} = +11.5 \text{ cm}$$

This time,

$$11.5 - 8.5 = \boxed{3 \text{ cm}}$$

which means that $\mathbf{H}_2$ is located 3 cm to the left of the last lens, or 1 cm *to the right* of the midpoint.

ANGULAR MAGNIFICATION

Before we come to *angular magnification,* consider the concept of *angular size.* Initially the object may be far away. The image formed on the retina is small. Then, as the object is brought closer, the image will grow larger. But there is a limit to how close the object can be brought and still remain in focus. This limit is a matter of age, but on the average, the *distance of most distinct vision* is 25 cm (Figure 1.4-7, top).

But there is a way to bring the object closer (and hence see it larger) and *still keep it in focus.* This is possible by placing a converging lens in front of the eye (Figure 1.4-7, center). A lens used that way is called a *magnifier.*

To find out how much the lens magnifies, again we make a *comparison.* Without the lens, the tangent of the angle subtended at the eye by the object is

$$\tan \theta = \frac{y}{-25} \qquad [1.4\text{-}16]$$

With the lens, the tangent is

$$\tan \theta' = \frac{y}{-f} \qquad [1.4\text{-}17]$$

The angular magnification (sometimes called "magnifying power"), therefore, is the ratio of these two tangents:

$$M_\theta = \frac{\tan \theta'}{\tan \theta} \qquad [1.4\text{-}18]$$

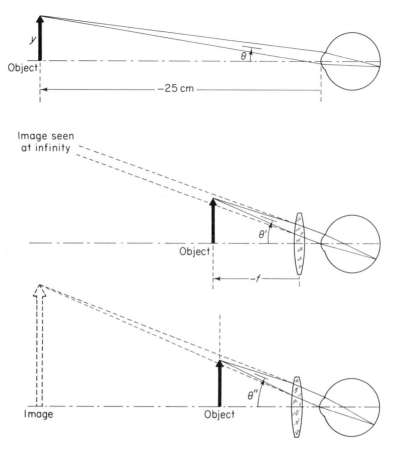

Figure 1.4-7   (*From top to bottom*) Object at the distance of most distinct vision, as viewed through magnifier and seen at infinity, and as viewed through magnifier and seen at −25 cm.

We insert Equations [1.4-17] and [1.4-16] in [1.4-18]; that gives

$$M_\theta = \frac{y/-f}{y/-25} = \frac{25}{f}$$ [1.4-19]

with f given in centimeters. If instead of the focal length of the lens, we use its *power,* we have

$$\boxed{M_\theta = \tfrac{1}{4}P}$$ [1.4-20]

This is known as the "quarter-power equation." It applies when the magnifying lens is held close to the eye and the image is seen at infinity, *without accommodation.*

Next, let the observer *accommodate.* Accommodation means that the image is seen at the distance of most distinct vision, −25 cm (Figure 1.4-7, bottom). The light entering the eye is now divergent, the same as when reading, and the object can be brought even closer, to within the focal length of the lens. This makes the object distance

$$s = \frac{s'f}{f - s'} = \frac{(-25)(f)}{f - (-25)}$$

so that, instead of $\tan \theta' = y/-f$, we have

$$\tan \theta'' = \frac{(y)(f + 25)}{-25f}$$

Inserting this expression in Equation [1.4-18] gives

$$M_\theta = \frac{y(f + 25)/-25f}{y/-25} = \frac{25}{f} + 1 \qquad [1.4\text{-}21]$$

or

$$\boxed{M_\theta = \tfrac{1}{4}P + 1} \qquad [1.4\text{-}22]$$

In this case the image is seen *with accommodation*.

Third, we consider that the magnifying lens is held some distance away from the eye, rather than next to it as we had assumed so far. That is a more complex problem, which is probably best solved by the vergence method.

As shown in Figure 1.4-8, the object is located at a distance a to the left of the lens, and the eye at a distance b to the right of the lens. The image, again, is seen at the distance of most distinct vision, $s' = -25$ cm.

Now consider the various vergences. At the lens, the entrance vergence is

$$\mathbf{V}_L = \frac{1}{a}$$

and the exit vergence, from Equation [1.3-7], is

$$\mathbf{V}'_L = \frac{1}{a} + P$$

At the eye, the entrance vergence is found from the change-of-vergence equation,

$$\mathbf{V}_E = \frac{P + 1/a}{1 - (P + 1/a)(b)} \qquad [1.4\text{-}23]$$

This vergence, no matter what the exit vergence $\mathbf{V}'_E$ *inside* the eye, in order to see an image at -25 cm, must always be

$$\mathbf{V}_E = \frac{1}{-0.25} = -4 \text{ diopters}$$

Finally, we consider the magnification. While generally, from Equation [1.3-17],

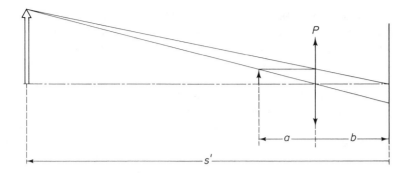

Figure 1.4-8 Deriving angular magnification with a lens held some distance in front of the eye.

$$M = \frac{\mathbf{V}}{\mathbf{V}'}$$

the total magnification (of the retinal image) is the product of the individual magnifications,

$$M_R = M_L M_E$$

where the subscript R refers to the retinal image, L to the lens, and E to the eye. We replace the two right-hand terms by the vergences just derived,

$$M_R = \frac{\mathbf{V}_L}{\mathbf{V}'_L} \frac{\mathbf{V}_E}{\mathbf{V}'_E}$$

$$= \frac{1/a}{P + 1/a} \frac{P + 1/a}{1 - (P + 1/a)(b)} \frac{1}{\mathbf{V}'_E}$$

$$= \frac{1/a}{1 - (P + 1/a)(b)} \frac{1}{\mathbf{V}'_E} \qquad [1.4\text{-}24]$$

If the object were seen without the lens, at the distance of most distinct vision, the magnification would be

$$M_E = \frac{\mathbf{V}_E}{\mathbf{V}'_E} = \frac{1/-0.25}{\mathbf{V}'_E} \qquad [1.4\text{-}25]$$

But if we compare the size of the image seen with the lens to the size of the image seen without the lens, we need to divide Equation [1.4-24] by [1.4-25]; that results in

$$M_\theta = \frac{\dfrac{1/a}{1 - (P + 1/a)(b)} \dfrac{1}{\cancel{\mathbf{V}'_E}}}{\dfrac{1/-0.25}{\cancel{\mathbf{V}'_E}}}$$

$$= (-0.25) \frac{1}{a - (aP + 1)(b)} \qquad [1.4\text{-}26]$$

where, needless to say, all distances are measured in meters.

Nodal points.   We have seen that the principal planes are where all refraction is assumed to occur. In contrast, the nodal points are where *no refraction occurs.* In a thin lens the nodal point is the center of the lens: light passing through the center does so without refraction. In a thick lens, the same as with the principal planes, the center separates into *two nodal points.*

Whenever the refractive indices on either side of the lens are the same, the nodal points, N_1 and N_2, coincide with the principal points, H_1 and H_2. This is shown in Figure 1.4-9. (If the refractive indices are different, the N points would move away from the H planes, toward the side of the *higher* index.) Thus, from knowing the locations of the N points we find the H planes (and these we like to know because from there the focal lengths and the object and image distances are measured). We find the N points using the *nodal slide method.*

A nodal slide is a particular kind of a lens support that can be swiveled back and forth about a vertical axis. While doing so, the image formed by the lens or lens system is watched for any sideways motion. Use collimated light and focus the light by the system on a ground-glass screen. Cautiously move the lens holders *on top of* the nodal slide in one direction and move the nodal slide *as a whole* an equal distance in the opposite direction. *Keep the image in focus all the time!* At the position where, on rotating the nodal slide, there is *no sideways motion of the image,* the axis of rotation goes through the second nodal point, N_2.

Measure the distance from the axis of rotation to the ground glass, that is, the distance from N_2 to F_2. With the same medium (air) on either side of the system, the nodal points coincide with the principal points. The distance from N_2 to F_2, therefore, is the *equivalent focal length,* **f** (of the system).

Determining Focal Length.   There are several ways of determining the focal length of a lens or combination of lenses.

1. Measure the object distance and the image distance, and calculate the (equivalent) focal length using Gauss' *thin-lens equation,*

$$f = \frac{ss'}{s - s'} \qquad\qquad [1.4\text{-}27]$$

2. *Bessel's method.* Make the distance between object and image longer than four times the estimated focal length of the lens. Slowly move the lens back and forth along the axis; you will find *two*

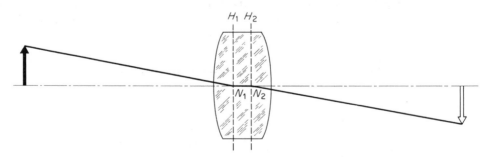

Figure 1.4-9   Nodal points. Note displacement, but no deviation, of the chief ray.

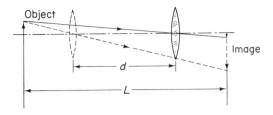

Figure 1.4-10   Bessel's method.

positions where the image is in focus. In one position the image is magnified and in the other it is reduced in size. As shown in Figure 1.4-10, L is the distance between object and image,

$$L = -s + s'$$

Solving for s and substituting for s' the thin-lens equation gives

$$s = \frac{sf}{s + f} - L$$

$$s(s + f) = sf - L(s + f)$$

$$s^2 + Ls + Lf = 0$$

Solving for s yields

$$s_1 = -\frac{L}{2} + \frac{1}{2}\sqrt{L^2 - 4Lf}$$

$$s_2 = -\frac{L}{2} - \frac{1}{2}\sqrt{L^2 - 4Lf}$$

We subtract s_2 from s_1; that gives d, the distance between the two positions:

$$s_1 - s_2 = d = \sqrt{L^2 - 4Lf}$$

$$d^2 = L^2 - 4Lf$$

and hence

$$f = \frac{L^2 - d^2}{4L} \qquad [1.4\text{-}28]$$

The advantage of this method is that only L and d, rather than the object distance and image distance, need be measured.

    3. *Autocollimation.* (Collimation means placing a source or target in the left-hand focal plane of a converging lens so as to obtain parallel light. Autocollimation means that the light, in addition, is reflected back along the same path and comes to a focus in the plane of the source. In a way, collimation is one-half of autocollimation.)

    Place a light source with an object–image screen attached to it near the left-hand end of the optical bench. At some distance from the source mount the lens to be measured and, next to it to the right, a plane mirror. Slide the lens, together with the mirror, back and forth along the axis until the image of the target is in focus on the screen. The distance between the lens and the screen is the focal length of the lens because light emerging parallel from the lens will be reflected parallel and form an image in the focal plane (Figure 1.4-11). This method is well suited for the student laboratory.

    4. *Abbe's method.* Form a real image on the right. Record the position of the object. Measure object size and image size. Determine the magnification, M_1. Then, leaving the lens in place, move the

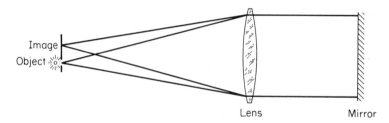

Figure 1.4-11   Autocollimation.

object a short distance A to the right (which makes A positive) and focus again, moving the image screen only. Determine the new magnification, M_2. The focal length is found from Abbe's equation:

$$f = \frac{AM_1M_2}{M_1 - M_2}$$

[1.4-29]

This method is preferred in industrial laboratories; it requires a microscope for the precise determination of the image size.

5. The focal length of a *diverging lens* cannot be determined directly as it can with a converging lens. Therefore, another approach is needed. In the *virtual-object method,* for example, we combine the minus lens with a sufficiently strong plus lens, but the two lenses need not be in contact. First form an image using the plus lens only. Note where the image is in focus; this is screen position 1 (Figure 1.4-12).

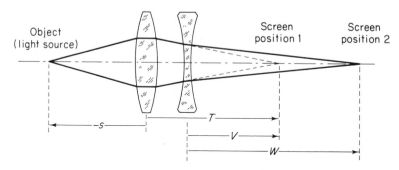

Figure 1.4-12   Determining the focal length of a diverging lens.

Then insert the minus lens. Move the screen to the right to bring the image back into focus (screen position 2). The earlier image (position 1) now acts as a virtual object for the minus lens. The image distance T, less the separation of the two lenses, therefore, becomes the object distance V. The focal length of the minus lens then follows from

$$f = \frac{VW}{V - W}$$

[1.4-30]

6. A *lens clock* actually measures the curvature of a surface, but for a given refractive index, it can be calibrated to show the surface *power.* The clock has two outer points a distance D apart, and a movable inner point. When the three points touch a plane surface, they form a straight line; on a

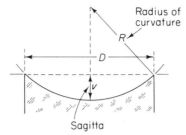

Figure 1.4-13   Deriving the vertex depth formula.

curved surface, by contrast, the center point either protrudes, or is pushed in, by a certain distance v, called the *vertex depth* or *sagitta* (Figure 1.4-13). From Pythagoras' theorem

$$(R - v)^2 + \left(\frac{D}{2}\right)^2 = R^2 \qquad [1.4\text{-}31]$$

$$2Rv = v^2 + \left(\frac{D}{2}\right)^2$$

Since $v \ll R$, v^2 may be neglected and

$$2Rv \approx \left(\frac{D}{2}\right)^2 \qquad [1.4\text{-}32]$$

Solving for R and substituting in the surface power equation then leads to

$$P \approx \frac{2(n' - n)v}{(D/2)^2}$$

which shows that, D being constant, P is directly proportional to v.

7. A *lensometer* is used to measure the power, and other characteristics, of a lens. A lensometer consists of a light source, a target that can be moved back and forth along the axis, a stationary lens, and a telescope focused for both object and image at infinity. If there is no other lens present in the path, the target is seen in focus (Figure 1.4-14, top).

However, if another lens is placed between stationary lens and telescope (usually at a location 50 mm from the stationary lens, at its right-hand focal point), the light entering the telescope is not parallel any longer, and the target will not be seen in focus (center). To make the light parallel again, the light to the left of a plus test lens must be divergent, and the light to the left of the stationary lens must be even more divergent than it was before; hence, the target must be moved to the *right* (bottom).

The opposite holds true for a minus lens. This time the target must be moved to the *left*, through a distance that is in essence Newton's object distance, N. Similarly, the distance that extends to the right from the test lens is Newton's image distance, N'. We then use Newton's equation,

$$NN' = f_1 f_2$$

and convert the focal lengths into power,

$$\frac{1}{N}\frac{1}{N'} = P^2$$

To see the target in focus, the light emerging from the test lens must be parallel and, hence,

$$N' = f_T$$

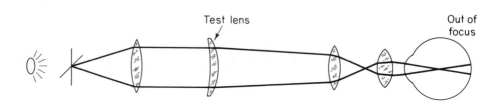

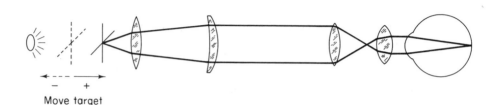

Figure 1.4-14   Ray diagram of a lensometer. (*Top*) Stationary lens, with target at F_1, produces parallel light. Telescope is focused for infinity. Target is seen in focus. (*Center*) Adding unknown lens (plus lens in the example shown) will defocus image. (*Bottom*) Moving target brings image back into focus.

Substituting this in the preceding equation and setting $1/f_T = P_T$ gives

$$P_T = NP_S^2 \qquad [1.4\text{-}33]$$

where, again, N is the distance through which the target must be moved to see it in focus.

8. In recent years, fully *automated lensometers* have become available. The lens is simply inserted into the instrument and within seconds the built-in optical system and subsequent microprocessor produce the result, often in the form of both a numerical display and a printout.

SUGGESTIONS FOR FURTHER READING

R. S. Longhurst, *Geometrical and Physical Optics,* 3rd edition (New York: Longman, Inc., 1974).

M. Katz, "The Human Eye as an Optical System," in T. D. Duane, *Clinical Ophthalmology,* Vol. 1, Chap. 33, pp. 1–52 (Philadelphia: Harper & Row, Publishers, Inc., 1982).

E. B. Mehr and A. N. Freid, *Low Vision Care* (New York: The Professional Press, 1975).

D. F. Horne, *Optical Production Technology,* 2nd edition (Bristol, England: Adam Hilger Ltd., 1983).

PROBLEMS

1.4-1. In the lens shown in Figure 1.4-15, find the object point conjugate to image point S'.

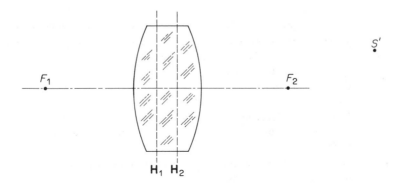

Figure 1.4-15

1.4-2. Two plano-convex lenses are placed a short distance apart, their plane sides facing each other. Trace two rays, coming from an off-axis pixel located in the left-hand focal plane of the first lens. Assume refraction at the principal plane only and find the conjugate pixel. Directly below draw the same *outer* surfaces with the space between the lenses filled with glass, that is, consider a rather thick biconvex lens. Trace two similar rays.

1.4-3. Draw the principal planes inside a thick equiconcave lens and show how a parallel ray, a focal ray, and a chief ray, all coming from the tip of an extended object, are (assumed to be) refracted at these planes.

1.4-4. Two lenses of equal power but opposite sign are mounted coaxially a short distance apart from each other.
  (a) If parallel light passes first through the positive lens and then through the negative lens, what will be the equivalent power of the combination?
  (b) What will it be if the light passes first through the negative lens?

1.4-5. Two lenses of −10.00 diopters power each are separated by a distance of 5 cm. What is the equivalent focal length of the system?

1.4-6. What must be the minimum separation of two thin lenses of +6.00 and −4.00 diopters power, respectively, so that the combination of both has an equivalent power of +5.00 diopters?

1.4-7. A light collector is often a combination of two identical plano-convex lenses, their focal length chosen so as to be slightly less than twice the desired equivalent focal length of the system. If the latter is to be 50 mm, then probably $f = 90$ mm is a suitable figure for the former. How far apart from each other should the lenses be placed?

1.4-8. Two thin lenses have a combined (approximate) power of +10.00 diopters. If they become separated by 20 cm, their equivalent power decreases to +6.25 diopters. What are the powers of the two lenses?

1.4-9. Two thin lenses in air have powers of +5.00 diopters and −10.00 diopters, respectively. If the combination is to be afocal, what must be the separation of the lenses?

1.4-10. If the first lens of a system of two has a power of +7.5 diopters, if the distance between the two lenses is 10 cm, and if the system is to be afocal, what is the power of the second lens?

1.4-11. An equiconvex lens has an index of refraction of 1.5 and radii of curvature of 10 cm. How thick a lens would have an infinite focal length?

1.4-12. Two positive lenses are placed 30 cm apart. If the second lens has twice the power of the first lens, and the system is afocal, what are the powers of the two lenses?

1.4-13. A 10.6-mm-thick lens has $+16.00$ diopters front surface power and -6.00 diopters back surface power. If the lens is made of spectacle crown ($n = 1.523$), what is its back *vertex* power?

1.4-14. Two thin lenses of $+5.00$ diopters each are separated by a distance of $6\frac{2}{3}$ cm. If the distance from the second lens to the focus is 10 cm, where is the second principal plane of the system?

1.4-15. Two plano-convex lenses of $+4.00$ diopters power each are placed coaxially and 6 cm apart, their plane sides facing each other. Determine:
   (a) The back vertex power.
   (b) The equivalent power.
   (c) The location of the second principal plane.

1.4-16. Continue with Problem 1.4-15 and now let the space between the lenses be filled with water ($n = \frac{4}{3}$). Determine the location of the second principal plane and compare the result with Problem 1.4-15c.

1.4-17. When a thin lens, held close to the nonaccommodating eye, magnifies 3 times, what is the refractive power of the lens?

1.4-18. If a $+8.50$-diopter magnifying glass is held close to the eye and the image is seen at the distance of most distinct vision:
   (a) What is the magnification?
   (b) How far from the lens is the object?

1.4-19. An object is held 4 cm in front of a $+20.00$-diopter lens and the lens 5 cm in front of the eye. With the image seen at the distance of most distinct vision, what is the angular magnification?

1.4-20. A $+50.00$-diopter lens is held 4 cm in front of the eye and an object is placed in front of the lens. As the object is moved back and forth along the axis, a distance is found where the image is seen to expand rapidly, each pixel filling the whole lens. At that distance, the angular magnification is infinite. What is this distance?

1.4-21. Nodal points are where rays, *without refraction,* intersect the optic axis. Using the same definition, find the nodal point of a single spherical refracting surface.

1.4-22. Two $+6.00$-diopter lenses are placed a short distance apart from each other. The nodal points of the system, therefore, are located symmetrically between the lenses. If then one of the $+6.00$ lenses is replaced by a $+10.00$ lens, how do the nodal points change?

1.4-23. The distance between object and screen is 100 cm. If a lens produces an image on the screen when placed at either of two positions 38 cm apart, what is the power of the lens?

1.4-24. Using Abbe's method, find the focal length of a lens from the following data: initial position of the target on the bench 12.0 cm; subsequent position 17.0 cm; target size 12 mm; image size first 6 mm, then 8 mm.

1.4-25. The outer points of a lens clock are 20 mm apart. If the center point is pushed in by 0.8 mm, a reading of $+9.25$ diopters results. What is the refractive index the lens clock is calibrated for?

1.4-26. If the stationary lens in a lensometer has a focal length of 50 mm, and the target must be moved 8 mm away from the observer to see it in focus, what is the power of the test lens?

Mirrors

IN MANY WAYS, MIRRORS AND LENSES are alike. Often, one can be replaced by the other. The only difference between them is that a mirror *returns* the light while a lens allows it to go on.

SPHERICAL MIRRORS

Perhaps, when dealing with mirrors, we foresee some difficulties when we apply the Cartesian sign convention because both a real object and a real image lie on the same side, in front of the mirror. At least, as before, the distance of a real *object* is negative. With an *image*, we note that axial *distances are measured in the direction of propagation of the light.* Hence, if the image is formed along the return path, to the left of the mirror, the image is *real* and the image distance is *positive.* If the image is formed to the right of the mirror, the image is virtual and the image distance is negative.

The same as with a single refracting surface, the radius of curvature of a mirror concave to the left is *negative* (Figure 1.5-1). Such a mirror acts like a converging lens and its focal length and power are *positive.* A convex mirror, by contrast, acts like a diverging lens and its focal length and power are negative.

Assume now that a real object is placed some distance to the left of a concave mirror. If the object distance chosen is longer than the focal length, the mirror, the same as a converging lens, forms a real image. As the light travels from the object to the mirror, it advances in the $+x$ direction and the velocity of the light, x/t, is positive:

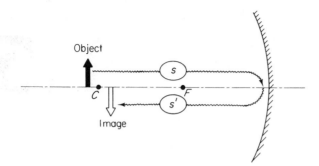

Figure 1.5-1   Concave mirror forming real image.

$$\frac{+x}{+t} = +v$$

If the light travels in the $-x$ direction, then, because time can only go forward ($+t$, rather than $-t$), the velocity of the light is negative:

$$\frac{-x'}{+t} = -v'$$

By definition, $v/v' = n'/n$, and therefore the index after reflection is negative also:

$$n' = -n$$

All that is needed, then, is to substitute $-n$ for n' in the surface power equation:

$$P = \frac{n' - n}{R} = \frac{(-n) - n}{R} = \frac{-2n}{R} \qquad [1.5\text{-}1]$$

where P is now the *reflective power* of the mirror, R is its radius of curvature, and n is the refractive index of the medium in front of the mirror. We substitute $P = n/f$, cancel n, and combine the result with Gauss' thin-lens equation solved for $1/f$. This gives

$$\boxed{\frac{1}{f} = \frac{1}{s'} - \frac{1}{s} = \frac{-2}{R}} \qquad [1.5\text{-}2]$$

which is the *spherical-mirror equation*. Although I have derived this equation for a concave mirror and for a real object and real image, it holds as well for any mirror and for any object and image distance.

Many characteristics of a mirror are equivalent to those of a lens. The *transverse magnification* produced by a mirror, for instance, is the same as that produced by a lens:

$$M = \frac{y'}{y} = \frac{s'}{s} \qquad [1.5\text{-}3]$$

In contrast to a lens, however, the refractive index in front of a mirror is (almost) irrelevant. True, a mirror in air has the same focal length as the same mirror under water. Only the *power* of the mirror changes, as we see from Equation [1.5-1].

Mirrors and vergence.   Vergences and changes of vergence are handled the same way with a mirror as with a lens.

Example 1

An object is located 2.5 cm to the left of a concave mirror of 10 cm radius of curvature. Find the image distance using the vergence method.

Solution.   First we determine the entrance vergence,

$$\mathbf{V} = \frac{1}{s} = \frac{1}{-0.025} = -40 \text{ m}^{-1}$$

and the power of the mirror,

$$P = \frac{1}{f} = \frac{-2}{R} = \frac{-2}{-0.1} = +20 \text{ m}^{-1}$$

Adding both gives the exit vergence,

$$\mathbf{V}' = (-40) + (+20) = -20 \text{ m}^{-1}$$

The image, as we see from the minus sign, is virtual and located

$$s' = \frac{1}{\mathbf{V}'} = \frac{1}{-20} = -0.05 = \boxed{-5 \text{ cm}}$$

to the *right* of the mirror.

Example 2

Consider the combination of a lens and a mirror. The object is located 50 cm to the left of a -2.00-diopter lens. A $+12.50$-diopter mirror is placed 15 cm to the right of the lens. Find the position of the image.

Solution.   In Figure 1.5-2 the vergences of the light coming from the object are shown *above the*

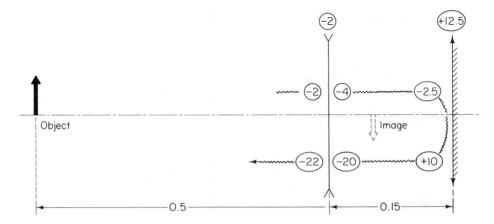

Figure 1.5-2   Schematic representation of light passing through a lens–mirror combination.

axis and the vergences of the light reflected back by the mirror *below the axis.* With the object 50 cm away, the entrance vergence at the lens is

$$\mathbf{V} = \frac{1}{-0.5} = -2 \text{ m}^{-1}$$

The exit vergence therefore is

$$\mathbf{V}' = (-2) + (-2) = -4 \text{ m}^{-1}$$

This means that the center of curvature of the wavefronts leaving the lens is

$$L = \frac{1}{\mathbf{V}'} = \frac{1}{-4} = -0.25 \text{ m}$$

to the left of the lens, or 40 cm to the left of the mirror. At the mirror, therefore, the entrance vergence is

$$\mathbf{V} = \frac{1}{-0.4} = -2.5 \text{ m}^{-1}$$

and the exit vergence is

$$\mathbf{V}' = (-2.5) + (+12.5) = +10 \text{ m}^{-1}$$

At this point a slight complication occurs. From the exit vergence $\mathbf{V}' = +10 \text{ m}^{-1}$ we conclude that 10 cm to the left of the mirror the light comes to a focus. But the light does not stop there; it continues, becomes *di*vergent, and passes once more through the lens (Figure 1.5-3).

For the dimensions given, the crossover point (where the light becomes divergent) lies 5 cm before the light reaches the lens. On entering the lens, therefore, the vergence is

$$\mathbf{V} = \frac{1}{-0.05} = -20 \text{ m}^{-1}$$

As before, the lens makes the light still more divergent:

$$\mathbf{V}' = (-20) + (-2) = -22 \text{ m}^{-1}$$

The final image, then, is located at a distance

$$s' = \frac{1}{-22} = (-)0.045 = \boxed{(-)4.5 \text{ cm}}$$

to the right of the lens.

I have put the minus sign in parentheses. True, the image is located on the "wrong" side

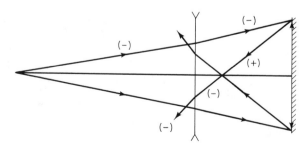

Figure 1.5-3 Lens–mirror combination. The plus and minus signs indicate whether the light is convergent or divergent.

considering the direction of the light, but this time the wrong side is to the *right* of the lens. The image is also virtual and inverted, an odd combination, due to reflection at the mirror.

Graphical construction.   In principle very similar to refraction, rays are deflected as follows:

The *parallel ray* (1), after reflection, goes through the focal point.
The *focal ray* (2) is reflected back parallel to the optic axis.
The *chief ray* (3) is returned in its own path.

The object in Figure 1.5-4, for example, lies *outside* the center of curvature; in other words, the object is farther away from the mirror than twice its focal length. The image, the same as with a lens, is real, inverted, and reduced in size. In Figure 1.5-5, the object is *inside the focal length*. In that case, the image is virtual, upright, and magnified. Convex mirrors produce only virtual, upright, and reduced images. The principal points of a mirror coincide at the vertex, and the nodal points coincide at the center of curvature.

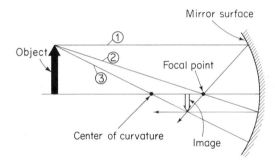

Figure 1.5-4 Image formed by a spherical mirror. Object *outside* the center of curvature: Image is real, inverted, and reduced in size.

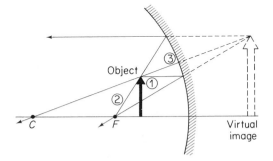

Figure 1.5-5 Object *inside* the focal length: Image is virtual, upright, and magnified.

Applications.   An interesting example of the reflection of light on a (convex) spherical mirror occurs in the *keratometer*. That is an instrument used to determine the radius of curvature, and therefore the power, of the cornea of the eye.

A keratometer consists of a back-lit target, usually two "mires" separated by a distance *a* (Figure 1.5-6). The mires produce two virtual images, seen on the cornea, which are separated by a distance *b*.

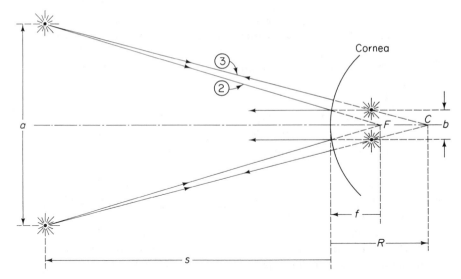

Figure 1.5-6 Principle of the keratometer.

To locate the images, use graphical construction. First draw the two chief rays, (3), one from each of the sources toward the center of curvature, *C*. By definition, the chief rays hit the surface at normal incidence, and hence are reflected back in themselves. Then draw the two focal rays, (2), one from each of the sources toward the focal point, *F*. By definition, focal rays are reflected back parallel to the axis, as shown. Intersection of the extensions of the rays defines the two images.

Now consider the triangle formed by the two sources and by *F*, and the triangle formed by the two images as seen on the cornea and also by *F*. Since the triangles are similar,

$$\frac{a}{s+f} = \frac{b}{f}$$

Cross-multiplication gives

$$af = bs + bf$$

$$f(a - b) = bs$$

The radius of curvature, *R*, of a mirror is twice its focal length, *f*. Therefore, the radius of curvature of the surface is

$$R \approx \frac{2bs}{a} \qquad \text{[1.5-4]}$$

In a given keratometer, the distances *a* and *s* are usually fixed and cannot be changed; thus, we only need to measure the separation, *b*, of the two reflections to find the radius of the cornea.

To convert the radius into (reflective) power, we use the surface power equation, $P = (n' - n)/R$, and set $n' - n$ arbitrarily equal to 0.3375. Consequently, a radius of 7.5 mm, for example, means a power of 45 diopters.

PLANE MIRRORS

A plane mirror has a radius of infinite length, $R = \infty$, and thus, from Equation [1.5-2],

$$\frac{1}{s'} - \frac{1}{s} = \frac{-2}{\infty}$$

so that

$$-s = -s' \qquad\qquad [1.5\text{-}5]$$

which means that *the image formed by a plane mirror is located at the same distance behind the mirror as the object is in front of it.* The image, furthermore, is *virtual;* it cannot be received on a screen because the light does not actually reach the image plane.

Consider in more detail the *orientation* of such an image. The object may be the letter R. If the mirror is standing upright, like a dresser mirror, with its surface parallel to the straight line in the R, the images are upright but right and left are reversed (Figure 1.5-7, left and right). Such images, Я, are "upward backward"; they are *wrong-reading.*

If the mirror surface is horizontal, like a reflecting pool, the images, Я, are "inverted forward," and also wrong-reading. An image formed by a single plane mirror is always wrong-reading.

If a mirror is tilted *(rotated)* through a given angle, a beam reflected by the mirror will be rotated through twice that angle. This fact plays a role in optical levers, galvanometers, sextants, and similar instruments.

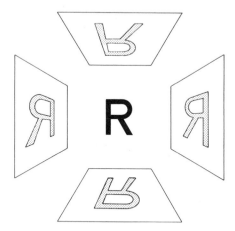

Figure 1.5-7 Looking at an object through a tunnel lined with four plane mirrors.

Multiple plane mirrors.   Two or more plane mirrors, when combined, cause multiple reflections, and therefore multiple virtual images. Consider *two* mirrors, and assume that all rays and the normals to the two mirrors lie in the same plane (Figure 1.5-8). The light may be incident on the first mirror at an angle α. If the two mirrors together subtend an angle γ, the light is deviated through δ, and

$$\delta = 2\gamma \qquad\qquad [1.5\text{-}6]$$

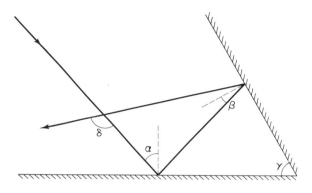

Figure 1.5-8   Reflection on two plane mirrors.

which means that the deviation is twice the angle subtended by the two mirrors, *independent of the angle of incidence*. (Solve Problem 1.5-21 to verify this statement.) When $\gamma = 90°$, $\delta = 180°$: In that case the incident ray and the emergent ray are antiparallel.

 Three plane mirrors, orthogonal to each other, form a *cube-corner retroreflector*. Such a reflector returns the light in the direction from where it came, even in three-dimensional space (Figure 1.5-9). Cube-corner reflectors are used in highway signs and geodetic instruments; they were placed on the moon to determine, very precisely, the moon's distance from the earth.

 A *kaleidoscope* consists of a tube containing two or three rectangular plane mirrors. The mirrors form either a V or a triangle, causing multiple images that lie in pie-shaped sectors or fill the entire field of view. With the angle subtended by two mirrors 60°, six images are produced; with an angle of 45°, eight.

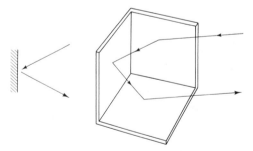

Figure 1.5-9   Conventional *specular* reflection (*left*) and cube corner reflector (*right*).

ASPHERICAL MIRRORS

Parallel light will come to a focus only if reflected at a *paraboloidal* mirror. A paraboloid is a figure of revolution of a parabola about its axis of symmetry. Paraboloidal mirrors are used widely, from astronomical reflectors to searchlights to microwave antennas.

 A parabola is defined as the location of points in a plane that are equidistant from a given point (the *focus F* of the parabola) and a given line (the *directrix r*). The line through *F*

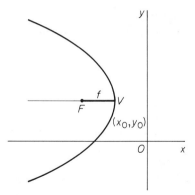

Figure 1.5-10   Parabola opening to the left.

which is perpendicular to *r* is the *axis* of the parabola. The point at which the axis intersects the parabola is the *vertex*. The function

$$y = x^2$$

for example, describes a parabola open at the top and symmetric about the *y* axis.

To turn the parabola so that its axis is horizontal, the same as the axis of most optical systems, interchange *x* and *y* and specify the focal length *f* and the coordinates x_0, y_0 of the vertex:

$$(y - y_0)^2 = -4f(x - x_0) \qquad [1.5\text{-}7]$$

A plot of that function is shown in Figure 1.5-10.

In contrast to a paraboloid, an *ellipsoidal* mirror has two foci. These are conjugate: Light originating at one focus is projected into the other focus (Figure 1.5-11).

With the center of the ellipse at x_0, y_0, the equation of an ellipse is

$$\frac{(x - x_0)^2}{a^2} + \frac{(y - y_0)^2}{b^2} = 1 \qquad [1.5\text{-}8]$$

where $2a$ is the length of the major axis (the *length* of the ellipse) and $2b$ the length of the minor axis (the *width*). For an ellipsoid we have

$$\frac{(x - x_0)^2}{a^2} + \frac{(y - y_0)^2}{b^2} + \frac{(z - z_0)^2}{c^2} = 1 \qquad [1.5\text{-}9]$$

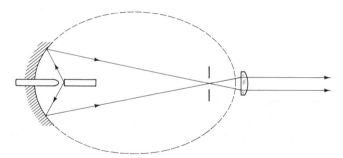

Figure 1.5-11   Ellipsoidal mirror together with collimating lens makes for efficient light collector. Note that the angle subtended by the mirror surface is much larger than that subtended by the lens (with the source then placed in the right-hand focus).

If $a = b = c$, the ellipsoid is a sphere.

Graphical construction and using Pythagoras' theorem show that the distance from the center of the ellipse to either of its foci is $\sqrt{a^2 - b^2}$. The *focal length* of an ellipse is approximately

$$f \approx \frac{b^2}{2a} \qquad [1.5\text{-}10]$$

Twice that distance, b^2/a, is the radius of curvature at either end.

Determining the Focal Length of a Mirror. 1. With a *concave* mirror, place an object–image screen (a transilluminated target) in front of the mirror at a distance such that the target is focused back into the plane of the target. The distance from the target to the mirror is then equal to the radius of curvature. One-half that distance is the focal length.

2. Focus a telescope for infinity and then aim at the (concave) mirror. A small object such as a pin is mounted, slightly off-axis, between mirror and telescope. Moving the object back and forth along the axis, find a position where the image is seen in focus. Light coming from the object and reflected by the mirror must now be parallel and the object be located in the mirror's focus.

3. A *convex* mirror, like a diverging lens, does not form a real image. Proceed as with a minus lens, first forming an image with the plus lens only. Focus on the mirror surface (dashed curve on the right in Figure 1.5-12). Then move the mirror to the left (solid curve) until an image is formed in the plane of the target. The distance between the two positions of the mirror is equal to the radius of curvature. The same procedure can be used with a concave mirror, moving the mirror *away* from the lens.

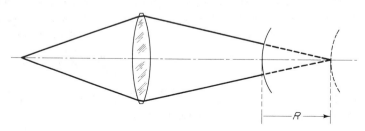

Figure 1.5-12   Determining the radius of curvature of a mirror.

SUGGESTIONS FOR FURTHER READING

R. E. HOPKINS, "Mirror and Prism Systems," in R. Kingslake, editor, *Applied Optics and Optical Engineering,* Vol. III, pp. 269–308 (New York: Academic Press, Inc., 1965).

W. B. ELMER, *Optical Design of Reflectors,* 2nd edition (New York: John Wiley & Sons, Inc., 1980).

T. B. GREENSLADE, JR., "Multiple Images in Plane Mirrors," *Phys. Teacher* **20** (1982), 29–32.

J. WALKER (Kaleidoscope) *Sci. Am.* **253** (Dec. 1985), 134–45.

PROBLEMS

1.5-1. An object is placed 15 mm in front of a concave mirror of 4 cm radius of curvature. What is the image distance?

1.5-2. With an object 2 m away, a concave mirror forms a real image 50 cm away. What is the radius of curvature of the mirror?

1.5-3. An object 2 cm high is placed 6 cm in front of a concave mirror. If the mirror has a radius of curvature of 16 cm, how high is the image?

1.5-4. How far from a concave mirror of 10 cm radius of curvature must a real object be placed so that its image is real and four times the size of the object?

1.5-5. If a concave mirror of 60 cm radius of curvature forms a real image twice as far away as the object, what is the object distance?

1.5-6. Which two object distances will produce an image, real or virtual, that is 25 cm away from a mirror of +12 diopters power?

1.5-7. Which two radii of curvature of a mirror will produce an image twice the size of an object 15 cm away from the mirror?

1.5-8. If an object is placed halfway between the focal point and the vertex of a concave mirror, how much will the image be magnified?

1.5-9. A rock is dropped next to the edge of a swimming pool 6 m in diameter, producing ripples that are reflected at the opposite side. How far from the point of impact will the ripples come together, forming a point *conjugate* to the point of impact?

1.5-10. A ray of light is reflected at a concave mirror of 40 cm radius of curvature (Figure 1.5-13). When proceeding in a medium of index 1.6, how large an angle does the reflected ray subtend with the axis?

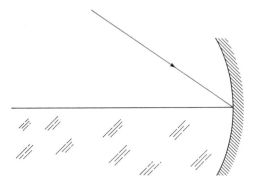

Figure 1.5-13

1.5-11. An object 16 mm high is placed 13 cm in front of a concave mirror of 26 cm radius of curvature. What is the angular size of the image?

1.5-12. A ray of light, proceeding parallel to the optic axis at a height of 20 mm, is incident on a mirror of −40 mm radius of curvature.
(a) What is the angle of incidence at the mirror's surface?
(b) What is the slope angle as the ray crosses the optic axis?

1.5-13. A small object is placed between the center of curvature and the focal plane of a concave mirror. By graphical construction find the orientation, type (real or virtual), and magnification of the image.

1.5-14. Assume that the object is placed in front of a *convex* mirror. As in Problem 1.5-13, determine the character of the image.

1.5-15. A −2.00-diopter lens is held in contact with a mirror of +80 cm focal length and of the same diameter as the lens. What is the power of the combination?

1.5-16. A symmetrical (equiconvex) lens has +10 diopters power and is made of glass of $n = 1.5$. If the back surface of the lens has a reflective coating, what is the (total) power of the combination?

1.5-17. If the distance between the mires in a keratometer is 42 mm and the conjugate image height 2 mm, and if the mires are 67.5 mm away from the eye, what is the power of the cornea?

1.5-18. When a polished steel ball 15 mm in diameter is observed in a keratometer, it gives a reading of +48.00 diopters. For what index is the keratometer calibrated?

1.5-19. How long must a (vertical) plane mirror be for a man who stands 1.82 m tall to see his full length?

1.5-20. A student, 1.60 m tall and with her eyes 150 cm above the floor, sees her full image in a vertical plane mirror $3\frac{1}{2}$ m away. How high above the floor, at most, can the mirror's lower rim be?

1.5-21. If two plane mirrors, as in Figure 1.5-8, subtend a certain angle and if light is incident on one mirror, prove that the light reflected from the other mirror is deviated through *twice* that angle.

1.5-22. Two plane mirrors subtend an angle of 35°. At what angle must light be incident on one mirror so that, after reflection at the other mirror, the light exactly retraces its path?

1.5-23. Two plane mirrors subtend a certain angle with each other. A ray of light is parallel to one of the mirrors, and after four reflections it exactly retraces its path. What angle do the mirrors subtend?

1.5-24. Two plane mirrors subtend an angle of 15°. A ray originates at a point on one of the mirrors. At what angle must this ray be incident on the other mirror so that after three reflections it is parallel to the first mirror?

1.5-25. If close to the optic axis a concave spherical mirror has the same radius of curvature as a concave paraboloidal mirror, how do the focal lengths of the two mirrors differ farther away from the axis?

1.5-26. What is the general equation of a parabola opening to the *right?*

1.5-27. A +6.00-diopter lens is placed 25 cm from a transilluminated target. If then a convex mirror, as in Figure 1.5-12, is set 40 cm from the lens, it returns the light in its original path. What is the power of the mirror?

1.5-28. A lens clock, held against a mirror, gives a reading of −2.50 diopters. If this mirror forms a real image at a distance one-half the object distance, what is the image distance?

1.6

Stops and Pupils

In any optical system the light passing through is limited in cross section, either by the finite diameter of the lenses or by additional diaphragms called *stops* (Figure 1.6-1). Stops, as well as *pupils*, are as important as lenses or mirrors. The exit pupil of a system, for instance, should be placed so as to coincide with the pupil of the observer's eye. Only then will the observer be comfortable looking through the system, an aspect just as necessary as seeing the image in focus.

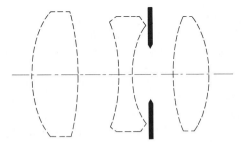

Figure 1.6-1 Aperture stop placed in a three-element lens system.

The slide projector.   When we set out to build a slide projector, perhaps we feel that all we have to do is take a light source, place the object (the *slide*) in front of it, and project a magnified image onto a distant screen (Figure 1.6-2, top). Most likely, the image will be disappointing. Even while it may be "in focus," probably only the center of the image will be bright; the corners will be dim or even be cut off entirely, an effect called *vignetting*

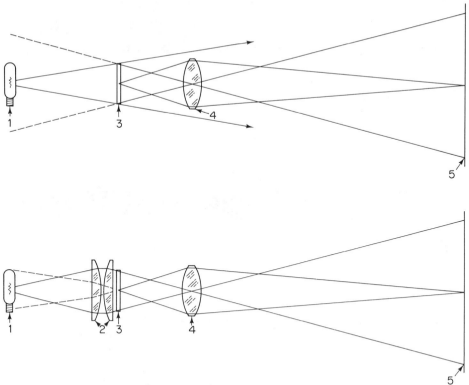

Figure 1.6-2 Wrong way (*top*) and good way (*bottom*) of building a slide projector. 1, light source; 2, condenser; 3, transparency; 4, projection lens; 5, screen.

(vig net' ting). Placing a sheet of ground glass between light source and transparency would help but that is hardly an elegant solution.

Obviously, if all of the transparency is to receive light that then goes on to the projection lens, a source of unreasonably large size would be required. Instead of trying to find such a source, we use a *condensing lens.* That lens is placed just ahead of the transparency; it should be slightly larger than the diagonal across the transparency, but since the condenser does not contribute directly to image formation, it need not be of high quality.

The purpose of the condenser is to project an image of the light source into, or close to, the projection lens (bottom). In other words, its purpose is to *fill the aperture of the projection lens.* The rays then continue and fill the screen. The projection lens, in turn, forms an image (of the slide). But with the light source relatively small, *its* image will be small too and hence the projection lens can be small also—which makes it less expensive to design and produce (but certainly it should be of high quality). The image, as it appears on the screen, is now fully and evenly illuminated, with no vignetting.

Clearly, the proper design of an optical system includes the correct choice of diameters of the various lenses, and of additional diaphragms. That is the subject of our discussion of the theory of *stops and pupils.*

STOPS

Aperture stop.   An *aperture stop* is an opening, usually circular, in an otherwise opaque screen. Sometimes, the aperture is simply the rim of a lens, or the edge of a mirror. In a camera the aperture stop is often an iris diaphragm (which can be varied in size). An aperture stop limits the cross section of the light that forms the image. Making the aperture smaller, as in Figure 1.6-3, limits the amount of light passing through and therefore makes the image dimmer (but it does not restrict the field of view); in addition, it increases the *depth of focus*.

The concept of depth of focus is tied to the thickness and graininess of photographic film. Because of these, we can, or need to, tolerate a certain amount of blur. A blur circle 30 μm in diameter or $1/1000$ the focal length of the lens, whichever is larger, is considered acceptable.

Assume that the light in Figure 1.6-4 comes from infinity. If the film were infinitesimally thin, the light would focus solely on that plane. But in reality, with light focusing on the rear face of the film, it causes a blur, of diameter b, on the front face. If D is the diameter of the aperture and Δf the depth of focus, then from similar triangles,

$$\frac{f + \Delta f}{D} = \frac{\Delta f}{b}$$

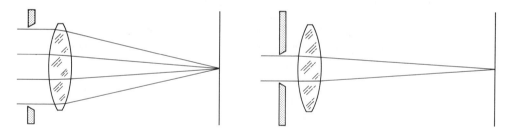

Figure 1.6-3   Aperture stop. Making the aperture stop smaller (*right*) limits the amount of light but does not restrict the field of view.

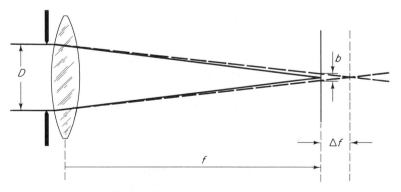

Figure 1.6-4   Defining depth of focus.

Because $\Delta f \ll f$, this simplifies to $f/D \approx \Delta f/b$, with the ratio f/D called the *f-stop number* of the system,

$$\text{\textit{f}-stop number} = \frac{\text{focal length}}{\text{diameter of aperture}} \qquad [1.6\text{-}1]$$

For example, for a camera lens of $f = 50$ mm and $D = 1.25$ cm the f-stop number is 50 mm/12.5 mm $= 4$, which customarily is written $f/4$. (Camera lenses often have f-stops chosen so that the next-higher number requires twice the exposure time.) For objects nearby and image distances, s', appreciably longer than f, the *effective f-number* is s'/D, rather than f/D.

Depending on the depth of focus, Δf or $\Delta s'$, there is a conjugate distance, Δs, in the object space (not shown). That distance is called the *depth of field*. The same as with the depth of focus, a smaller aperture will increase the depth of field. But to reach the largest possible depth of field, it would not be wise to focus at infinity because some depth of field, "beyond infinity," would be wasted. Instead, we focus at the *hyperfocal distance;* the depth of field then extends from one-half that distance out to infinity. As a rule of thumb, use the focal length of the lens, in mm, and divide it by the f-stop setting: that gives the hyperfocal distance, in meters.

Field stop.   As the name suggests, a *field stop* limits the *field of view.* As the stop is made smaller, the field of view becomes smaller too (as in Figure 1.6-5) but the amount of light admitted to the image space remains the same, the opposite of what happens with an aperture stop. A field stop is often located in the image plane (as shown), but it may also be located in the object plane (as in a slide projector, where the opaque mount of the slide limits the field).

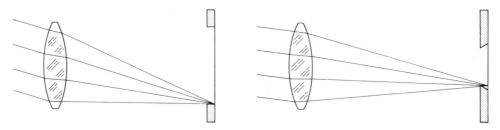

Figure 1.6-5 Field stop. Making the field stop smaller (*right*) restricts the field of view but does not limit the light.

PUPILS

A stop is something tangible and real. A pupil is more conceptual. A pupil may be a real physical stop; or it may be a real or even a virtual image of a stop. A pupil is defined as the cross section of a bundle where light from all parts of the object passes through, *completely mixed,* with no preferential spatial separation.

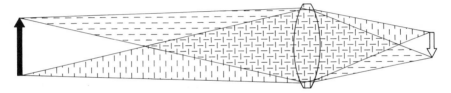

Figure 1.6-6   Lens acting as a pupil.

Examples.

1. Assume that we have an extended object and that a lens forms an image of this object as shown in Figure 1.6-6. There is only one cross section common to all rays: *that cross section is the lens.* Of course, we could draw an infinite number of other cross sections through the beam, closer to the object or closer to the image. But these cross sections would all contain light that comes predominantly from certain parts of the object, more from the tip of the arrow or more from its foot, and thus they are not pupils.

2. If we place a stop between an (extended) object and a lens, that *stop will become the pupil* (Figure 1.6-7). Rays from all points in the object pass through the stop, completely mixed, with no spatial separation. If we partially occlude the stop with the point of a pencil and at the same time look at the image, the point can hardly be seen (except that the image becomes dimmer). In this case, the lens does *not* act as a pupil: rays from the tip of the arrow pass through the lower part of the lens, and rays from the foot pass through the upper part. If we hold the pencil close to the lens, the point *would* show as a shadow superimposed on the image.

3. So far, I have referred to *a* pupil. More specifically, the stop in Figure 1.6-7 is the *entrance pupil. Its image,* as formed by the lens, *is the exit pupil.* Note that there are two processes of image formation that go on side by side: (a) the lens transforms an object into its

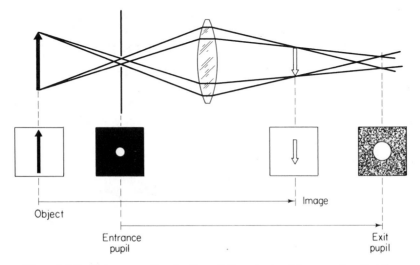

Figure 1.6-7   Entrance pupil and exit pupil. Long horizontal arrows show that there are two independent processes of image formation.

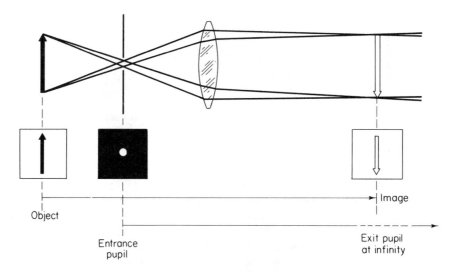

Figure 1.6-8   System telecentric on the image side.

image, and (b) the lens transforms the entrance pupil into the exit pupil. Object and image are conjugate, and entrance pupil and exit pupil are conjugate also.

4. We could *move* the entrance pupil, that is, we could move the stop, back and forth along the axis. If we move the stop closer to the lens, its image, the exit pupil, moves away. If the entrance pupil is moved into the left-hand focal plane, the exit pupil moves out to infinity on the right. The system is then called *telecentric on the image side* (Figure 1.6-8). Conversely, if the stop is moved into the right-hand focal plane, the system is *telecentric on the object side*. Such systems are useful in precision measurements, whenever the (transverse) size of an object is to be compared with a scale that cannot be brought into contact with the object.

5. If the stop is moved even closer to the lens so that it comes to *within the focal length*, the entrance pupil and exit pupil lie on the same side (Figure 1.6-9). The exit pupil, like the image formed by a magnifying glass, has become a virtual, magnified image of the stop.

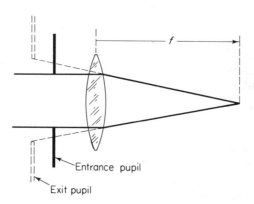

Figure 1.6-9   Entrance pupil located within first focal length of lens.

Lab Experiment. Consider a filter one-half of which is green, the other half red. Hold the filter in front of a diffuse light source and, using a converging lens, form an image of the filter on a screen. The image will show both colors clearly separate. Then take a white card, hold the card first next to the screen, and then move it slowly toward the lens: the colors will fuse into white. This shows that the *lens* is the pupil (*where all rays mix*).

Place a stop halfway between the light source and the lens. Hold a pencil point close to the lens: it will show as a shadow superimposed on the image (of the filter). But when held close to the stop, the point's image can hardly be seen at all—which means that now the *stop* has become the pupil.

Move the screen farther away from the lens to where you see an image of the stop. Hold the pencil point inside the stop. An image of the stop with the point inside can easily be seen on the screen: While the stop as such is the *entrance pupil,* the stop's image is the *exit pupil.*

Numerical Example

A stop 8 mm in diameter is placed halfway between an extended object and a large-diameter lens of 9 cm focal length. The lens projects an image of the object onto a screen 14 cm away. What is the diameter of the exit pupil?

Solution. Consider again Figure 1.6-7. From the known focal length and the image distance, determine the object distance,

$$s = \frac{s'f}{f - s'} = \frac{(14)(9)}{9 - 14} = -25.2 \text{ cm}$$

One-half that distance is the distance of the stop (from the lens),

$$\frac{-25.2 \text{ cm}}{2} = -12.6 \text{ cm}$$

With the diameters given, we have little doubt that it is the stop, rather than the lens, that is the entrance pupil. The image of the stop, hence, is the exit pupil. It is formed

$$s'' = \frac{(-12.6)(9)}{(-12.6) + 9} = 31.5 \text{ cm}$$

to the right of the lens. The diameters of the two pupils are proportional to their distances,

$$\frac{D_{EP}}{D_{XP}} = \frac{-12.6}{31.5}$$

Therefore, the diameter of the exit pupil is

$$D_{XP} = \frac{(0.8)(31.5)}{-12.6} = \boxed{2 \text{ cm}}$$

Pupils of combinations of lenses. When a stop is placed ahead of a lens, that stop becomes the entrance pupil. But if another lens is placed ahead of the stop, the light does not reach the stop directly. Instead, the light "sees" the stop the same way we see newsprint through a magnifying glass. In other words, the light comes to an *image* of the stop, rather than to the stop itself, and *this image,* by definition, *is the entrance pupil.* Therefore, the entrance pupil of a combination of lenses is the image of the stop formed by all lenses *preceding* the stop. Conversely, the exit pupil of a combination of lenses is the image of the stop formed by all lenses *following* the stop.

If we are not sure whether it is the stop, or its image, or one of the lenses that limits the bundle, the general rule is that the aperture that subtends the *smallest* angle at the axial point object is the entrance pupil. We also note that the exit pupil may be located to the right of the entrance pupil or to the left of it, or it may coincide with it.

Example

Two lenses, a +8.00-diopter lens and a minus lens of unknown power, are mounted coaxially and 8 cm apart. The system is afocal, that is, light entering the system parallel at one side emerges parallel at the other. If a stop 15 mm in diameter is placed halfway between the lenses:
(a) Where is the entrance pupil?
(b) Where is the exit pupil?
(c) What are their diameters?

Solution.   (a) For the system to be afocal, the focal points of the two lenses must coincide (at F in Figure 1.6-10). Therefore, if the first lens has +8.00 diopters power, or $(1/8)(100) = 12.5$ cm focal length, and if the two lenses are 8 cm apart, the second lens must have $-(12.5 - 8) = -4.5$ cm focal length.

The *entrance pupil* is the image of the stop formed by the first lens. This time, contrary to our usual custom, the object (that is, the stop) lies to the right of the lens, the rays that form the image proceed from right to left, and the f_2 focal length is on the left. Hence, the image of the stop (that is, the entrance pupil) lies

$$s' = \frac{sf_2}{s + f_2} = \frac{(4)(-12.5)}{(4) + (-12.5)} = \boxed{+5.88 \text{ cm}}$$

to the right of the (first) lens.

(b) The *exit pupil* is the image of the stop formed by the second lens. It is

$$s' = \frac{(-4)(-4.5)}{(-4) + (-4.5)} = \boxed{-2.12 \text{ cm}}$$

to the left of the (second) lens. In other words, the two pupils coincide—which is typical of an afocal system.

(c) However, the pupil *sizes* are different. The diameter of the entrance pupil is

$$D_{\text{EP}} = \frac{D_{\text{STOP}}\, s'}{s} = \frac{(15)(58.8)}{40} = \boxed{22 \text{ mm}}$$

and the diameter of the exit pupil

$$D_{\text{XP}} = \frac{(15)(21.2)}{40} = \boxed{8 \text{ mm}}$$

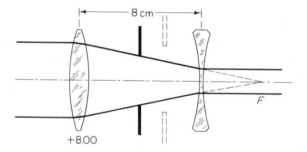

Figure 1.6-10

Finally, we consider a *combination of lenses and stops*. First look at the aperture stop, placed some distance to the right of the first lens (Figure 1.6-11a). Trace a parallel ray and a chief ray. As before, the image of the aperture stop, formed by the lens preceding it, is the entrance pupil, EP. Conversely, the image of the stop, formed by the lens following it, is the exit pupil, XP (b). In the case shown, the stop is placed outside the focal lengths; that makes the pupils *real*.

Next look at the field stop. By definition, the image of a field stop, formed by the lens preceding it, is the *entrance window,* EW (c). The image of the stop, formed by the lens following it, is the *exit window,* XW (d). Those are appropriate terms: If we are inside a room and look out, it is the window, and its position relative to the observer, that limits the field of view.

Now we are ready to locate the rays limiting the bundle. First, there is a cone of light that comes from the center of the entrance pupil, touches the rim of the entrance window, and goes on to the center of the aperture stop. From there it proceeds to the rim of the field stop, the center of the exit pupil, and the rim of the exit window (full lines in Figure 1.6-11e). The other bundle, limited by dashed lines, begins at the rim of the entrance pupil and goes on to the center of the entrance window. It then proceeds to the rim of the aperture stop, the center of the field stop, the rim of the exit pupil, and finally the center of the exit window. The two bundles are completely intertwined yet independent of each other. In particular, we note that

1. *Aperture stops relate to pupils.* The image of an aperture stop, formed in the direction of propagation of the light, is the *exit pupil;* formed in the opposite direction, it is the *entrance pupil.*

2. *Field stops relate to windows.* The image of a field stop, formed in the direction of propagation of the light, is the *exit window;* formed in the opposite direction, it is the *entrance window.*

3. A chief ray (solid lines, Figure 1.6-11e) always enters the system through the center of the entrance pupil. And because entrance pupil and exit pupil are conjugate, the chief ray leaves the system through the center of the exit pupil.

4. At the exit pupil, the bundle has the *least diameter* because an aperture stop is by necessity farther away from the last lens than a field stop. That has important consequences: In many optical systems, such as in a telescope or microscope, the entrance pupil is the objective lens (no separate stop is needed). The exit pupil then is the image of this lens formed by the eyepiece. This image is where the density of the light is highest and where, as we had seen earlier, the pupil of the observer's eye should be placed to assure comfortable viewing and avoid vignetting.

5. Finally, the light source, or in place of it a condenser lens, should be large enough to fill the entrance pupil. A source or lens that does not fill the entrance pupil cannot fill the aperture stop either, and if it does not, the amount of light reaching the image will be less than what it could be. This is why, in designing an optical system, we need to make sure that the lenses and other elements not only have the right focal lengths, but also the right *diameters.*

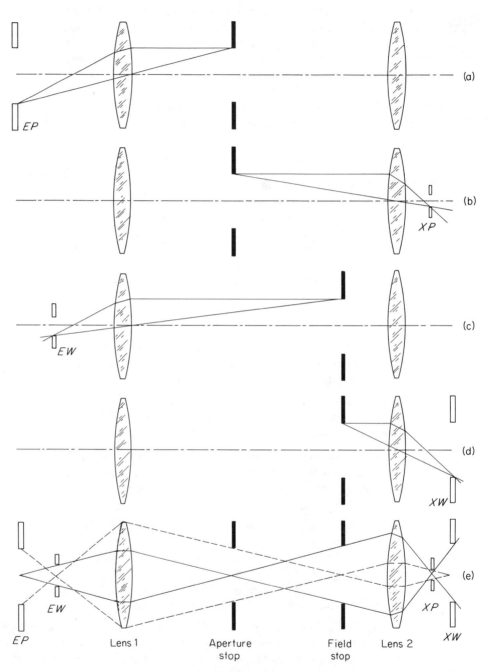

(a)

(b)

(c)

(d)

(e)

EP

EW

XP

XW

EW

EP

Lens 1

Aperture
stop

Field
stop

Lens 2

XP

XW

Figure 1.6-11 Entrance pupil, exit pupil, entrance window, and exit window of a system composed of two lenses and two stops.

SUGGESTION FOR FURTHER READING

F. A. Jenkins and H. E. White, *Fundamentals of Optics*, 4th edition, pp. 115–129 (New York: McGraw-Hill Book Company, 1976).

PROBLEMS

1.6-1. If a camera lens of 50 mm focal length is set at $f/8$ and focused at an object 15 cm away, what is the effective f-stop number?

1.6-2. If a camera lens of 50 mm focal length is set at $f/4$ and focused at an object 30 cm away, what is the effective f-stop number?

1.6-3. Compare the depth of focus, as a percentage of focal length, of a 30-mm-focal-length wide-angle lens with the depth of focus of a 135-mm-focal-length telephoto lens, both set at $f/4$.

1.6-4. When figuring the depth of field of the eye, we assume that the observer will not notice a blur if the entrance vergence changes by no more than $0.25\ \text{m}^{-1}$. Hence, to see clearly out to infinity, what is:
   (a) The hyperfocal distance at which the eye should focus?
   (b) The nearest distance seen in focus?

1.6-5. If a pinhole camera has the shape of a cube, how large a field of view, measured horizontally, will it cover?

1.6-6. Continue with Problem 1.6-5 and determine the size of the field if the pinhole camera is built around a cube made of glass of $n = 1.52$.

1.6-7. A photographic slide, taken on 35-mm film, has an area $24 \times 36\ \text{mm}^2$. What is the angular field, measured diagonally, that can be projected through a lens of 122.7 mm focal length?

1.6-8. A narrow stop is placed two focal lengths away on the right-hand side of a converging lens. Where is:
   (a) The entrance pupil of the system?
   (b) The exit pupil?

1.6-9. A stop, placed a short distance to the rear of a converging lens, acts as the exit pupil. By ray tracing find the entrance pupil.

1.6-10. A thin lens of 50 mm focal length has a diameter of 4 cm. A stop 2 cm in diameter is placed 3 cm to the left of the lens, and an axial point object is located 20 cm to the left of the stop.
   (a) Which of the two, the stop or the lens, limits the bundle?
   (b) Where is the exit pupil located?

1.6-11. An object is placed 25 cm in front of a 60-mm-focal-length lens and a stop 2 cm behind the lens. If the lens is 50 mm in diameter, and the stop 20 mm, determine:
   (a) The position of the entrance pupil relative to the lens.
   (b) The diameter of the entrance pupil.

1.6-12. A lens of 50 mm focal length and 36 mm diameter is used as a magnifier. If a stop 12 mm in diameter is placed 10 mm to the right of the lens:
   (a) How far from the lens is the entrance pupil?
   (b) What is the diameter of the entrance pupil?

1.6-13. A first lens of 3 cm focal length and a second lens of 5 cm focal length are mounted 4 cm apart from one another and a narrow stop is placed halfway between them. By graphical ray tracing find the positions of the entrance pupil and the exit pupil.

1.6-14. Two thin lenses, 5 cm in diameter each and of focal lengths $+10$ cm and $+6$ cm, respectively, are placed 4 cm apart. An aperture stop 2 cm in diameter is set halfway between the lenses. Find the diameters of the entrance pupil and the exit pupil.

1.6-15. A first lens of $+4$ cm focal length and 2 cm diameter is combined with a second lens of -2 cm focal length and 1 cm diameter, the two lenses separated by 2 cm. By graphical ray tracing locate the:

(a) Entrance pupil.

(b) Exit pupil.

(c) Field stop.

(d) Entrance window.

1.6-16. A lens of $+5$ cm focal length is mounted 6 cm in front of another lens but of -7 cm focal length. When a stop 5 mm in diameter is placed halfway between the two lenses, what is:

(a) The location and diameter of the entrance pupil?

(b) The location and diameter of the exit pupil?

2.1

Ray Tracing

So far, we have considered the individual elements of a system: lenses, mirrors, and stops. Now we consider these elements as parts of a whole. We follow the light as it passes through the system, and we look at limitations and at ways to compensate for these limitations. That is mainly a matter of *ray tracing*. Ray tracing can be practiced to different degrees of precision and generality. *Paraxial ray tracing* is restricted to rays that lie close to the optic axis. *Trigonometric ray tracing* is more precise; still, it applies only to rays that lie in a meridional plane (in the plane of the drawing, for example, as in Figure 2.1-1). The tracing of *skew rays* is most general; such rays proceed in three-dimensional space, outside meridional planes. The most modern approach is by computer *(computer-aided lens design),* a subject to be discussed in detail in Chapter 2.4.

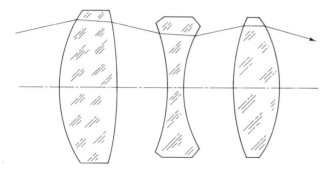

Figure 2.1-1 Tracing a ray through a three-lens system.

TRIGONOMETRIC RAY TRACING

The Q-U method.   In contrast to paraxial ray tracing, *trigonometric ray tracing* is inherently rigorous; it applies to any rays, even those that subtend with the optic axis reasonably *large angles, U*. As shown in Figure 2.1-2, we let an arbitrary ray be incident on a single spherical surface. The ray comes from an axial point object, located at a distance $-L$ from the vertex, and subtends with the axis an angle $+U$. After refraction, the ray intersects the axis at a distance $+L'$, subtending an angle $-U'$. The surface has a radius of curvature $+R$; it separates a medium of index n to the left of the surface from a medium of index n' to the right.

For trigonometric ray tracing we consider the perpendicular *vertex-ray distance, Q*, rather than the *height* of the ray above the vertex. From the construction it follows that

$$\sin +U = \frac{+Q}{-L} \qquad\qquad [2.1\text{-}1]$$

$$\sin +I = \frac{(+a) + (+Q)}{+R}$$

and

$$\sin +U = \frac{+a}{+R}$$

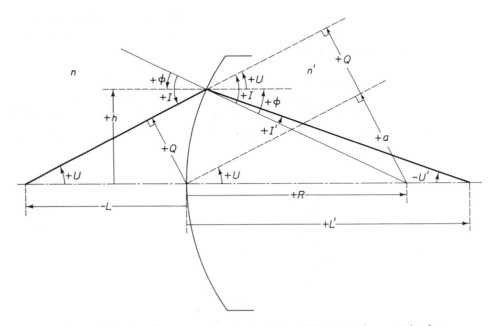

Figure 2.1-2   Ray diagram used for the derivation of trigonometric ray tracing formulas.

Rearranging the last two equations, eliminating a, and solving for Q yields

$$Q = R \sin I - R \sin U \qquad [2.1\text{-}2]$$

Following Equation [1.3-1], we replace the radius, R, of the surface by the reciprocal of its curvature, $R = 1/C$; thus,

$$QC = \sin I - \sin U \qquad [2.1\text{-}3]$$

The same equation holds after refraction where now all quantities (except C) are primed:

$$Q'C = \sin I' - \sin U' \qquad [2.1\text{-}4]$$

The difference between I and U is another angle, ϕ,

$$I - U = \phi$$

and since $+I'$ and $-U'$ are interior angles in a triangle whose opposite exterior angle is $+\phi$,

$$I' - U' = \phi$$

Combining these two equations by eliminating ϕ gives

$$I - U = I' - U' \qquad [2.1\text{-}5]$$

The angles of incidence and refraction, I and I', are connected through Snell's law,

$$n \sin I = n' \sin I'$$

which, solved for $\sin I'$, becomes

$$\sin I' = \frac{n}{n'} \sin I \qquad [2.1\text{-}6]$$

Next we consider the *transfer* of the refracted ray from one surface to the next, where it becomes the incident ray, the slope angle losing its prime in the process, $U_1' \rightarrow U_2$.

We replace the L in Equation [2.1-1] by the axial distance, d, between the two vertices. Then from Figure 2.1-3,

$$\sin U_1' = \frac{Q_1' - Q_2}{-d}$$

$$-d \sin U_1' = Q_1' - Q_2$$

and

$$Q_2 = Q_1' + d \sin U_1' \qquad [2.1\text{-}7]$$

For rays proceeding *parallel* to the optic axis, the vertex-ray distance Q is equal to the height of the ray, $Q = h$.

If one of the surfaces is *plane,* its curvature is zero, $C = 0$, the slope angle equals the angle of incidence, $U = I$, and Equation [2.1-3] becomes indeterminate, $Q = 0/0$. In that case, we use Snell's law to find U',

$$\sin U' = \frac{n}{n'} \sin U \qquad [2.1\text{-}8]$$

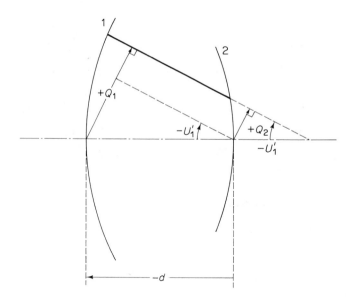

Figure 2.1-3 Transfer of ray from surface 1 to surface 2.

From the construction in Figure 2.1-4 we find that

$$\cos(-U) = \frac{+Q}{+h} \quad \text{and} \quad \cos(-U') = \frac{+Q'}{+h}$$

Combining these two equations by eliminating h and solving for Q' gives

$$Q' = \frac{Q \cos U'}{\cos U} \qquad [2.1\text{-}9]$$

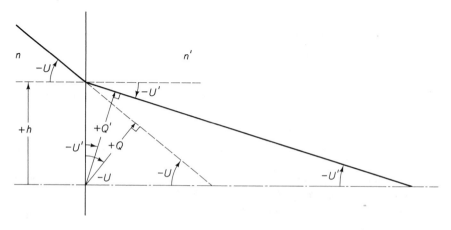

Figure 2.1-4 Ray trace through a plane surface.

In conclusion, we have four sets of equations, the *opening equations* which start the ray into the system,

$$\text{initial } Q = -L \sin U \qquad\qquad [2.1\text{-}1]$$

$$QC = \sin I - \sin U \qquad\qquad [2.1\text{-}3]$$

the *refraction equations,*

$$\sin I' = \frac{n}{n'} \sin I \qquad\qquad [2.1\text{-}6]$$

$$I - U = I' - U' \qquad\qquad [2.1\text{-}5]$$

$$Q'C = \sin I' - \sin U' \qquad\qquad [2.1\text{-}4]$$

the *transfer equation,*

$$Q_2 = Q'_1 + d \sin U'_1 \qquad\qquad [2.1\text{-}7]$$

and the *closing equation,* in analogy with Eq. [2.1-1],

$$\text{final } L' = -\frac{\text{final } Q'}{\text{final } \sin U'} \qquad\qquad [2.1\text{-}10]$$

where L' is the *axial intercept,* the distance from the vertex to the point where the refracted ray crosses the optic axis.

Example

Trace a ray through a combination of two lenses in contact, a *doublet* such as that shown in Figure 2.1-5. The first lens of the doublet is equiconvex; its surfaces have radii of curvature of $+50$ mm and -50 mm, respectively. The second lens is a negative meniscus; its surfaces have radii of -50 mm and -87 mm. Lens 1 has an axial thickness of 20 mm and is made of glass of index 1.50. Lens 2 has an axial thickness of 5 mm and an index of 1.80. The ray enters the system parallel to the axis at a height of 22 mm. How far from the back vertex does the ray intersect the axis?

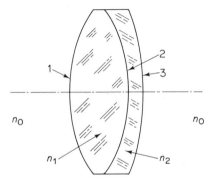

Figure 2.1-5 Example of a doublet. To scale.

Solution.   The first step in solving a problem of this kind is to set up a table, listing the various surfaces from left to right, and entering the data given below the surfaces, or between them:

	Surface 1		Surface 2		Surface 3
Radius, R	+50		−50		−87
Curvature, C	+0.02		−0.02		−0.0115
Thickness, d		20		5	
Index, n		1.5		1.8	
Q	+22				

Since the incident ray is parallel to the axis, its height is equal to Q_1,

$$h_1 = Q_1 = +22 \text{ mm}$$

and the initial slope angle is zero,

$$U_1 = 0$$

First we determine the angle of incidence, I_1, using Equation [2.1-3],

$$\sin I_1 = \sin U_1 + Q_1 C_1$$

$$= (0) + (+22)(+0.02) = +0.44$$

$$I_1 = +26.1°$$

The angle of refraction, I_1', is found from Snell's law,

$$\sin I_1' = \left(\frac{1.0}{1.5}\right)(+0.44) = +0.2933$$

$$I_1' = +17.1°$$

Knowing U_1, I_1, and I_1', we determine the slope of the ray after refraction, using Equation [2.1-5] solved for U_1',

$$U_1' = U_1 + I_1' - I_1$$

$$= (0) + (+17.1°) - (+26.1°) = -9.0°$$

Knowing U_1', we find Q_1', using Equation [2.1-4],

$$Q_1' = \frac{\sin I_1' - \sin U_1'}{C_1}$$

$$= \frac{(+0.2933) - (-0.1572)}{+0.02} = +22.5$$

By now our table looks like this:

	Surface 1		Surface 2		Surface 3
Radius, R	$+50$		-50		-87
Curvature, C	$+0.02$		-0.02		-0.0115
Thickness, d		20		5	
Index, n		1.5		1.8	
Q	$+22$				
Q'	$+22.5$				
$\sin U$	0				
U	0				
$\sin U'$	-0.1572				
U'	$-9.0°$				
$\sin I$	$+0.44$				
I	$+26.1°$				
$\sin I'$	$+0.2933$				
I'	$+17.1°$				

To transfer the ray to the next surface, we use Equation [2.1-7]:

$$Q_2 = Q'_1 + d_{12} \sin U'_1$$

$$= (+22.5) + (20)(-0.1572) = +19.4$$

Using once more Equation [2.1-3] and setting $U'_1 = U_2$, we determine the angle of incidence at the second surface,

$$\sin I_2 = \sin U_2 + Q_2 C_2$$

$$= (-0.1572) + (+19.4)(-0.02) = -0.5448$$

$$I_2 = -33.0°$$

Again using Snell's law, we find the angle of refraction:

$$\sin I'_2 = \left(\frac{1.5}{1.8}\right)(-0.5448) = -0.454$$

$$I'_2 = -27.0°$$

The slope angle after refraction at the second surface, from Equation [2.1-5], is

$$U'_2 = (-9.0°) + (-27.0°) - (-33.0°) = -3.0°$$

and again from Equation [2.1-4],

$$Q'_2 = \frac{\sin I'_2 - \sin U'_2}{C_2}$$

$$= \frac{(-0.454) - (-0.05296)}{-0.02} = +20.1$$

For the transfer from surface 2 to 3, we have

$$Q_3 = Q_2' + d_{23} \sin U_2'$$

$$= (+20.1) + (5)(-0.05296) = +19.8$$

and for the third surface, where $U_2' \rightarrow U_3$,

$$\sin I_3 = \sin U_3 + Q_3 C_3$$

$$= (-0.05296) + (+19.8)(-0.0115) = -0.2804$$

$$I_3 = -16.3°$$

$$\sin I_3' = \left(\frac{1.8}{1.0}\right)(-0.2804) = -0.5047$$

$$I_3' = -30.3°$$

$$U_3' = (-3.0°) + (-30.3°) - (-16.3°) = -17.1°$$

$$Q_3' = \frac{\sin I_3' - \sin U_3'}{C_3}$$

$$= \frac{(-0.5047) - (-0.2934)}{-0.0115} = +18.4$$

Entering these data makes our table complete (Table 2.1-1).

Finally, to find the axial intercept, we use Equation [2.1-10]:

$$L' = -\frac{Q_3'}{\sin U_3'}$$

$$= -\frac{+18.4}{(-0.2934)} = \boxed{+62.7 \text{ mm}}$$

TABLE 2.1-1   RAY TRACE THROUGH CONVERGING DOUBLET

	Surface 1		Surface 2		Surface 3
Radius, R	+50		−50		−87
Curvature, C	+0.02		−0.02		−0.0115
Thickness, d		20		5	
Index, n		1.5		1.8	
Q	+22		+19.4		+19.8
Q'	+22.5		+20.1		+18.4
$\sin U$	0		−0.1572		−0.05296
U	0		−9.0°		−3.0°
$\sin U'$	−0.1572		−0.05296		−0.2934
U'	−9.0°		−3.0°		−17.1°
$\sin I$	+0.44		−0.5448		−0.2804
I	+26.1°		−33.0°		−16.3°
$\sin I'$	+0.2933		−0.454		−0.5047
I'	+17.1°		−27.0°		−30.3°

which means that a ray, entering the system at a height of 22 mm, will be refracted such that it intersects the axis 62.7 mm from the back vertex of the last lens.

SKEW RAYS

Rays that pass through an optical system are either *meridional rays* or *skew rays*. Meridional rays lie in planes containing the optic axis. As a meridional ray proceeds through the system, it will remain in its meridional plane as long as the system is centered, that is, as long as the vertices and centers of curvature all lie on a common straight line, the optic axis. Tracing a meridional ray is comparatively easy; it is a matter of plane geometry.

Skew rays pass through a system *outside* meridional planes. Tracing a skew ray is much more difficult; it is a matter of solid geometry.

For the tracing of skew rays, and in lens design in general, it is best to use x and y as the coordinates perpendicular *(normal)* to the optic axis and to use z for distances along the axis, as shown in Figure 2.1-6.

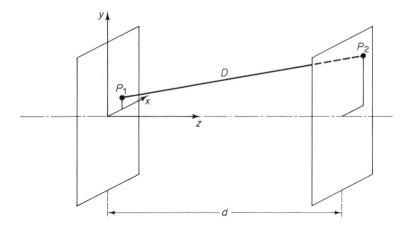

Figure 2.1-6   Tracing a skew ray from one surface to the next.

Assume that the ray comes from a point P_1, which has the coordinates (x_1, y_1, z_1), and proceeds to a point P_2, which has the coordinates (x_2, y_2, z_2). The distance between the two points is

$$D = \sqrt{(x_2 - x_1)^2 + (y_2 - y_1)^2 + (z_2 - z_1)^2} \qquad [2.1\text{-}11]$$

This ray is a *vector;* its three components are the projections of the ray on the three coordinate axes: $(x_2 - x_1)$, $(y_2 - y_1)$, and $(z_2 - z_1)$. The ray subtends with the axes a set of three angles whose cosines are the *direction cosines, X, Y,* and *Z,*

$$X = \frac{x_2 - x_1}{D}$$

$$Y = \frac{y_2 - y_1}{D} \qquad [2.1\text{-}12]$$

$$Z = \frac{z_2 - z_1}{D}$$

We note that an arbitrary ray in space is completely specified by its direction cosines and by the coordinates of the point at which it intersects a given surface. For example, if point P_1 (x_1, y_1, z_1) is given and we know the cosines X, Y, and Z, we can find the coordinates of point P_2 using Equations [2.1-12],

$$x_2 = x_1 + DX$$
$$y_2 = y_1 + DY \qquad [2.1\text{-}13]$$

and if $z_1 = 0$,

$$z_2 = DZ$$

The sum of the squares of the direction cosines is unity,

$$X^2 + Y^2 + Z^2 = 1 \qquad [2.1\text{-}14]$$

which shows that two of the angles are independent variables and one is dependent. If both x and X are zero, the ray is a meridional ray and arccos Z is the slope of the ray, U.

Now assume that the second surface is part of a sphere of radius R, with its center, C, on the optic axis (Figure 2.1-7). To find the point where the ray intersects the sphere, we use the *vertex depth formula,* noting that the half-diameter, $D/2$, that occurs in the formula is actually h, the (total) distance of P_2 from the axis,

$$(R - v)^2 + h^2 = R^2$$
$$v = R - \sqrt{R^2 - h^2} \qquad [2.1\text{-}15]$$

At P_2 we erect a *surface normal,* again a vector. Its direction cosines (which we identify by a degree sign, °), follow Equations [2.1-12], except that we change subscripts 1 to 2 (because we are at point P_2) and distance D to radius R,

$$X° = -\frac{x_2}{R}$$

$$Y° = -\frac{y_2}{R} \qquad [2.1\text{-}16]$$

$$Z° = \frac{z_C - z_2}{R}$$

Next we determine the *angle of incidence I,* at P_2. In general, the cosine of an angle subtended by two lines in space is given by the direction cosines of the two lines,

$$\cos I = XX° + YY° + ZZ° \qquad [2.1\text{-}17]$$

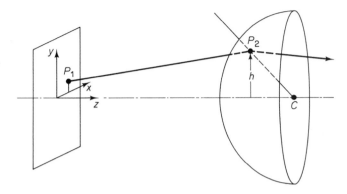

Figure 2.1-7 Continuing the ray through a spherical surface.

Substituting for $X°$, $Y°$, and $Z°$ the terms in Equations [2.1-16], we obtain

$$\cos I = (X)\left(-\frac{x_2}{R}\right) + (Y)\left(-\frac{y_2}{R}\right) + (Z)\left(\frac{z_C - z_2}{R}\right)$$

$$= -\frac{1}{R}\left[x_2 X + y_2 Y - (z_C - z_2)Z\right] \qquad [2.1\text{-}18]$$

Then we use *Snell's law* to find the angle of refraction, I'. Ordinarily, Snell's law is written

$$n \sin I = n' \sin I'$$

but since Equation [2.1-18] refers to $\cos I$, it is more practical to convert Snell's law to cosine terms. To do so, we square both sides,

$$n^2 \sin^2 I = n'^2 \sin^2 I'$$

use the identity

$$\sin^2 I + \cos^2 I = 1$$

solve for $\sin^2 I$, and substitute in the preceding equation:

$$n^2(1 - \cos^2 I) = n'^2(1 - \cos^2 I')$$

$$n^2 - n^2 \cos^2 I = n'^2 - n'^2 \cos^2 I'$$

$$n'^2 \cos^2 I' = n'^2 - n^2 + n^2 \cos^2 I$$

$$\cos^2 I' = 1 - \left(\frac{n}{n'}\right)^2 + \left(\frac{n}{n'}\cos I\right)^2$$

$$\cos I' = \sqrt{1 - \left(\frac{n}{n'}\right)^2 + \left(\frac{n}{n'}\cos I\right)^2} \qquad [2.1\text{-}19]$$

Now comes the essential point. Since both the incident ray and the surface normal are vectors, we anticipate that the refracted ray is a vector as well. To show that these three

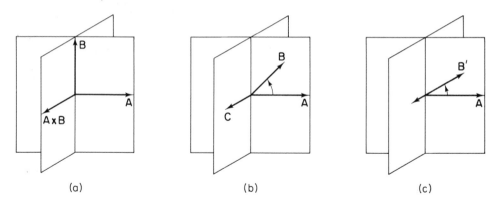

Figure 2.1-8   (*From left to right*) As the angle subtended by **A** and **B** decreases, the vector cross product **A** × **B** also decreases.

vectors all lie in the same plane, we use *vector cross products.* Generally, the cross product of two vectors is another vector, pointing in a direction *normal* to the plane defined by the two initial vectors. The *direction* of **A** × **B**, for example, is found from the right-hand rule: Rotate **A** toward **B** with the fingers of the right hand; the thumb will point in the direction **A** × **B** (Figure 2.1-8, left).

If **A** and **B** are unit vectors, the resultant **A** × **B**, likewise, is a unit vector. But if **A** and **B** subtend an angle less than 90°, the resultant, **C**, becomes less also (Figure 2.1-8, center and right). In the example of light passing through a surface, **A** is the surface normal, **N**°; **B** is the incident ray, **R**; and **B**′ is the refracted ray, **R**′. This makes the angle subtended by **N**° and **R** the angle of incidence, I, and the angle subtended by **N**° and **R**′ the angle of refraction, I',

$$\mathbf{N}° \times \mathbf{R} = \sin I$$

$$\mathbf{N}° \times \mathbf{R}' = \sin I'$$

Substituting these terms in Snell's law,

$$n \sin I = n' \sin I'$$

gives

$$\boxed{n(\mathbf{N}° \times \mathbf{R}) = n'(\mathbf{N}° \times \mathbf{R}')}$$ [2.1-20]

which is the *vector form of Snell's law.*

The cross products, though, do not mean that the light actually goes where the resultant vectors point to; the resultants merely connect the two sides of an equation. Their ratios, however, are significant. The ratio $(\mathbf{N}° \times \mathbf{R})/(\mathbf{N}° \times \mathbf{R}')$ is equal to n'/n, and it also is equal to $\sin I/\sin I'$. Consequently, we can redraw Figure 2.1-8, retaining **R**, **R**′, and **N**° and eliminating the cross-product resultants and the plane that contains them. The three remaining vectors, **R**, **R**′, and **N**°, in other words, lie in the same plane, the *plane of incidence,* as indeed they should (Figure 2.1-9).

From the vector form of Snell's law, Equation [2.1-20], from the definition of the angle

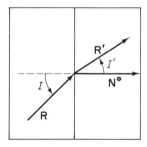

Figure 2.1-9 Eliminating the resultant leads again to the conventional form of Snell's law.

of incidence, Equation [2.1-17], and from a similar definition of the angle of refraction it follows that

$$n(X - X^\circ \cos I) = n'(X' - X^\circ \cos I') \qquad [2.1\text{-}21]$$

$$nX - nX^\circ \cos I = n'X' - n'X^\circ \cos I'$$

$$n'X' = nX - nX^\circ \cos I + n'X^\circ \cos I'$$

$$X' = \left(\frac{n}{n'}\right)X - \left(\frac{1}{n'}\right)(n \cos I - n' \cos I')(X^\circ)$$

Substituting Equation [2.1-16a], $X^\circ = -x_2/R$, and repeating the process for Y' and Z', we obtain the *direction cosines of the refracted ray:*

$$
\boxed{
\begin{aligned}
X' &= \frac{n}{n'} X - \frac{1}{n'R}(n' \cos I' - n \cos I)x_2 \\[2mm]
Y' &= \frac{n}{n'} Y - \frac{1}{n'R}(n' \cos I' - n \cos I)y_2 \\[2mm]
Z' &= \frac{n}{n'} Z - \frac{1}{n'R}(n' \cos I' - n \cos I)(z_2 - z_C)
\end{aligned}
}
\qquad [2.1\text{-}22]
$$

To sum up the pertinent steps: Presumably we know the coordinates of two points of intersection, P_1 and P_2. From the coordinates of these points we find the distance between them:

$$D = \sqrt{(x_2 - x_1)^2 + (y_2 - y_1)^2 + (z_2 - z_1)^2} \qquad [2.1\text{-}11]$$

Then we calculate the direction cosines:

$$X = \frac{x_2 - x_1}{D}$$

$$Y = \frac{y_2 - y_1}{D} \qquad [2.1\text{-}12]$$

$$Z = \frac{z_2 - z_1}{D}$$

If the second surface is part of a sphere, we need to know its radius of curvature, R, and the location of the center, C. This gives us the angle of incidence, I:

$$\cos I = -\frac{1}{R} [x_2 X + y_2 Y - (z_C - z_2)Z] \qquad [2.1\text{-}18]$$

Knowing $\cos I$ and the refractive indices on both sides of the surface, n and n', we find the angle of refraction, I':

$$\cos I' = \sqrt{1 - \left(\frac{n}{n'}\right)^2 + \left(\frac{n}{n'} \cos I\right)^2} \qquad [2.1\text{-}19]$$

With I' known, we determine the refraction cosines:

$$X' = \frac{n}{n'} X - \frac{1}{n'R} (n' \cos I' - n \cos I)x_2$$

$$Y' = \frac{n}{n'} Y - \frac{1}{n'R} (n' \cos I' - n \cos I)y_2 \qquad [2.1\text{-}22]$$

$$Z' = \frac{n}{n'} Z - \frac{1}{n'R} (n' \cos I' - n \cos I)(z_2 - z_C)$$

At the next surface, the refracted ray becomes the incident ray and the process is repeated until the ray reaches the image plane.

Example

Trace a ray through a spherical surface that has a radius of curvature of $+26$ mm and separates air ($n = 1.0$) on the left from glass ($n' = 1.8$) on the right. The ray originates at an object placed 100 mm in front of the surface, at a point P_1 that has the coordinates (2, 2, 0); it enters the surface at a point P_2 that has the coordinates $x_2 = 6$, $y_2 = 8$. Determine the direction cosines before and after refraction and the angles of incidence and refraction.

Solution.   First we determine the distance from P_1 to P_2. So far we know only the distance measured parallel to the axis (100 mm) to the *tangent plane,* the plane tangent to the surface at the vertex. The (axial) distance from there to P_2 is found from the vertex depth formula, but for using that formula we need to know h, the distance of P_2 from the axis:

$$h = \sqrt{x_2^2 + y_2^2} = \sqrt{6^2 + 8^2} = 10$$

The additional axial distance follows from Equation [2.1-15],

$$v = R - \sqrt{R^2 - h^2} = 26 - \sqrt{26^2 - 10^2} = 2$$

and the total distance, from P_1 to P_2, from Equation [2.1-11]:

$$D = \sqrt{(x_2 - x_1)^2 + (y_2 - y_1)^2 + (z_2 - z_1)^2}$$
$$= \sqrt{(6 - 2)^2 + (8 - 2)^2 + (102 - 0)^2} = 102.25$$

Now we determine the direction cosines of the incident ray, using Equations [2.1-12]:

$$X = \frac{x_2 - x_1}{D} = \frac{6 - 2}{102.25} = \boxed{0.039}$$

$$Y = \frac{8 - 2}{102.25} = \boxed{0.059}$$

$$Z = \frac{102 - 0}{102.25} = \boxed{0.998}$$

The cosine of the angle of incidence is found from Equation [2.1-18]:

$$\cos I = -\frac{1}{R} [x_2 X + y_2 Y - (z_C - z_2)Z]$$

Note that the center of curvature, z_C, is located at the z-coordinate of the vertex plus R, $z_C = 100 + 26 = 126$; thus

$$\cos I = -\frac{1}{26} [(6)(0.039) + (8)(0.059) - (126 - 102)(0.998)] = 0.894$$

Next we determine the cosine of the angle of refraction, using Equation [2.1-19]:

$$\cos I' = \sqrt{1 - \left(\frac{n}{n'}\right)^2 + \left(\frac{n}{n'} \cos I\right)^2}$$

$$= \sqrt{1 - \left(\frac{1}{1.8}\right)^2 + \left(\frac{1}{1.8} 0.894\right)^2} = 0.9685$$

and the direction cosines of the refracted ray, using Equations [2.1-22]:

$$X' = \frac{n}{n'} X - \frac{1}{n'R} (n' \cos I' - n \cos I)x_2$$

$$= \frac{n}{n'} X - (K)(x_2)$$

$$= \frac{1}{1.8} (0.039) - \frac{1}{(1.8)(26)} [(1.8)(0.9685) - 0.894](6)$$

$$= \boxed{-0.087}$$

$$Y' = \frac{n}{n'} Y - (K)(y_2)$$

$$= \frac{1}{1.8} (0.059) - (K)(8)$$

$$= \boxed{-0.112}$$

$$Z' = \frac{n}{n'} Z - (K)(z_2 - z_C)$$

$$= \frac{1}{1.8} (0.998) - (K)(102 - 126)$$

$$= \boxed{+0.990}$$

The angle of incidence is easily found from its cosine:

$$I = \arccos 0.894 = \boxed{+26.6°}$$

and the angle of refraction either from extending Equation [2.1-19]:

$$I' = \arccos 0.9685 = \boxed{+14.4°}$$

or from Snell's law:

$$\sin I' = \frac{n}{n'} \sin I = \frac{1}{1.8} \sin 26.6°$$

$$I' = \boxed{+14.4°}$$

This completes our example.

Ray tracing through a new design often requires tracing more than 100 rays, some of them meridional rays but probably most of them skew rays. Tracing that many rays is a formidable task, all but impossible without the help of a high-speed computer. Lens design then becomes a matter of *computer-aided lens design,* the topic to be discussed in Chapter 2.4.

SUGGESTIONS FOR FURTHER READING

R. KINGSLAKE, *Lens Design Fundamentals* (New York: Academic Press, Inc., 1978).

R. KINGSLAKE, *Optical System Design* (New York: Academic Press, Inc., 1983).

D. C. O'SHEA, *Elements of Modern Optical Design* (New York: John Wiley & Sons, Inc., 1985).

Military Standardization Handbook, *Optical Design,* MIL-HDBK-141 (Washington, DC: Defense Supply Agency, 1962).

PROBLEMS

2.1-1. A parallel bundle of light proceeds inside a solid rod of $n = 1.5$ until it reaches the back surface, ground on the rod with a radius of curvature of -24 mm. How wide a beam, limited only by total internal reflection, can emerge from the back surface?

2.1-2. Continue with Problem 2.1-1 and determine the (total) angular spread of the beam after it has passed through the surface.

2.1-3. A ray originates at an axial point object 200 mm in front of a convex surface of 24 mm radius of curvature, ground on a block of glass of $n = 1.72$. The ray enters the surface at a height of 12 mm. Determine the distance, Q, of the ray from the vertex.

2.1-4. Continue with Problem 2.1-3 and find the axial intercept of the refracted ray.

2.1-5. A meridional ray subtends with the $+z$ axis an angle of $+30°$ and another meridional ray subtends an angle of $+45°$. Determine the direction cosines of the two rays and the angle between them.

2.1-6. Determine the direction cosines of a ray that proceeds from a point P_1 (5, 2, 0) to a point P_2 (-3, -2, 1). Check by Equation [2.1-14].

2.1-7. A ray of light comes from a point (2, 4, 0) and at another point (4, 7, 10) enters a thick, plane-parallel plate of glass of $n = 1.6$, placed normal to the optic axis. What is the angle of incidence at the plate?

2.1-8. Continue with Problem 2.1-7. Determine the angle of refraction (both by the skew ray formula and by Snell's law) and the direction cosines of the refracted ray.

2.1-9. Consider a surface separating air ($n = 1.0$) from glass ($n' = 1.72$). The surface has a radius of curvature $R = +16$ mm; its vertex is 24 mm from the object. A ray originates at a point (2, 2, 0) and is incident on the surface at $x_2 = 3$, $y_2 = 4$. Find the direction cosines.

2.1-10. Continue with Problem 2.1-9 and determine:

(a) The angle of incidence and

(b) The angle of refraction, using both the skew ray formula and Snell's law.

2.2

Aberrations

AGAIN, OUR THREE-ELEMENT SYSTEM SERVES as a guide to further exploration. If a few sample rays, as in Figure 2.2-1, are traced through the first lens, we find that they do not come together as we would like them to. Instead, the peripheral rays intersect the optic axis closer to the lens, the central rays intersect farther away. This is an example of an *aberration*.

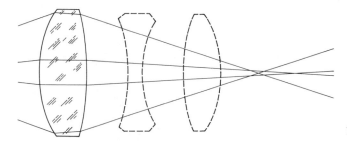

Figure 2.2-1 Spherical aberration.

Aberrations are due to inherent shortcomings of a lens, even a lens made of the best glass and free from manufacturing and other defects. Some aberrations occur already with monochromatic light; they are *monochromatic aberrations* (1 through 5 in the list below). Other aberrations occur only with light that contains at least two wavelengths; these are *chromatic aberrations* (6). We distinguish the following:

1. Spherical aberration
2. Coma
3. Oblique astigmatism
4. Curvature of field
5. Distortion
6. Chromatic aberration

In addition, spherical aberration and chromatic aberration both have a longitudinal (axial) and a transverse (lateral) variety, each with its own causes and corrections.

We have seen earlier that in *paraxial ray tracing* the angles that the rays subtend with the optic axis are assumed to be small (no larger than a few degrees). If so, the sines of the angles can be set equal to the angles themselves (in radians). In *trigonometric ray tracing*, the angles can be larger but then the sine of an angle must be represented by a series expansion:

$$\sin x = x - \frac{x^3}{3!} + \frac{x^5}{5!} - \frac{x^7}{7!} + \cdots \qquad [2.2\text{-}1]$$

with x again given in radians. The factorials, 3! for example, are $3! = 1 \cdot 2 \cdot 3$, and so on.

If only the first term, x, is retained (that is, if we assume that $\sin x = x$), we have an example of paraxial or *first-order optics*. If the next term, $x^3/3!$, is retained also (and we assume that $\sin x = x - x^3/3!$), we have *third-order optics*.*

As we have seen in Figure 2.2-1, aberrations can well be considered a matter of misaligned rays, and then may be called *ray aberrations*. But aberrations may as well be described in terms of wavefronts, and then be called *wave aberrations*.

For example, the dashed curve in Figure 2.2-2 may represent a reference sphere. It converges toward point P. The actual wavefront, however, has a different shape, and may converge at P'. The wavefront could also be tilted (such that the point of convergence moves off axis). In either case, the wavefronts are spherical and the wavefront normals come together, but at the wrong point. That happens in curvature of field and distortion. On the other hand, the wavefronts may be distorted and *not* converge toward a point. That happens in spherical aberration, coma, and astigmatism.

SPHERICAL ABERRATION

Spherical aberration is the phenomenon wherein rays passing through different zones of a surface come to different foci. We have already seen that rays passing through a lens farther away from the axis are refracted more, and come to a focus closer to the lens, than paraxial

---

*Aberrations based on third-order theory are often called *von Seidel aberrations*, named after Ludwig Philipp von Seidel (1821–1896), German mathematician and astronomer, professor at the University of Munich. After working on divergent and convergent series, von Seidel became interested in stellar photometry, probability theory, and the method of least squares, studied the trigonometry of skew rays, and became the first to establish a rigorous theory of monochromatic third-order aberrations which led to the construction of much improved astronomical telescopes. L. Seidel, "Ueber die Entwicklung der Glieder 3ter Ordnung, welche den Weg eines ausserhalb der Ebene der Axe gelegenen Lichtstrahles durch ein System brechender Medien bestimmen," *Astron. Nachr.* **43** (1856), 289–304, 305–20, 321–32.

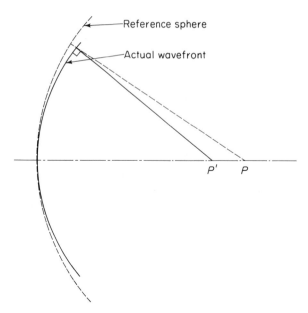

Reference sphere

Actual wavefront

P' P

Figure 2.2-2 Illustrating the concept of wave aberration.

rays (Figure 2.2-1). That is called *positive longitudinal spherical aberration,* and a lens acting this way is said to be *undercorrected.* An image pixel that is formed by paraxial rays, therefore, will be surrounded by a diffuse halo formed by peripheral rays, and vice versa—which accounts for the blur of the image.

Consider first in more detail the spherical aberration of a single surface. The distance from the surface to the point where *paraxial* rays come to a focus is the conventional, paraxial image distance, s'. This distance is found from Gauss' formula (page 32),

$$\frac{n}{s} + \frac{n' - n}{R} = \frac{n'}{s'} \qquad [2.2\text{-}2]$$

The distance where *peripheral* rays come to a focus is found from trigonometric ray tracing (pages 91–95). The result, from Equation [2.1-10], is the axial intercept, L',

$$L' = -\frac{Q'}{\sin U'} \qquad [2.2\text{-}3]$$

The difference between the two distances is a measure of the (longitudinal) spherical aberration,

$$\text{LSA} = s' - L' \qquad [2.2\text{-}4]$$

Example

Determine the spherical aberration of a single surface that has a radius of curvature of $+50$ mm and is ground on a block of glass of index 1.50. Compare two rays, both parallel to the optic axis, but one of them next to the axis and the other 22 mm away. (That is the same ray we have traced earlier in the example on page 95.)

Solution. The paraxial focal length is found from Gauss' formula or, more directly, from Equation [1.3-5] solved for f,

$$f = \frac{n'R}{n' - n} = \frac{(1.5)(+50)}{1.5 - 1.0} = +150 \text{ mm}$$

For the peripheral ray we determine first the angle of incidence at the surface, using Equation [2.1-2],

$$Q = R \sin I - R \sin U$$

Since $\sin U = 0$, this simplifies to

$$\sin I = \frac{Q}{R} = \frac{22}{+50} = +0.44$$

The angle of incidence, therefore, is

$$I = +26.1°$$

The angle of refraction is found from Snell's law,

$$\sin I' = \frac{n}{n'} \sin I = \frac{1.0}{1.5} (+0.44) = +0.2933$$

$$I' = +17.1°$$

Next we determine the ray slope after refraction, using Equation [2.1-5] solved for U',

$$U' = U - I + I' = (0) - (+26.1°) + (+17.1°) = -9°$$

Knowing U', we find Q', again using Equation [2.1-2],

$$Q' = R(\sin I' - \sin U')$$

$$= (+50)[(+0.293) - (-0.157)] = +22.5$$

The axial intercept is then found from Equation [2.2-3],

$$L' = \frac{Q'}{\sin U'} = -\frac{+22.5}{-0.157} = +143 \text{ mm}$$

The difference between f and L' is the longitudinal spherical aberration, at the heights specified,

$$\text{LSA} = 150 - 143 = \boxed{+7 \text{ mm}}$$

Farther away from the optic axis, generally, the spherical aberration becomes worse. If we plot the focal length, or the length of the axial intercept, as a function of height, we obtain a curve characteristic of the type and magnitude of the aberration. In Figure 2.2-3, for example, the spherical aberration is positive. If the curve were to lean to the right, the aberration would be negative and the lens be *overcorrected*.

Besides longitudinal (axial) spherical aberration, LSA, there is also a *transverse* (lateral) spherical aberration, TSA. It is defined as the height above, or below, the axis at which a peripheral ray intersects the plane of the paraxial focus (see Figure 2.2-4); in a way, the

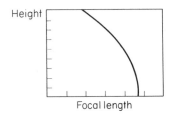

Figure 2.2-3 Positive spherical aberration.

TSA is a more meaningful measure of image blur than the LSA. TSA and LSA are connected as

$$\text{TSA} = \text{LSA} \tan U' \qquad\qquad [2.2\text{-}5]$$

From Figure 2.2-4 we see that the light, approaching the focus, at first converges and then diverges. The cross section where the beam contracts to a minimum is called the *circle of least confusion,* a term that will come up again later when we discuss cylinder lenses.

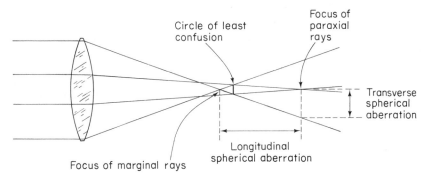

Figure 2.2-4 Longitudinal and transverse spherical aberration.

Correction for spherical aberration.   With a single spherical lens, spherical aberration cannot be eliminated completely. But it can be minimized. That is possible by making the two surfaces of the lens contribute equally, similar to setting a prism to the angle of minimum deviation (Figure 2.2-5). The lens, in short, must have the right shape, called *bending* (Figure 2.2-6).

To further reduce spherical aberration, the radii should be as long as practical, which means using *high-index* glass. Complete elimination of spherical aberration is possible only by making either or both surfaces *aspherical,* or by using a *gradient-index lens* whose index of refraction is higher in the center than in the periphery (see page 134). Finally, in a lens

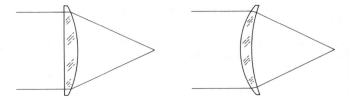

Figure 2.2-5 If one surface contributes nothing to the refraction (*left*), spherical aberration is worse than if both surfaces contribute about equally (*right*).

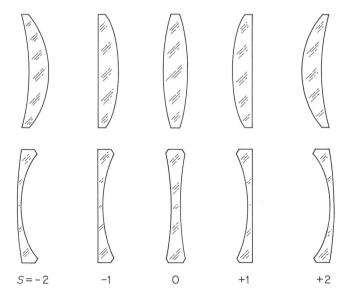

$S = -2$ -1 0 $+1$ $+2$

Figure 2.2-6 Lenses of different bending but (within each row) equal focal length. Coddington shape factor shown below.

system, the undercorrection of one lens can be compensated for by the overcorrection of another lens.

The shape of a lens at which it has the least amount of spherical aberration is called its *best form.* To find the best form, we need to consider two factors. One is the *Coddington shape factor,* * S. It describes the degree of bending as a function of the two radii of curvature, R_1 and R_2:

$$S = \frac{R_2 + R_1}{R_2 - R_1}$$
[2.2-6]

For example, when $S = 0$, the lens is symmetric, either equiconvex or equiconcave. When $S = -1$, the first surface is plane and the second surface is either convex or concave. When $S = +1$, the second surface is plane. When S is less than -1, or larger than $+1$, the lens is of the meniscus type. If the lens is turned around so that the light is incident on the other side, the shape factor retains its absolute value but the sign changes.

Whereas S is determined by the physical shape of the lens, the *Coddington position factor, P,* depends on the object and image distances actually used:

$$P = \frac{s' + s}{s' - s}$$
[2.2-7]

---

*Henry Coddington (born around 1800, died 1845), English mathematician and cleric. Coddington had many interests, spoke several languages, was a good musician, draftsman, and botanist. He wrote two books on optics, the more important one the two-volume *A System of Optics.* Part I, *A Treatise on the Reflexion and Refraction of Light* (Cambridge: Cambridge University Press, 1829), contains a thorough investigation of reflection and refraction. In volume 2, Coddington discusses the eye and the theory and construction of various types of eyepieces, telescopes, and microscopes.

When $P = -1$, the incident light is parallel. When $P = 0$, object distance and image distance are equal, except for the sign; and when $P = +1$, the emergent light is parallel. Like the shape factor, P has no dimensions.

Now, to find the best form, we first determine the position factor. Substituting in Equation [2.2-7] the Gaussian thin-lens equation, solved for s', gives

$$P = \frac{sf + s(s + f)}{sf - s(s + f)} = -\left(\frac{2f}{s} + 1\right) \qquad [2.2\text{-}8]$$

and solved for s,

$$P = \frac{s'(f - s') + s'f}{s'(f - s') - s'f} = 1 - \frac{2f}{s'} \qquad [2.2\text{-}9]$$

Then from the lens-makers formula,

$$\frac{1}{f} = (n - 1)\left(\frac{1}{R_1} - \frac{1}{R_2}\right) = \frac{n - 1}{R_1} - \frac{n - 1}{R_2}$$

$$R_1 = f(n - 1) - \frac{fR_1(n - 1)}{R_2} \qquad [2.2\text{-}10]$$

Next we take the shape factor, Equation [2.2-6], and solve for R_2:

$$R_2 = R_1 \frac{S + 1}{S - 1} \qquad [2.2\text{-}11]$$

Substituting Equation [2.2-11] in [2.2-10] gives

$$R_1 = f(n - 1) - \frac{f(n - 1)(S - 1)}{S + 1}$$

Multiplying the $f(n - 1)$ term by $(S + 1)/(S + 1)$ and simplifying leads to

$$R_1 = \frac{2f(n - 1)}{S + 1} \qquad [2.2\text{-}12]$$

and similarly,

$$R_2 = \frac{2f(n - 1)}{S - 1} \qquad [2.2\text{-}13]$$

Differentiation then shows that *minimum spherical aberration* occurs whenever

$$\boxed{S = -\frac{2(n^2 - 1)}{n + 2} P} \qquad [2.2\text{-}14]$$

Example

Determine the radii of curvature of a lens of $f = +10$ cm, $n = 1.5$, which for parallel incident light has minimum spherical aberration.

Solution.   First, determine the position factor:

$$P = \frac{s' + s}{s' - s} = \frac{(+10) + (-\infty)}{(+10) - (-\infty)} = -1$$

Substitute $n = 1.5$ and $P = -1$ in Equation [2.2-14]:

$$S = -\frac{(2)(1.5^2 - 1)}{1.5 + 2}(-1) = +0.714$$

Then from Equations [2.2-12] and [2.2-13],

$$R_1 = \frac{2f(n - 1)}{S + 1} = \frac{(2)(10)(1.5 - 1)}{(+0.714) + 1} = \boxed{+5.83 \text{ cm}}$$

$$R_2 = \frac{2f(n - 1)}{S + 1} = \frac{(2)(10)(1.5 - 1)}{(+0.714) - 1} = \boxed{-35 \text{ cm}}$$

Dividing R_1 by R_2 shows that to attain minimum spherical aberration for parallel incident light, the radii of curvature of a lens should be related as $+1 : -6$. Such a lens is nearly plano-convex, the convex side facing the light. If the lens were placed halfway between object and image, the best form would be symmetric.

COMA

Now we consider points that lie *off the optic axis*. Look at Figure 2.2-7, top, and compare it with our earlier Figure 2.2-1. This time the object is a circular opening of a certain diameter, D. The lens forms an image of diameter D'. Note that rays 1, which pass through the periphery of the lens, form an image *larger* than rays 2 which pass through the center. The image, therefore, is blurred, the result of *coma*.*

If we consider only a single object point (as on the top right in Figure 2.2-7), rays 2 form a well-defined image point, but rays 1 will give a larger, more diffuse patch. The result is a *comatic flare*. The diffuse tail of the flare most often lies in a direction away from the axis ("outward" coma), as shown, but depending on the type of lens used, it can also extend toward the axis ("inward" coma).

To correct for coma, we need to make the different images coincide, as shown in Figure 2.2-7, bottom. This time the light comes from an extended source, placed far to the left of the opening. Again some of the light proceeds along the axis. Other light is more oblique, subtending an angle U with the axis. This angle relates to the path difference p as

$$\sin U = \frac{p}{D}$$

---

*The term *coma* comes from the Greek κόμη, long hair, referring to the long tail of a comet. Coma is easy to see. Hold a magnifying glass in the path of sunlight and tilt it: The image of the sun will elongate into the cometlike shape characteristic of coma.

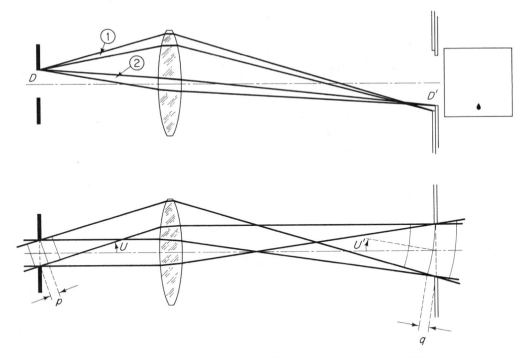

Figure 2.2-7   Different magnifications in coma (*top*) and deriving the sine condition for its elimination (*bottom*).

At the image, the conjugate angle is U' and

$$\sin U' = \frac{q}{D'}$$

For the different images to coincide, elemental lengths of p must equal conjugate lengths of q and, consequently,

$$D \sin U = D' \sin U' \qquad\qquad [2.2\text{-}15]$$

which is *Abbe's sine condition.* Since the ratio D'/D is the transverse magnification, this means that if all details of the image have the same magnification and

$$M_T = \frac{\sin U}{\sin U'} \qquad\qquad [2.2\text{-}16]$$

no matter what the ray slopes, there is no coma.

Correction for coma is possible by bending, or by using a combination of lenses symmetric about a central stop. A system free of both spherical aberration and coma is called an *aplanat.*

OBLIQUE ASTIGMATISM

Oblique astigmatism, like coma, is another off-axis aberration.* Its main characteristic is that for oblique rays (that come from off-axis points), a focal *point* is drawn out into a complex, three-dimensional focal figure called a *conoid.* The conoid is limited by two line foci, oriented at right angles to each other. These foci lie on (hypothetical) surfaces, one called the *tangential surface, T,* and the other the *radial surface, R* (Figure 2.2-8). Right at the axis, the two surfaces touch: for axial rays there is no astigmatism.

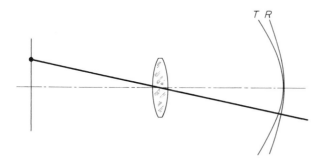

Figure 2.2-8 Tangential focal surface (*T*) and radial focal surface (*R*) as they occur in oblique astigmatism.

In positive astigmatism, the line focus closer to the lens is oriented *tangential* (tangent to a circle drawn around the optic axis) and the line focus farther away from the lens is oriented *radial.* Consequently, a point object, because of astigmatism, will be imaged as a tangential line at *T* and as a radial line at *R* (Figure 2.2-9a).

If the object were a radial line, its image would be blurred at the tangential focus, *T,* and sharp at the radial focus, *R* (b). In contrast, if the object were a tangential line, its image would be sharp at *T,* and blurred at *R* (c). If the object were a combination of radial and tangential lines, such as a spoked wheel, the rim of the wheel (which is the sum of short tangential lines) will be in focus at *T* and the spokes (which are radial lines) at *R* (d).

To understand why these lines are oriented tangential and radial, respectively, consider the following. Let a bundle of light originate at an axial point source. The light spreads out as a right-circular cone. If and when this cone falls off-axis on a lens surface, it forms an ellipse. The major axis of the ellipse is always radial, the minor axis tangential. But rays that lie in the meridian of the major axis of the ellipse encounter a lens whose projected diameter is *shorter,* and its power higher. Such light comes to a focus *closer* to the lens, forming the tangential focus. Rays in the plane of the minor axis form the radial focus.

Cylinder lenses. Cylinder lenses have properties that remind us of astigmatism. But *oblique* astigmatism is an off-axis aberration. Cylinder lenses have *axial astigmatism.* While oblique astigmatism is a nuisance, axial astigmatism is often induced on purpose; it is used in the correction of visual deficiencies and in anamorphic (wide-screen) motion picture cameras and projectors.

Consider the cylinder lens shown in Figure 2.2-10. Assume that two pieces of cardboard are held

---

*The term astigmatism comes from the Greek, $\overset{\text{,}}{\alpha}$ = alpha privative, meaning "not," and $\sigma\tau\acute{\iota}\gamma\mu\alpha$ = point, meaning that a point object is no longer imaged as a point. A system free of astigmatism is called an *anastigmat.*

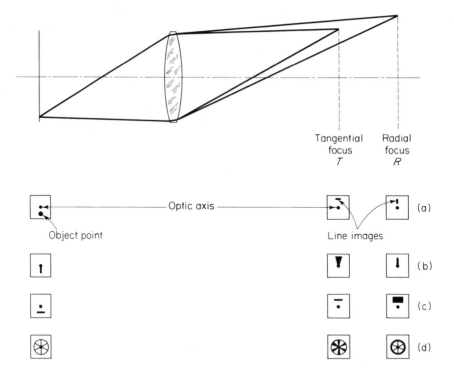

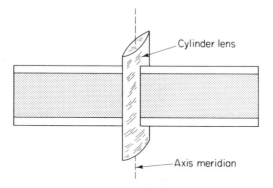

Figure 2.2-9   Astigmatic images of a sequence of objects.

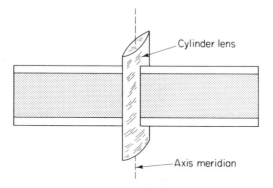

Cylinder lens

Axis meridian

Figure 2.2-10   Biconvex cylinder. Ribbon of light in the axis meridian.

to either side of the lens, with the light grazing along their surfaces; on the cardboard we see a *ribbon* of light. In the orientation shown, the light continues straight through the lens with *no refraction,* because the thickness of the lens, in this *axis meridian,* is constant; the lens acts merely as a plane-parallel slab of glass.

But imagine that the cardboard is turned through 90°, as in Figure 2.2-11. The ribbon of light then lies in the *power meridian,* 90° away from the axis meridian. In that orientation the light is refracted, and comes to a focus the same as with a conventional converging lens. A diverging, or *negative cylinder,* not shown, is comparable to a minus lens. Both types of cylinder lenses are *simple cylinders;* they have power in their power meridians but no power in their axis meridians.

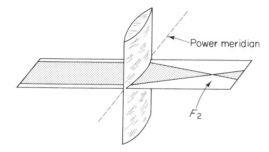

Figure 2.2-11 Cylinder with ribbon of light in the power meridian.

A *spherocylinder* is a combination of a conventional lens and a simple cylinder. Consider the two lenses illustrated in Figure 2.2-12. The spherical lens alone, in the meridian shown in the upper diagram, projects a point object into a point image; the cylinder, in this meridian, *does not contribute.*

In the power meridian, however, *both lenses contribute,* causing the light to come to a focus *closer* to the lens (bottom). Obviously, the light does not stop at the foci; at the first focus to the right of the lenses (lower diagram) it continues, and by the time it has reached the second focus, it has spread out into a *line,* oriented in the direction of the power meridian. On the other hand (upper diagram), when the light is at the first focus, it has *not yet* converged into a point; it still is a line. (Note that the first line focus is always parallel to the meridian of lesser power.) In reality, the two separate lenses shown in Figure 2.2-12 are fused into one, a *spherocylinder.* A spherocylinder, in other words, has one surface that is spherical and another that is cylindrical.

The two lines that limit the conoid are a certain distance apart, called the *interval of Sturm.** Within this interval there is one cross section where the bundle of light is circular; this is the *circle of least confusion.*

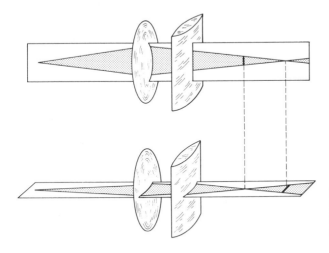

Figure 2.2-12 Spherocylinder (schematic), with ribbons of light in the axis meridian (*top*) and in the power meridian (*bottom*).

---

*Charles-François Sturm (1803–1855). Born in Switzerland, Sturm became a French citizen, mathematician and physicist, and professor of analysis and mechanics at the École Polytechnique in Paris. He is known for the theorem that bears his name (referring to the number of roots of an algebraic equation) and for his contributions to projective geometry, propagation of sound in water, and the optics of cylinder lenses. Much of Sturm's scientific work is contained in lecture notes, published posthumously as two books, *Cours d'analyse de l'École Polytechnique* (Paris, 1857–1859) and *Cours de mécanique de l'École Polytechnique* (Paris, 1861).

A *toric surface* has two radii of curvature (different in the two principal meridians), but in contrast to a simple cylinder, even the axis meridian has some power other than zero. Toric surfaces are used in ophthalmic lenses (which are menisci) whenever an additional cylinder component is needed. Usually, only one surface is toric, but *bitoric* lenses can be made also.

Example

A spherocylinder 70 mm in diameter produces two line foci whose distances from the lens are 12.5 cm and 20 cm, respectively. Determine the position and size of the circle of least confusion.

Solution.   First change the two focal lengths into powers:

$$P = \frac{1}{0.125} = +8 \text{ diopters}$$

$$P' = \frac{1}{0.2} = +5 \text{ diopters}$$

The circle of least confusion is located at the "dioptric midpoint,"

$$\frac{8 + 5}{2} = +6.5 \text{ diopters}$$

Thus it is

$$\frac{1}{6.5} = 0.1538 \approx \boxed{15.4 \text{ cm}}$$

from the lens, always a little closer to the left-hand focus (Figure 2.2-13).

The diameter d of the circle of least confusion is found from similar triangles:

$$\frac{d}{200 - 154} = \frac{70}{200}$$

$$d = (200 - 154)\frac{70}{200} = \boxed{16.1 \text{ mm}}$$

CURVATURE OF FIELD

Closely related to oblique astigmatism, curvature of field is another off-axis aberration. This time, though, the tangential and the radial surfaces coincide, forming the *Petzval surface.**
Shown schematically in Figure 2.2-14, curvature of field is especially objectionable in cam-

---

*Josef Max Petzval (1807–1891), Hungarian mathematician and professor of mathematics at the University of Vienna. Petzval is known for his eloquent, colorful lectures; he wrote two volumes on the integration of linear differential equations and extended Gaussian optics beyond the paraxial approximation by including higher powers. For diversion he liked fencing and other forms of physical activity. Soon after Louis Jacques Maudé Daguerre (1787–1851) had announced, in 1839, his process of photography, Petzval set out to calculate, with the help of eight artillery men familiar with arithmetic and loaned to him by Archduke Ludwig of Austria, a portrait camera lens, of much larger aperture (to allow shorter exposures) and much better correction than any other lens known at that time (1840). Unfortunately, when in 1859 burglars broke into his country home on the Kahlenberg near Vienna looking for valuables, they destroyed part of the manuscript and only a short version remains: J. Petzval, *Bericht über die Ergebnisse einiger dioptrischer Untersuchungen* (Pest, 1843).

Figure 2.2-13

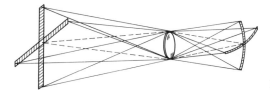

Figure 2.2-14   Curvature of field.

eras, photographic enlargers, and slide projectors, where the image plane is expected to be flat.

Correction for curvature of field, called "field flattening," is possible by using a combination of at least two lenses that meet the *Petzval condition:*

$$n_1 f_1 + n_2 f_2 = 0 \qquad\qquad [2.2\text{-}17]$$

Example

The Petzval condition can be met even if the two lenses have the same index of refraction. Consider a combination of two lenses that have the same power but opposite signs, and set them some distance d apart. Assume that $P_1 = +10\ \mathrm{m}^{-1}$, $P_2 = -10\ \mathrm{m}^{-1}$, $n = 1.55$, and $d = 25$ mm. This satisfies the Petzval condition,

$$\frac{1.55}{+10} + \frac{1.55}{-10} = 0$$

but is there any power left? (Work Problem 2.2-15 to find out.)

Obviously, it would be better to make the plus lens of glass of higher index than the minus lens. But this is just the opposite of what is needed for correction of chromatic aberration. It is only by the use of newer types of high-index low-dispersion glass that the conditions for achromatism and flatness of field can be met at the same time.

DISTORTION

Distortion is the transverse counterpart of curvature of field. Like curvature, distortion does not cause any image blur; instead, distortion refers to a sideways (radial) displacement of image points, either toward or away from the optic axis; in other words, it refers to a change of magnification.

We distinguish two types of distortion, *pincushion distortion* and *barrel distortion*. In pincushion distortion, the transverse magnification increases with increasing obliquity of the

Figure 2.2-15   Undistorted image (*left*), pincushion distortion (*center*), and barrel distortion (*right*).

rays; thus, the corners of a square, for example, are drawn out like a pillow. In barrel distortion the magnification decreases and the corners retract (Figure 2.2-15).

Look at a sheet of graph paper from a distance of about 50 cm. Hold a high plus lens first fairly close to the paper and then farther away from it. You will have no difficulty distinguishing the two types of distortion.

Distortion is readily explained as follows. The lens in Figure 2.2-16 forms an image of object 1–2 on the screen on the right. While point 1 is imaged (correctly) into 1′, point 2, because of astigmatism and curvature of field, is imaged into 2′, which lies to the left of the screen. But when the rays forming 2′ have reached the screen, they have expanded into a blur circle with the chief ray at its center, *C*. If a circular aperture is placed to the right of the lens, the blur circle contracts and its center moves to *D* (solid line), which is farther away from the axis. This results in pincushion distortion.

On the other hand, if the stop is placed to the left of the lens, *C* moves to *D′*, which is closer to the axis; this results in barrel distortion. To eliminate distortion, the stop should be placed *between* two lenses. Systems free of both curvature of field and distortion are called *orthoscopic*.

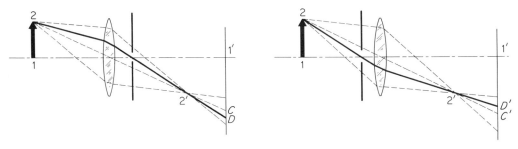

Figure 2.2-16   Rays causing pincushion distortion (*left*) and barrel distortion (*right*).

CHROMATIC ABERRATION

Since the index of refraction of matter varies with wavelength, a single lens has different powers for different colors: *Blue* light comes to a focus *closer* to the lens than red light (Figure 2.2-17). The horizontal distance between the two images is called *longitudinal chromatic aberration,* or "*longitudinal color.*"

In addition, the images produced by different colors are of different sizes; they have different transverse (lateral) magnifications. This is called *lateral chromatic aberration,* or "*lateral color.*" An image formed in the "blue focus" is closer to the lens and smaller; its details are surrounded by a red halo. The "red focus" is farther away from the lens and larger; its details have a blue halo.

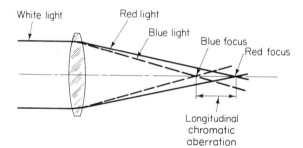

Figure 2.2-17   Longitudinal chromatic aberration.

Correction for chromatic aberration.   There are several ways of correcting for chromatic aberration. The best known is to use two lenses in contact, one made of crown, the other of flint (Figure 2.2-18). The crown lens is given more plus power than necessary. Its dispersion is moderate. Flint, on the other hand, has high dispersion. The two dispersions are then made equal but of opposite sign so that they cancel. Therefore, because of the higher dispersion of flint, the flint component can have less minus power than the crown has plus power, the combination has excess plus power, and the result is a *positive achromat.* *

Consider a combination of two lenses in contact, to be corrected for blue and red. We know that the approximate power of a combination is the sum of the powers of the elements:

$$\mathbf{P} = P_1 + P_2 \qquad\qquad [2.2\text{-}18]$$

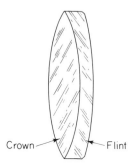

Figure 2.2-18   Achromatic doublet.

*This method of correcting for chromatic aberration goes back to 1729, when Chester Moor-Hall (1704–1771), British justice of the peace and amateur astronomer, designed the first achromatic contact doublet. Apparently, Moor-Hall had discovered that glass containing lead oxide, of the type used to make fine table glassware, had higher dispersion than window glass. To keep his discovery secret, he ordered one element for his doublet from a certain lens-maker in London, the other from another. As it happened, both men subcontracted the work to a third optician, who, on finding that both lenses were for the same customer and had one radius in common, placed them in contact and saw that the image was free of color. News of this discovery slowly spread to other opticians, among them John Dollond (1706–1761), whose son Peter (1739–1820) urged him to apply for a patent so that he could collect royalties. Naturally, the other London opticians objected and took the case to court, producing Moor-Hall as a witness. The court agreed that indeed Moor-Hall was the inventor, but in a much-quoted decision, the judge, Lord Camden, ruled in favor of Dollond, saying: "It is not the person who locked up his invention in his scritoire that ought to profit by a patent for such invention, but he who brought it forth for the benefit of the public."

We substitute for P the lens-makers formula, once for the crown element and once for the flint,

$$\underbrace{P = (n - 1)\left(\frac{1}{R_1} - \frac{1}{R_2}\right)}_{\text{crown}} + \underbrace{(n - 1)\left(\frac{1}{R_1} - \frac{1}{R_2}\right)}_{\text{flint}}$$

For convenience we replace the terms containing the R's by A's, writing

$$P = (n_1 - 1)A_1 + (n_2 - 1)A_2 \qquad [2.2\text{-}19]$$

In order to make the combination achromatic, its power at the two colors must be the same. Denoting blue by the subscript F and red by C, we obtain

$$(n_{1F} - 1)A_1 + (n_{2F} - 1)A_2 = (n_{1C} - 1)A_1 + (n_{2C} - 1)A_2$$

Multiplying out and canceling gives

$$\frac{A_1}{A_2} = \frac{n_{2F} - n_{2C}}{n_{1F} - n_{1C}} \qquad [2.2\text{-}20]$$

We repeat the process for yellow (subscript d),

$$P_{1d} = (n_{1d} - 1)A_1 \qquad \text{and} \qquad P_{2d} = (n_{2d} - 1)A_2$$

Dividing one by the other gives

$$\frac{A_1}{A_2} = \frac{P_{1d}(n_{2d} - 1)}{P_{2d}(n_{1d} - 1)} \qquad [2.2\text{-}21]$$

Then setting Equation [2.2-21] equal to [2.2-20] and solving for P_{1d}/P_{2d} yields

$$\frac{P_{1d}}{P_{2d}} = \frac{(n_{2F} - n_{2C})/(n_{2d} - 1)}{(n_{1F} - n_{1C})/(n_{1d} - 1)}$$

Both the numerator and the denominator in the right-hand term are the inverse of Abbe's number; therefore,

$$\frac{P_1}{P_2} = -\frac{\nu_1}{\nu_2} \qquad [2.2\text{-}22]$$

Note the minus sign. It means, since Abbe's number can only be positive, that if one of the lenses is positive, the other must be negative. Finally, we substitute the values of P_1 and P_2, respectively, from Equation [2.2-18] and obtain

$$\boxed{P_1 = P\left(\frac{\nu_1}{\nu_1 - \nu_2}\right) \qquad \text{and} \qquad P_2 = -P\left(\frac{\nu_2}{\nu_1 - \nu_2}\right)} \qquad [2.2\text{-}23]$$

Example

Design a crown-flint doublet of $+10$ cm focal length, achromatic for blue and red, using the refractive indices listed in Chapter 1.2, page 23.

Solution.   First, we determine Abbe's numbers. We find for crown

$$\nu_1 = \frac{n_d - 1}{n_F - n_C} = \frac{1.5230 - 1.0}{1.5293 - 1.5204} = 58.7640$$

and for flint

$$\nu_2 = \frac{1.7200 - 1.0}{1.7378 - 1.7130} = 29.0323$$

Inserting these figures in Equations [2.2-23] gives

$$P_1 = (+10)\left(\frac{58.7640}{58.7640 - 29.0323}\right) = +19.7648 \text{ m}^{-1}$$

and

$$P_2 = -(+10)\left(\frac{29.0323}{58.7640 - 29.0323}\right) = -9.7648 \text{ m}^{-1}$$

The combined power of the two lenses is $+10$ m^{-1}, which serves as a check on our calculations so far.

Knowing the focal lengths required of the two lenses, we are ready to choose their radii. For reasons of economy, the converging lens is made symmetric, equiconvex. Also, the two lenses are to be in contact. Thus $R_1 = -R_2 = -R_3$. For the first lens, from the lens-makers formula,

$$P = (n_{\text{lens}} - 1)\left(\frac{1}{R_1} - \frac{1}{R_2}\right)$$

$$+19.7648 = (1.523 - 1.000)\left(\frac{2}{R_1}\right) = \frac{1.046}{R_1}$$

and thus

$$R_1 = 0.0529 = \boxed{+5.29 \text{ cm}}$$

The next two surfaces have radii of

$$R_2 = R_3 = \boxed{-5.29 \text{ cm}}$$

and for the last surface,

$$-9.7648 = (0.7200)\left(\frac{1}{-0.0529} - \frac{1}{R_4}\right)$$

$$R_4 = -\frac{0.7200}{3.8458} = -0.1872 = \boxed{-18.72 \text{ cm}}$$

Another method of making a system achromatic is to use two positive lenses, made of the same type of glass and separated by a distance equal to one-half the sum of their focal lengths. To see why this approach works, we start out from the equivalent power equation, Equation [1.4-9],

$$\mathbf{P} = P_1 + P_2 - P_1 P_2 d$$

Then, following Equation [2.2-19],

$$\mathbf{P} = (n - 1)A_1 + (n - 1)A_2 - (n - 1)A_1(n - 1)A_2d$$
$$= (n - 1)(A_1 + A_2) - (n - 1)^2 A_1 A_2 d$$

For the combination to be achromatic, $\mathbf{P}$ must stay constant at different wavelengths; thus by differentiation,

$$\frac{d\mathbf{P}}{dn} = A_1 + A_2 - 2(n - 1)A_1 A_2 d = 0$$

We multiply by $(n - 1)$ and substitute for $(n - 1)A$ the corresponding P:

$$P_1 + P_2 = 2P_1 P_2 d$$
$$d = \frac{P_1 + P_2}{2P_1 P_2}$$

which is equal to

$$\boxed{d = \frac{1}{2}(f_1 + f_2)} \qquad [2.2\text{-}24]$$

Spaced doublets of this type are used in eyepieces (the Huygens and Ramsden eyepiece), discussed in Chapter 2.5.

A system corrected for chromatic aberration is called an *achromat,* but more precisely, if it is corrected for two colors, it should be called a *dichromat,* and if corrected for three colors, a *trichromat.**

Concluding remarks.   Are some aberrations more important than others? That depends on the application. A high-speed camera lens must be well corrected for spherical aberration; distortion is not as critical. For a lens used in reproducing topographic

TABLE 2.2-1   SUMMARY OF ABERRATIONS

Aberration	Character	Correction
1. Spherical aberration	Monochromatic, on- and off-axis, image blur	Bending, high index, aspherics, gradient index, doublet
2. Coma	Monochromatic, off-axis only, blur	Bending, spaced doublet with central stop
3. Oblique astigmatism	Monochromatic, off-axis, blur	Spaced doublet with stop
4. Curvature of field	Monochromatic, off-axis	Spaced doublet
5. Distortion	Monochromatic, off-axis	Spaced doublet with stop
6. Chromatic aberration	Heterochromatic, on- and off-axis, blur	Contact doublet, spaced doublet

---

*Microscope objectives corrected for three colors are sometimes called "apochromats" and systems corrected for four colors "superachromats." See M. Herzberger and N. R. McClure, "The Design of Superachromatic Lenses," *Appl. Opt.* **2** (1963), 553–60.

maps it is just the opposite. In any case, there is no way to eliminate all aberrations at the same time; in fact, eliminating some of them will usually make others come up worse. All that we can do is try to *balance* them. A summary of the various aberrations and what to do about them is found in Table 2.2-1.

SUGGESTIONS FOR FURTHER READING

W. H. Price, "The Photographic Lens," *Sci. Am.* **235** (Aug. 1976), 72–83.

A. Nussbaum and R. A. Phillips, *Contemporary Optics for Scientists and Engineers* (Englewood Cliffs, NJ: Prentice-Hall, Inc., 1976).

R. Kingslake, *Lens Design Fundamentals* (New York: Academic Press, Inc., 1978).

PROBLEMS

2.2-1. If peripheral rays are traced through a $+5.00$-diopter lens of 39 mm diameter, it is found that these rays come to a focus 5 mm ahead of the paraxial rays. What is the diameter of the blur circle in the paraxial focus?

2.2-2. If a lens of 50 mm diameter and 25.4 cm focal length has a longitudinal spherical aberration of 4 mm, what is its transverse spherical aberration?

2.2-3. A thin meniscus of index 1.60 has radii $R_1 = +15$ cm and $R_2 = +30$ cm. Determine:
   (a) The Coddington shape factor.
   (b) The position factor for an object 1.0 m away.

2.2-4. A lens of index 1.6 forms of an object 40 cm away an image 8 cm away. The front surface of the lens has a radius of curvature of $+120$ mm. Find the shape and position factors.

2.2-5. A lens made of flint of index 1.72 has a focal length of $+5$ cm. For parallel incident light and for the lens to have minimum spherical aberration, determine the position and shape factors and the two radii of curvature necessary.

2.2-6. A lens made of crown ($n = 1.523$) forms of an object 7 mm away an image 16 mm away. What is the best form to minimize spherical aberration?

2.2-7. If a target such as that shown in Figure 2.2-19 is imaged through a system not corrected for astigmatism, what changes will occur in the two principal foci?

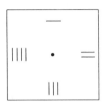

Figure 2.2-19

2.2-8. If the object consists of a series of dots, arranged in a circle concentric with the optic axis, what does the image look like:
   (a) In the tangential focus?
   (b) In the radial focus?

2.2-9. When a conventional (spherical) lens is tilted about its horizontal axis, a point object 20 cm in front of the lens gives two line images, one 25 cm and the other 40 cm behind the lens. How much cylinder has been induced?

2.2-10. The letters ⱻHⱢ serve as the object for a plus spherocylinder, axis vertical. Show what the image looks like in the two principal image planes.

2.2-11. A large-diameter spherocylinder forms one line focus at a distance of 21.7 cm and another line focus at a distance of 23.8 cm. To what diameter must an iris diaphragm close to the lens be stopped down in order to produce a circle of least confusion 0.8 mm in diameter?

2.2-12. Collimated light is focused by a spherocylinder into two line images 20 cm and 25 cm, respectively, away from the lens. If then a −3.00-diopter sphere is placed in contact with the spherocylinder, how long will the interval of Sturm be?

2.2-13. A spherocylinder forms with parallel light a circle of least confusion 6 mm in diameter on a screen located 30 cm from the lens. If a point source, placed 60 cm from the lens, gives a *line* image on the screen, what is the diameter of the lens?

2.2-14. A conventional thin lens forms of a pinhole 40 cm to the left of the lens a real image 20 cm to the right of the lens. On adding a cylinder to the lens, a horizontal line image is formed 10 cm to the right of the two lenses. What is the power and the orientation of the cylinder?

2.2-15. To meet the Petzval condition for the elimination of curvature of field, two lenses, of +10.00 and −10.00 diopters power and both of $n = 1.55$, are placed 25 mm apart. What is the power of the combination?

2.2-16. Derive a general equation for the equivalent focal length of a system of two spaced lenses of equal power but opposite sign. Check by using the same data as in Problem 2.2-15.

2.2-17. If a lens, made of glass of index 1.5, has +9.00 diopters power, what is the power of a lens of the same shape but made of glass of index 1.6?

2.2-18. A certain lens has, for yellow light, an index of refraction of 1.6500 and a focal length of 62 cm. For red light, the focal length becomes 62.5 cm. What is the refractive index for that light?

2.2-19. A collimated bundle of white light is passing through a single converging lens. How does the image change because of chromatic aberration when the image screen is moved slightly out of focus, once in a direction toward the lens and then farther away from it?

2.2-20. A +6.1-diopter crown lens is to be combined with a flint lens to make a contact doublet achromatic for blue and red. Using the V-numbers from Table 1.2-2, determine the power of the flint lens.

2.2-21. What are the powers of two thin lenses, one made of crown, the other of dense flint, that must be placed in contact to obtain a +6.00-diopter doublet achromatic for blue and red? Use the refractive indices from Table 1.2-2.

2.2-22. Design an achromatic doublet of 40 cm focal length, following the example in the text but using as the first lens a negative flint meniscus and as the second lens a convex-plane crown element, the plane surface being last.

2.2-23. An achromatic doublet consists of two positive lenses separated by 8 cm. The first lens has a focal length of 12 cm. What is the focal length:
   (a) Of the second lens?
   (b) Of the whole system?

2.2-24. Using two positive lenses, one of them of $f = 12$ cm, design an achromatic doublet of $f = 8$ cm. Find the focal length of the second lens and the distance between both.

2.3

Gradient-Index, Fiber, and Integrated Optics

GRADIENT-INDEX OPTICS, GRIN optics for short, has become a prominent part of modern optics. The term *gradient index* refers to the fact that lenses and other optical elements can be produced whose refractive index varies as a function of space. For example, circular, plane-parallel plates can be made whose refractive index is higher in the center than in the periphery. Such a plate will act as a positive, converging lens. If the plate in addition has curved surfaces, like an ordinary lens, it is equivalent to a combination of lenses and can be corrected for various aberrations, using no more than a single element. Other applications of GRIN materials extend to optical waveguides (*fiber optics*) and to miniaturized systems (*integrated optics*).

Atmospheric refraction. Gradient-index phenomena occur in everyday life. The atmosphere, for example, has a refractive index that decreases with height; higher up its density is less. This causes light to proceed in a curved path, an effect known as *regular atmospheric refraction* (Figure 2.3-1, top). A similar effect, called *looming,* is seen when looking across a body of cold water; it makes objects on the surface appear to be lifted up (center). The opposite is the *mirage;* it occurs when looking across an expanse of hot desert. Here, the air directly above the ground is hotter than the air higher up, its refractive index is lower, and distant objects appear below the horizon as though reflected in a pool of water (bottom). *Random atmospheric refraction* is due to turbulence; it causes the twinkling, or scintillation, of the stars.

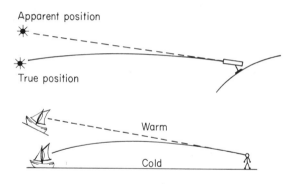

Apparent position

True position

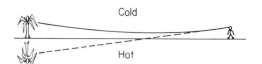

Warm

Cold

Cold

Hot

Figure 2.3-1 Regular atmospheric refraction (*top*), looming (*center*), and mirage (*bottom*). Dashed lines indicate apparent positions, full lines true positions.

THEORY OF REFRACTIVE GRADIENTS

Imagine a medium whose index of refraction is lowest at the top and increases from there toward the bottom. Let the light, as shown in Figure 2.3-2, be incident from the left. Outside the gradient field the light has a wavelength λ_0. At the top of the beam's cross section, the index is n and the wavelength λ. At the bottom, the index is n' and the wavelength λ'. If there are N wavefronts within the volume shown, the upper arc has a length

$$\Delta L = N\lambda = N\frac{\lambda_0}{n} \qquad [2.3\text{-}1]$$

and the lower arc

$$\Delta L' = N\lambda' = N\frac{\lambda_0}{n'}$$

The arc lengths depend on the radius of curvature, R, of the rays and on the angle of deviation, δ, measured in radians:

$$\Delta L = R\delta \qquad [2.3\text{-}2]$$

and

$$\Delta L' = (R - \Delta y)\delta \qquad [2.3\text{-}3]$$

Subtracting Equation [2.3-3] from [2.3-2] gives

$$\Delta L - \Delta L' = \Delta y\delta \qquad [2.3\text{-}4]$$

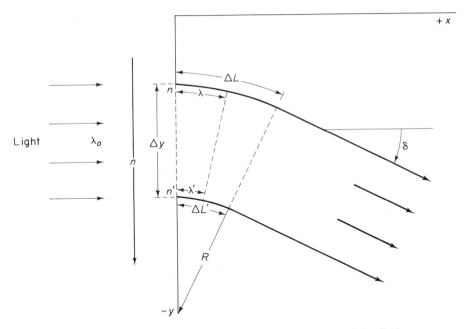

Figure 2.3-2   Deflection of light passing through a gradient-index field.

Since

$$\Delta L - \Delta L' = N\frac{\lambda_0}{n} - N\frac{\lambda_0}{n'} = N\lambda_0\left(\frac{n' - n}{nn'}\right)$$

and since from Equation [2.3-1]

$$N\lambda_0 = n\Delta L$$

$$\Delta L - \Delta L' = n\Delta L\left(\frac{n' - n}{nn'}\right) \qquad [2.3\text{-}5]$$

Combining Equations [2.3-4] and [2.3-5] by eliminating $\Delta L - \Delta L'$, and solving for δ yields

$$\delta = \frac{1}{\Delta y}\left(1 - \frac{n}{n'}\right)\Delta L$$

For light proceeding horizontally, in the $+x$ direction, it is sufficient to resolve δ into two components, δ_y and δ_z. The y component, in a first approximation, is given by

$$\delta_y = \int \frac{1}{n}\frac{\partial n}{\partial y}\, dL \qquad [2.3\text{-}6]$$

where the integral is taken over the length, L, traversed by the light. If this length is relatively short and the medium homogeneous in that direction,

$$\boxed{\delta_y = \frac{1}{n}\frac{\partial n}{\partial y}L} \qquad [2.3\text{-}7]$$

The angle of deflection, hence, is a function of the gradient $\partial n/\partial y$, of the index n, and of the length of path traversed L. Inserting Equation [2.3-7] into [2.3-2] gives

$$\Delta L = R\, \frac{1}{n}\, \frac{\partial n}{\partial y}\, L$$

and setting $\Delta L = L$ and solving for R,

$$R = \frac{n}{\partial n/\partial y} \qquad\qquad [2.3\text{-}8]$$

Thus *the steeper the gradient, the shorter the radius of curvature through which the light is bent.*

Example

Assume that a cuvette is filled with a salt solution whose refractive index varies from $n_1 = 1.4$ at the top to $n_2 = 1.6$ at the bottom. If the distance between top and bottom is 5 cm and the path through the solution 1 cm:
(a) Determine the *angle of deflection* of the light.
(b) Determine the *radius of curvature* through which the light is bent.
(c) Determine the *apex angle of a prism* (made of glass of $n = 1.5$) which would give the same deflection.

Solution.   (a) The angle of deflection is found from Equation [2.3-7], using for n the average refractive index, $(1.4 + 1.6)/2 = 1.5$:

$$\delta = \frac{1}{n}\, \frac{\partial n}{\partial y}\, L = \frac{1}{1.5}\, \frac{0.2}{0.05}\, (0.01) = 0.02667 \text{ rad} = \boxed{1.53°}$$

(b) The radius of curvature follows from Equation [2.3-8],

$$R = \frac{1.5}{0.2/5} = \boxed{37.5 \text{ cm}}$$

(c) Note that the 1.53° angle refers only to deflection within the solution. Outside, after passing through the rear surface of the cuvette, the angle, from Snell's law, becomes

$$I' = \arcsin(1.5)(\sin 1.53°) = 2.29°$$

The apex angle of a (thin) prism that gives the same deflection is found from Equation [1.2-10],

$$A = \frac{\delta}{n-1} = \frac{2.29°}{1.5-1} = \boxed{4.58°}$$

GRADIENT-INDEX LENSES

In a conventional lens, the refractive index is the same throughout. Refraction takes place only at the surfaces of the lens. In a *gradient-index lens,* or GRIN lens, refraction takes place also within the lens. That has major advantages. A single GRIN lens can be corrected to specifications that could not be met before. Hard-to-grind aspherics can be replaced by (spherical) GRIN lenses. And in complex optical systems the number of elements can be reduced, without sacrificing performance, by replacing a number of homogeneous lenses by a lesser number of GRIN lenses.

GRIN lenses come in three forms. One form has a *radial* gradient (of cylindrical symmetry), which means that the index of refraction varies as a function of distance *from* the optic axis.* The endfaces of such lenses can either be plane or can be ground to curved surfaces to give additional power.

Another type of GRIN lens contains an *axial* gradient. Here the refractive index varies as a function of distance *along* the axis: the surfaces of constant index are plane and normal to the axis. Axial refractive gradients are particularly useful for the correction of spherical aberration, replacing aspheric surfaces.

In Figure 2.3-3, top, we have an example of spherical aberration, similar to the one we have seen earlier in Chapter 2.2: the marginal rays come to a focus closer to the lens than the paraxial rays. Now consider a GRIN-lens blank containing an axial gradient (center left). The refractive index is highest near the front surface of the lens. If the front surface is ground convex (center right), some of the peripheral high-index material is removed and the marginal rays are bent less. With the gradient profile chosen right, all rays passing through come together at the same point, eliminating spherical aberration (bottom).

The third type is the *spherical* GRIN lens, where the index of refraction varies symmetrically about a point: The surfaces of constant index are *spheres*. An example of such a lens is the crystalline lens of the human eye (page 172).

A persistent problem in GRIN optics is how to produce the gradients. There are several methods available, neutron irradiation, chemical vapor deposition, polymerization, and ion stuffing, but the most promising, it seems, is *ion exchange*. Usually the process starts with an aluminosilicate glass. The sodium ions in such glasses are only loosely bound to the glass structure by weak interactions with the silicon and oxygen atoms. The Na^+ ions are then exchanged by other ions known to increase the refractive index, most notably silver, Ag^+. A lens blank is suspended in a bath of molten AgCl at a temperature of about 500°C for some 40 h. During that time, while the bath is continuously being stirred, the Ag^+ ions diffuse into the glass, replacing the Na^+ ions and raising the index. Theoretically, the index difference can be as high as 0.15, but in practice Δn is limited to about 0.05.

---

*GRIN lenses of radial symmetry were made as early as 1905 by Robert Williams Wood (1868–1955), professor of experimental physics at Johns Hopkins University. Wood used a mixture of glycerol and gelatin, and let it gel between two parallel plates, forming a cylinder a few centimeters in diameter. When the cylinder is soaked in water, the water slowly diffuses into the jelly, replacing (part of) the glycerol, and lowering the index. The cylinder is then cut into several slices, each slice representing a GRIN lens of radial symmetry—which forms an image just as a conventional (converging) lens does. Wood made notable contributions also to UV and IR spectroscopy, and was a superb lecturer with a flair for showmanship. For some of the spectroscopic experiments he did at his farm on Long Island, he used a grating at the end of a long section of sewer pipe; he cleaned the pipe of cobwebs by letting a cat run through it. Wood wrote two books, one of them a collection of nonsense poetry, *How to Tell the Birds from the Flowers* (New York: Dover Publications, Inc., 1959). When he sent a copy of it to President Theodore Roosevelt, the President asked for more of his writings, so Wood sent him a copy of his other book, *Physical Optics*. Even today, *Physical Optics* is worth reading for its wealth of experimental detail.

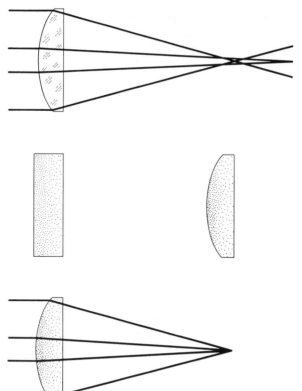

Figure 2.3-3 Eliminating spherical aberration by an axial GRIN lens. [For details, see D. T. Moore and D. P. Ryan, "Gradient Index Optical Lenses," *Phys. Teacher* **15** (1977), 409–13.]

EXPERIMENTAL GRADIENT-INDEX OPTICS

Under this heading we discuss various ways of evaluating the optical characteristics of GRIN elements, but for now we mention only *schlieren methods;* other ways, such as interferometry and moiré techniques, will come up later.

Töpler's method.   In Töpler's schlieren method* light from a point source, (1) in Figure 2.3-4, is focused by two lenses (2 and 4) onto a knife edge (5). In addition, the GRIN field (3) is imaged by a third lens (6), which is often the objective lens of a camera, onto a screen or photographic film (7). Sometimes, lenses 2 and 4 are replaced by concave first-surface mirrors.

---

*Robert Hooke in 1672 was the first to use schlieren techniques [according to J. Rienitz, "Schlieren experiment 300 years ago," *Nature (London)* **254** (1975), 293–95]. Léon Foucault applied them to testing large astronomical objectives [L. Foucault, "Mémoire sur la construction des télescopes en verre argenté," *Ann. Observatoire Imp. Paris* **5** (1859), 197–237], and August Joseph Ignaz Töpler, German scientist (1836–1912), applied them to the study of optical inhomogeneities in gases [A. Töpler, *Beobachtungen nach einer neuen optischen Methode* (Bonn: Max. Cohen und Sohn, 1864)]. The classic paper on schlieren optics is H. Schardin, "Die Schlierenverfahren und ihre Anwendungen," *Ergeb. exakt. Naturwiss.* **20** (1942), 303–439.

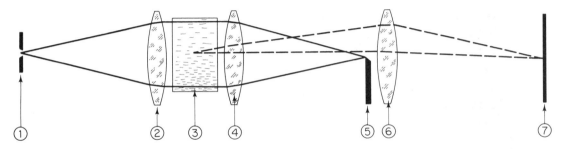

Figure 2.3-4   Töpler's method. 1, Light source; 2 + 4, schlieren head; 3, gradient-index field; 5, knife edge; 6, camera lens; 7, image plane.

If the sampling field is homogeneous and of uniform refractive index, the light passing through is blocked out by the knife edge and no light reaches the screen (solid lines). But if there are inhomogeneities (refractive gradients) present, these gradients deflect part of the light, which bypasses the knife edge, and the gradients show on the screen as bright streaks, real images of the gradients (dashed lines).

Ronchi grid method.   The Ronchi method has the advantage that it is quantitative; it shows the magnitude of the gradients. The Ronchi grid (see page 277) is placed somewhere between the objective and the eyepiece of a microscope (Figure 2.3-5). The exact

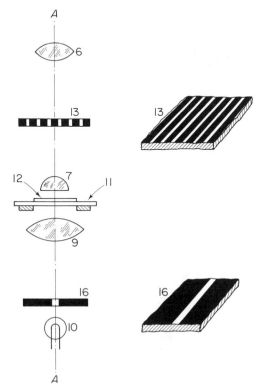

Figure 2.3-5   Ronchi-grid gradient-index microscope. *A-A*, Optic axis, 6, eyepiece; 7, objective lens; 9, condenser; 10, light source; 11 + 12, specimen; 13, Ronchi grid; 16, rotatable slit diaphragm. [From J. R. Meyer-Arendt, *Optical System for Microscopes or Similar Instruments*, U.S. Patent 2,977,847, Apr. 4, 1961.]

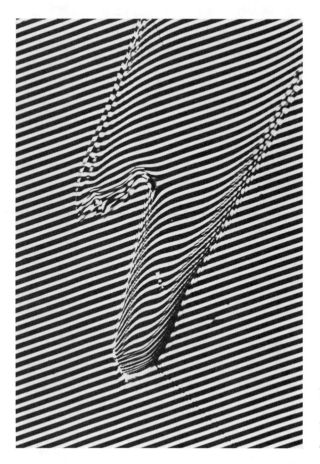

Figure 2.3-6 Photomicrograph of a thin layer of colorless cement spilled across a glass slide. Thickness gradients near the boundary of the layer cause characteristic distortions of the line pattern.

position is not critical. With the grid close to the eyepiece, the grid's shadow is more distinct but the sensitivity is less. With the grid near the objective, the opposite is true. When the slit (number 16 in Figure 2.3-5) is turned parallel to the grid, the image seen in the microscope has a series of parallel lines superimposed on it. If there are *no* gradients present, these lines are straight. But *with* gradients present, the shadow lines become distorted, as illustrated in Figure 2.3-6.

Color schlieren methods.   The magnitude of the gradients can also be represented by *color*. Early color schlieren systems were built in the form of prism spectrographs, with the light passing first through the prism and then through the gradient field. A second slit takes the place of the knife edge in Töpler's method. However, a slit significantly reduces both the amount of light and the resolution. Much better than a prism is a *wedge-type interference filter* and to use no slit at all (Figure 2.3-7).

Three processes take place:

1. *Without* index gradients, the filter merely casts a shadow on the film in the camera. This shadow provides a "background spectrum."

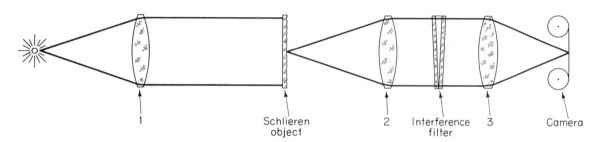

Figure 1 arrows and labels:

1   Schlieren   2   Interference   3   Camera
            object   filter

Figure 2.3-7   Color schlieren system using wedge-type interference filter. 1, Collimating lens; 2 + 3, schlieren head.

2. *With* index gradients, the two lenses 2 and 3 form a real image of the gradients, as in Töpler's method.

3. *In addition,* the gradients, which in effect are prisms, deflect the light through different parts of the filter. The result is that these gradients take on different colors, highly saturated and pleasing to look at.*

FIBER OPTICS

Optical fibers or, as they are called today, *fiber guides,* are of two types, *step-index fibers* and *GRIN fibers.* Step-index fibers were developed first.† They consist of a transparent core of glass or plastic of a given refractive index surrounded by a cladding of lower index. Most of the light travels inside the core and is contained there by total internal reflection.

Assume that n_0 in Figure 2.3-8 is the refractive index of the medium outside the fiber, n_1 the index of the core of the fiber, and n_2 the index of the cladding. To keep the light inside the core (and prevent leakage and "crosstalk"), the angle of incidence at the core–cladding interface must not fall below the minimum angle,

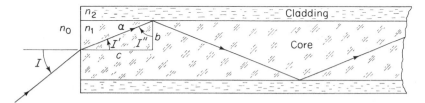

Figure 2.3-8   Propagation of light in a step-index fiber.

*See, for example, J. R. Meyer-Arendt, "Microscopy as a Spatial Filtering Process," in R. Barer and V. E. Cosslett, editors, *Advances in Optical and Electron Microscopy,* Vol. 8, p. 14 and facing page (London: Academic Press, Inc., 1982).

†At night, when we look at a water fountain illuminated from below, the light seems to follow the curved streams of water. Actually, the light follows a zigzag path, because of total internal reflection. The first to show this effect, using a stream of water flowing from a tank, was John Tyndall, "On Some Phenomena Connected with the Motion of Liquids," *Proc. Roy. Inst.* **1** (1854), 446–48. J. L. Baird extended the principle to fibers to convey an image, *An Improved Method of and Means for Producing Optical Images,* Brit. Patent 285,738, Feb. 15, 1928.

$$\sin I'' = \frac{n_2}{n_1} = \frac{c}{a} \qquad\qquad [2.3\text{-}9]$$

Since

$$a^2 = b^2 + c^2$$

$$1 = \frac{b^2}{a^2} + \frac{c^2}{a^2}$$

$$\left(\frac{b}{a}\right)^2 = 1 - \left(\frac{c}{a}\right)^2 = 1 - \left(\frac{n_2}{n_1}\right)^2$$

$$\frac{b}{a} = \sqrt{1 - \left(\frac{n_2}{n_1}\right)^2} = \sin I'$$

Then, from Snell's law,

$$n_0 \sin I = n_1 \sin I' = n_1\sqrt{1 - \left(\frac{n_2}{n_1}\right)^2} = \sqrt{n_1^2 - n_2^2} \qquad [2.3\text{-}10]$$

where $n_0 \sin I$ is the *numerical aperture,* NA, of the fiber.

In a GRIN fiber, the refractive index is highest along the axis; it decreases outward from there, as a function of radius. Assume, for simplicity, that the fiber material comes in discrete layers, as in Figure 2.3-9. Index n_1 is the highest. For some reason, for example when the fiber is bent, the light may deviate from the axis, making an angle I_1 with the normal. At the boundary n_1/n_2 the light is refracted, following Snell's law,

$$n_1 \sin I_1 = n_2 \sin I_2$$

At the *n*th boundary, at distance R from the axis, Snell's law becomes

$$n_1 \sin I_1 = n(R) \sin I(R)$$

and therefore,

$$n(R) \sin I(R) = \text{constant} \qquad\qquad [2.3\text{-}11]$$

This means that no matter whether the medium is layered or continuous, as n decreases, I increases and the light is bent toward the axis. The ray never even touches the

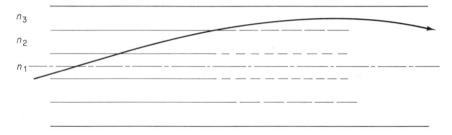

Figure 2.3-9   Refraction of light in a GRIN fiber.

surface. That is different from a step-index fiber, which relies on total internal reflection. The advantage of a GRIN fiber is that all rays inside have the same optical path length and a pulse of light injected at one end retains its shape when it emerges at the other.

Applications.   The most elementary application of fiber optics is the *transmission of light,* either to illuminate hard-to-reach places or to conduct light out of such places, perhaps to a photocell located at a more accessible site. In an instrument-laden cockpit, for instance, numerous individual light bulbs have given way to *light pipes* and a single light bulb that can easily be replaced when needed.

More interesting is the *transmission of images.* The best known example is the *flexible fiberscope.* As shown in Figure 2.3-10, some of the fibers conduct light into the cavity to be examined while the others carry the image back to the observer. The image-conducting fibers, up to 140,000 of them, are by necessity very thin, often no more than 10 μm in diameter each, and the entire fiber bundle may be no more than a few millimeters thick. Fiberscopes are used extensively in medicine and engineering; they make it possible to inspect just about any cavity in the human body, from the respiratory to the digestive tract, and to look inside the heart while it beats.

Of increasing interest is the use of fiber guides for *communications.* Compared to electrical conductors, optical fibers are lighter in weight, less expensive, equally flexible, not subject to electrical interference, and more secure to interception. Most important, fibers can now be made which have losses as low as 0.2 decibel (dB) per kilometer. This is a remarkable achievement considering that little more than a decade ago the best fibers had losses in excess of 1000 dB km^{-1}, and 20 dB km^{-1} was thought to be the limit.

Example

Both the *bel** and the *decibel* are comparative units. One bel means that the power in one channel, or at one time, is 10 times that in another channel, or at another time; 2 bel means 100$\times$, 3 bel 1000$\times$, and so on. For practical use the unit bel is too large, hence the decibel, dB,

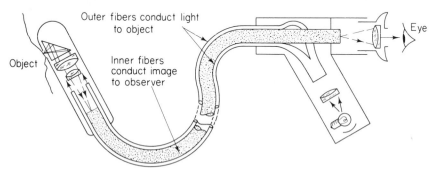

Figure 2.3-10   Flexible fiberscope.

---

*Named after Alexander Graham Bell (1847–1922), American scientist. Working at the Boston School for the Deaf, Bell reasoned that if sound could be made into fluctuating electric currents, these currents could be carried through wires over a distance and converted back into sound. When he accidentally spilled battery acid on his pants and, with his audio equipment nearby, called out to his assistant, near the other end of the line on another floor, "Watson, please come here. I want you," these famous words became the first telephone communication.

1 bel = 10 dB. The difference, in dB, between two powers, ϕ, therefore, is $10 \log_{10}(\phi_2/\phi_1)$. If one-half the initial power in a fiber is lost, $\phi_2 = \frac{1}{2}\phi_1$:

$$10 \log_{10} \frac{\frac{1}{2}\phi_1}{\phi_1} = 10 \log_{10} 0.5 = \boxed{-3 \text{ dB}}$$

the minus sign indicating the *loss*. Over a distance of 1 km, this is not much.

INTEGRATED OPTICS

In recent years, fiber optics has developed into complex systems of miniature dimensions. This is the field of *integrated optics*. There are two types of integrated waveguides: *planar* guides, which are relatively wide, and *strip* guides, which are more narrow, more similar to fibers. Both types are exceedingly thin, of the order of a wavelength of light, and either type can be made in the form of step-index guides or as GRIN guides, like fibers.

The light travels (mostly) inside the guide. But how does the light enter? Shining light head-on into the narrow side of a thin film would cause so much scatter that very little light would enter the film. Instead, *beam couplers* are used, which are either small prisms placed on top of the film, or gratings made interferometrically in a layer of photoresist.

A variety of (passive) elements can be made by changing the thickness of the film. In a thinner film the effective velocity of the light is higher (because its zigzag path is stretched out longer); in a thicker film it is less. A *thin-film prism* is made by adding, through a mask with a triangular opening, another layer of transparent material. When the light reaches the base of the triangular-shaped thicker film, it is delayed and, as in a conventional prism, it deviates from its initial direction. The same result can be obtained by a refractive gradient.

Thin-film lenses are of two varieties, *Luneburg lenses* and *geodesic lenses*. A Luneburg lens looks like a flat circular mound; it combines a thicker film and the GRIN feature, the index being highest in the center of the mound and decreasing toward the periphery (Figure 2.3-11, left). Light incident on the lens from *any* direction within the plane of the

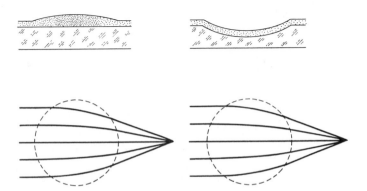

Figure 2.3-11   Thin-film Luneburg lens (*left*) and geodesic lens (*right*); ray trace for collimated light shown below.

film comes to a focus, interestingly enough without aberrations except curvature of field. A geodesic lens has the form of a dome or depression in a step-index film of uniform thickness (right). It works on the principle that rays follow the shortest path between two points on a surface—the same as with a conventional lens.

Active elements in integrated optics include light modulators, light switches, beam deflectors, and scanners. Modulators can be made to operate on the amplitude, phase, frequency, or state of polarization of the light. Beam deflectors and scanners change the direction of the light in response to an electric signal.

Manufacturing such elements is a rapidly evolving art. Earlier planar guides were simply films such as tantalum oxide or lithium niobate, coated on a substrate by vacuum deposition. Modern strip guides are made by diffusion techniques, ion implantation, proton bombardment, or material removal by electron or laser beam writing. Particularly interesting is the fabrication of *monolithic integrated optics* where all functions, from the generation of the light to guiding to modulation to detection, are performed in a single crystal, most commonly gallium arsenide, GaAs.

Gallium arsenide is a material well suited also for the construction of lasers (see Chapter 5.5, page 413). Such lasers serve as tiny, highly efficient light sources that together with other passive and active elements can be incorporated into small chips, often no larger than a few millimeters in size. Figure 2.3-12 gives an example.

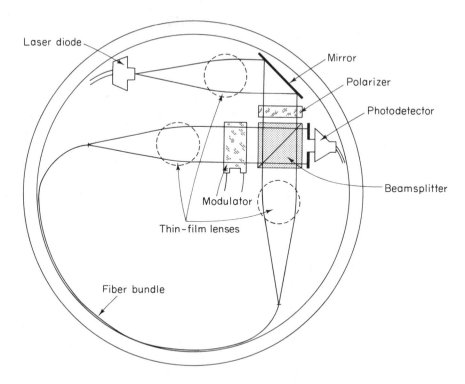

Figure 2.3-12 Schematic representation of integrated circuit used for fiber-optics gyroscope. For details on gyros, see Chapter 6.1, page 437.

SUGGESTIONS FOR FURTHER READING

Ch. K. Kao, *Optical Fiber Systems: Technology, Design, and Applications* (New York: McGraw-Hill Book Company, 1982).

D. Marcuse, *Light Transmission Optics*, 2nd edition (New York: Van Nostrand Reinhold Company, Inc., 1982).

M. Young, *Optics and Lasers*, 3rd edition, pp. 195–244 (New York: Springer-Verlag New York, Inc., 1986).

Y. Suematsu and K.-I. Iga, *Introduction to Optical Fiber Communications* (New York: John Wiley & Sons, Inc., 1982).

J. Gowar, *Optical Communication Systems* (Englewood Cliffs, NJ: Prentice-Hall, Inc., 1984).

R. G. Hunsperger, *Integrated Optics: Theory and Technology*, 2nd edition (New York: Springer-Verlag New York, Inc., 1984).

PROBLEMS

2.3-1. A slab of transparent material measures 3 mm $\times$ 10 mm $\times$ 60 mm. If the refractive index of the material varies linearly in the lengthwise direction from 1.51 to 1.55, how much will a beam of collimated light, incident normally on the widest face, be deflected at a distance of 2 m from the slab?

2.3-2. A palm tree is seen across 10 km of desert that has become so hot that directly above the ground the refractive index of the air is 1.000290, whereas 5 m higher up it is 1.000292. By how much does the tree appear to be displaced?

2.3-3. A plane-parallel GRIN plate of radial symmetry is 40 mm in diameter and 10 mm thick. Its refractive index varies from 1.6 in the center to 1.5 in the periphery. What is its focal length?

2.3-4. A plane-parallel plate, 50 mm in diameter and 10 mm thick, is made to have a peripheral refractive index $n_2 = 1.50$ and a focal length of 120 cm. What should be the refractive index in the center?

2.3-5. Imagine a solid sphere of gradient-index material whose index is highest in the center and 1.4 at the periphery.
   (a) Assume that light is injected tangentially into the sphere such that it travels inside, next to the surface. What is the *lowest-possible index* in the center to accomplish that?
   (b) What must be the *diameter* of the sphere?

2.3-6. Some GRIN material is made in the form of a doughnut. The inside diameter (the diameter of the hole in the doughnut) is 34 cm and the outside diameter 46 cm; the cross section of the ring, therefore, is 6 cm in diameter.
   (a) If the average index is 1.6, what radial gradient is needed to keep a beam of light traveling along exactly in the center of the ring?
   (b) What should be the highest, and the lowest, refractive index of the material?

2.3-7. Schlieren systems can be *calibrated* by placing a small, low-power lens in the schlieren field and replacing the knife edge by a Ronchi grid. What will the image look like and how can it be used for calibration?

2.3-8. Continue with Problem 2.3-7 and describe how the image will change if the grid is replaced by a wedge-type color filter.

2.3-9. If a step-index fiber has a core of index 1.55 and a cladding of 1.53:
(a) What is its numerical aperture?
(b) What is the maximum angle (of incidence) at which light can enter the fiber?

2.3-10. Parallel light enters a fiber of diameter $D = 0.1$ mm (Figure 2.3-13). Determine the least radius R through which the fiber may be bent, assuming that the core has an index of 1.54, the cladding 1.52.

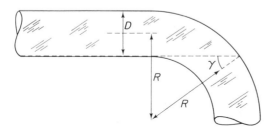

Figure 2.3-13

2.3-11. Assume that a series of object points lie on a circle concentric with a thin-film Luneburg lens. If the circle has a radius 1.5 times the focal length of the lens, where do we find the conjugate image points?

2.3-12. Three individual fibers enter an integrated circuit, and three other fibers leave it. A switching element can be designed such that any incoming fiber is connected to any outgoing fiber, but no incoming fiber is connected to more than one fiber, nor is any fiber *not* connected to any other fiber. How many permutations are possible?

2.4

Computer-Aided Lens Design

LENS DESIGN TODAY is a matter of high-speed computation. Whereas the design of a complete optical system is beyond the scope of this introduction, at least we show how things are done in principle. As a computer language that is both easy to learn, yet powerful enough to use, we choose BASIC. With this tool in hand, we discuss several applications: numerical data processing, computer graphics, and tracing a ray through a plane surface, through a spherical surface, and through a system of lenses such as that illustrated in Figure 2.4-1.

APPROACHING THE PROBLEM

By the turn of the century, ray tracing was a matter of actually tracing rays. Some time later, it became a numerical process done on a desk calculator. Then came card-programmed calculators. Today, ray tracing has become so complex, especially the tracing of skew rays, that a high-speed electronic computer is a virtual necessity.*

---

*The first electronic computer was built by John Vincent Atanasoff (1903–   ), American theoretical physicist and professor of physics and mathematics at Iowa State University, and by Clifford E. Berry, one of his graduate students, at about the same time that Alan Turing in England built the Colossus, and John W. Mauchly and J. Presper Eckert of the University of Pennsylvania built the Electronic Numerical Integrator and Calculator. Atanasoff, working on his thesis "The dielectric constant of helium," spent weeks of tedious computation with a desk calculator (to find numerical solutions to the Schrödinger equation) when, in the winter of 1937, he hit upon the idea of using electronic logic circuits (rather than mechanical relays) and a regenerative binary memory; in this way, he could separate one from the other.

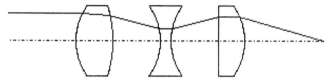

Figure 2.4-1   Triplet drawn by computer graphics.

The starting point is generally to ask for what purpose the new system is to be used. Is it a telescope objective, a camera lens, a slide projector? What should be its focal length, the size of the field, the aperture (f-number), and perhaps the zoom range? In addition, there may be other features, either required or desired: the resolution of the system (such as 150 lines/mm at the edge of a field of $6°$), the contrast transfer, and the freedom from various aberrations, especially distortion and chromatic aberration.

Perhaps at this point we may already guess what type of system might be best, and how many elements it should have. This step, in fact, is the most creative: To the experienced lens designer it is the most enjoyable, and to the beginner probably the most baffling.

Once we have chosen the type of system, we consider the powers and the spacing of the individual elements. So far, we assume that the lenses are thin. We also determine where to place the entrance pupil and exit pupil and how large these should be, and where to place stops, if any.

The next step is to correct for aberrations. We convert the powers of the lenses to curvatures of the surfaces, writing the lens-makers formula in the form

$$P = (n - 1)(C_1 - C_2) \tag{2.4-1}$$

The last term in parentheses tells us why by *bending* a lens we can change its shape without affecting its power.

By bending we can minimize certain aberrations, mainly spherical aberration and coma. That applies, moreover, not only to individual lenses but also to the system as a whole. Suppose that we make a small change in the curvature, C_1, of the first surface. That must be accompanied by a change in some other curvature (generally, but not always, the last) to hold the focal length constant. If this leads to an improvement, we continue in the same direction until the trend reverses. At that point we change the curvature of the second surface, C_2, repeating the process over and over by *iteration,* at each step tracing a series of rays to see how that will affect the image.

By now we need something to determine the quality of the image. That is often the *merit function,* defined as the sum of the squares of the various aberration coefficients. The smaller the merit function, the better the lens. (The coefficients are squared to prevent large coefficients of opposite sign from producing a small merit function.) Another measure of quality is the *spot diagram.* A spot diagram results when the entrance pupil of the system is divided into a checkerboard of small equal areas, and a ray is traced from a given point in the object through every one of them to the image plane. The more closely together the rays intersect the image plane, the better the lens. Still another method is the *modulation transfer function,* to be discussed in detail in Chapter 2.6.

Modern lens design, however, does more than trace rays and draw spot diagrams. These programs change variables within some preset limits, making the computer respond to the results of its own computations; that is called *optimization.* Interestingly enough, some lens systems have not only one but several best solutions, and sometimes these solutions can be reached by starting from different initial designs.

Clearly, all of this is rather complex. There can be no doubt that *computer-aided lens design* is a matter of intuition as much as a sophisticated and expensive art. Frequently, it requires at least a midsized computer, and costly software. Still, some aspects of the design process can be tried on a personal computer, as we will discuss in this chapter.

PROGRAMMING IN BASIC

A program written in BASIC* contains a sequence of instructions given to a computer, called *statements*. Each statement has a *line number,* followed by the instruction and by constants and variables to be operated on. *Commands* have no line numbers.

To begin a new program, type **NEW** and [ENTER]. To CLear the Screen, type **CLS** and [ENTER]. To LOAD a program (from the disk into the computer's memory), use **LOAD"** or, on the IBM Personal Computer, press **F3**; add file name and [ENTER]. To display a program (on the screen) that is currently in the memory, use **LIST** or press **F1** and [ENTER]. To RUN the program, use **RUN** or press **F2**. To SAVE the program, make sure the cursor is at the end of the program, then use **SAVE"** or press **F4**; add filename and [ENTER]. End the program by **END**.

To **edit** a program, type the number of the line to be changed wherever there is space and the new version and [ENTER]. The revised line will take the place of the old line. For REMarks such as headings or commentary, use **REM**, followed by the heading. The remarks will appear when the program is LISTed (displayed) but they are not acted on by the computer.

INPUT.   A string of text that asks for a response is set between quotation marks, followed by a comma and by the variable:

```
10   INPUT   "WHAT IS THE RADIUS?", R
```

When the program is RUN, the string appears on the screen:

```
WHAT IS THE RADIUS?
```

The system then waits for the response to be typed in, followed by [ENTER].

---

*The acronym BASIC stands for **B**eginner's **A**ll-purpose **S**ymbolic **I**nstruction **C**ode, a computer language developed in 1963–1964 by John George Kemeny (1926–) and Thomas Eugene Kurtz (1928–), both of Dartmouth College. A Hungarian-born mathematician, Kemeny came to the United States in 1940. After working under John von Neumann and later for Albert Einstein, he graduated in 1949 with a Ph.D. in mathematics from Princeton University. In 1953 Kemeny joined Dartmouth College as professor of mathematics and philosophy and chairman of the mathematics department; in 1970 he became president of the College. A native of Ilinois, Kurtz graduated in 1956 with a Ph.D. in statistics, also from Princeton University; in 1966 he became professor of mathematics at Dartmouth.

PRINT.   PRINT displays data on the screen. When used as a statement (with a line number), such as

```
20  PRINT X  [ENTER]
```

it causes the numeric value assigned to X to be displayed. If text is enclosed between quotation marks, as in

```
30  PRINT "FIRST ZONE"  [ENTER]
```

the words FIRST ZONE appear on the screen. PRINT by itself (with a line number) produces a blank line.

Constants and variables.   A *variable* may change during the run of a program, but a *constant* does not. A variable can be any letter, or group of letters, or it can be a letter followed by a single digit (in effect a subscript). Assume that a number such as 123.45 is to be added to 67, subtracted from 678, multiplied by 3.4, and divided by 5.6. This is done best by replacing 123.45 by A:

```
40  A = 123.45

50  PRINT A + 67; 678 - A; A * 3.4; A / 5.6
```

Line 40 is an *assignment* statement; it assigns a number to A. With only one variable in use, we could write

```
60  A = A + 2
```

This may look like an algebraic equation but it is not. As an equation, $A = A + 2$ makes no sense. In BASIC, it means that the new value of A, to the left of the equal sign, is set equal to the old value of A, to the right of the equal sign, plus 2. The new value will then replace the old value.

Arithmetic operations.   Arithmetic operations are performed in order of precedence, first exponentiation, $X = A^B$, then multiplication and division, $X = A * B$ and $X = A / B$, and then addition and subtraction, $X = A + B$ and $X = A - B$. For exponentiation (input) use the caret, $\wedge$, as in $5 \wedge 3 = 125$.

Operations in parentheses are carried out first. Otherwise, operations of equal priority are performed from left to right:

$$[(6 + 4)/(4 * 3 - 2)] \wedge 2 = [10/10] \wedge 2 = 1 \qquad [2.4\text{-}2]$$

An equation with built-up fractions such as

$$\frac{1}{s} + \frac{1}{f} = \frac{1}{s'}$$

must first be rewritten. Solving for s', for instance, gives

$$s' = \frac{1}{\dfrac{1}{s} + \dfrac{1}{f}}$$

or, as a program line:

$$120 \quad S' = 1 / ((1 / S) + (1 / F)) \qquad [2.4\text{-}3]$$

The terms FOR/NEXT set up a *loop* within the program. They let us use repeatedly the same set of instructions, each time, though, with a different set of data. FOR begins the loop, NEXT ends it. As an example, consider the quadratic equation

$$y = 3x^2 + 4x - 7 \qquad [2.4\text{-}4]$$

To evaluate (find) y, using for x any integer (whole number) from 1 to 20, proceed as shown in Program 2.4-1, QUADRA.

Graphics.   It is often helpful to show an optical system in *graphical form.* To draw a straight line between two points, go to the Graphics Mode, **Screen 1**, and use the **LINE** command with the coordinates (in parentheses). For example,

```
LINE (50,130) - (270,130)  [ENTER]
```

draws a line between the points (50,130) and (270,130). Adding ,1 or ,2 draws the line in color. Adding ,,,&H2222 draws a dotted line, ,,,&H7777 a dashed line, and ,,,&H2727 a line of alternating dots and dashes, as I use it to represent the optic axis.

To draw a spherical surface, use **CIRCLE**. For example,

```
CIRCLE (110,140), 40  [ENTER]
```

draws a circle with the center at (110,140) and with a radius of 40 (pixels). The two symbols **U** and **L**, as in

```
CIRCLE (150,100), 75,, U, L  [ENTER]
```

are the **U**pper limit and **L**ower limit (of the circumference). For a positive single surface (concave to the right), or a negative single surface (concave to the left), I have found it practical to use the following limits:

Radius of curvature	Positive surface		Negative surface	
	U	L	U	L
50	2.26	4	5.4	0.86
75	2.6	3.7	5.7	0.56
100	2.71	3.57	5.85	0.43
150	2.85	3.43	6	0.29

To draw a lens, place the optic axis at a height of 100. Draw two surfaces. Connect the upper ends, and the lower ends, of the surfaces by short horizontal lines. With lenses 64 (pixels) in diameter, these lines should be at heights of 68 and 132, respectively. An example is shown by Program 2.4-2, GRAFEX. When the program is RUN, it will draw the triplet seen on the opening page of this chapter (Figure 2.4-1).

PROGRAM 2.4-1 EVALUATING A QUADRATIC EQUATION

```
10      REM   Evaluating a Quadratic Equation
20      REM   Program QUADRA
30      PRINT
40      PRINT  "Evaluating the Equation Y = 3 X ∧ 2 + 4 X - 7"
50      PRINT,  "IF X=", "THEN Y="
60      FOR X = 1 TO 20
70         Y  =  3 * X * X  +  4 * X  -  7
80        PRINT, X, Y
90      NEXT
100     END
```

PROGRAM 2.4-2 COMPUTER-GENERATED TRIPLET

```
10      REM       Computer-Generated Triplet
20      REM       Program  GRAFEX
30      SCREEN 1
40      CIRCLE (150,100), 75, 1, 2.6, 3.7
50      CIRCLE (-40,100), 150, 1, 6, .29
60      CIRCLE (80,100), 75, 1, 5.7, .56
70      CIRCLE (240,100), 75, 1, 2.6, 3.7
80      LINE (210,68) - (210,132), 1
90      CIRCLE (160,100), 75, 1, 5.7, .56
100     LINE (10,100) - (310,100),,, &H2727
110     LINE (85,68) - (105,68), 1
120     LINE (145,68) - (175,68), 1
130     LINE (210,68) - (224,68), 1
140     LINE (85,132) - (105,132), 1
150     LINE (145,132) - (175,132), 1
160     LINE (210,132) - (224,132), 1
170     LINE (10,75) - (81,75)
180     LINE (81,75) - (107,77)
190     LINE (107,77) - (154,89)
200     LINE (154,89) - (166,89)
210     LINE (166,89) - (210,78)
220     LINE (210,78) - (230,78)
230     LINE (230,78) - (310,100)
240     END
```

TRACING A RAY THROUGH A PLANE SURFACE

We follow the ray using Snell's law:

$$n \sin I = n' \sin I' \qquad [2.4\text{-}5]$$

In BASIC, the sine of an angle R is found using **SIN(R)**, where the angle must be listed in parentheses and given in radians. To convert degrees to radians, set

$$PI = 3.141593$$

and multiply the angle in degrees, D, by PI/180:

$$R = D * PI/180 \qquad\qquad [2.4\text{-}6]$$

To conform to BASIC's notation, we replace n by N1, n' by N2, and I' by I2. We convert the argument (the angle) in sin I into radians using Equation [2.4-6], and solve for sin I', replacing sin I' by A,

$$A = N1 * SIN (I * PI / 180) / N2 \qquad\qquad [2.4\text{-}7]$$

Now we need to find the *inverse sine*. But BASIC has no inverse trigonometric functions except for the arc tangent, **ATN(T)**. ATN(T) returns the angle (in radians) whose tangent is T. The arc sine, however, can be derived from the arc tangent:

$$\text{If} \quad \sin \theta = A \quad \text{then} \quad \tan \theta = \frac{A}{\sqrt{1 - A^2}}$$

Take the inverse trigonometric functions of these two expressions and set them equal:

$$\theta = \arcsin A = \arctan \frac{A}{\sqrt{1 - A^2}} \qquad\qquad [2.4\text{-}8]$$

Written in BASIC, and substituting B for the angle in radians, that becomes

$$\theta_{(\text{in radians})} = B = ATN(A/SQR(1 - A * A)) \qquad\qquad [2.4\text{-}9]$$

Then, to convert radians to degrees, multiply the angle in radians by 180/PI:

$$D = B * 180/PI \qquad\qquad [2.4\text{-}10]$$

An example of determining the inverse sine is shown in Program 2.4-3, ARCSIN.

PROGRAM 2.4-3   DETERMINING THE INVERSE SINE

```
10      REM  Determining the Inverse Sine
20      REM  Program ARCSIN
30      PRINT
40      INPUT "What is the sine of the angle?    ", A
50      B  =  ATN (A / SQR (1 - A * A))
60      PI =  3.141593
70      D  =  B * 180 / PI
80      PRINT
90      PRINT "Arc sin"; A; " = "; B; "radian  = "; D; "degrees"
100     PRINT
110     END
```

Now we are ready to write the actual program, 2.4-4. The INPUT to Program 2.4-4,

PROGRAM 2.4-4   TRACING A RAY THROUGH A PLANE SURFACE

```
10      REM  Tracing a Ray Through a Plane Surface
20      REM  Program SNELLA
30      CLS
40      INPUT "Enter Angle of Incidence I:    ", I
50      INPUT "Enter Refractive Index N1:    ", N1
60      INPUT "Enter Refractive Index N2:    ", N2
70      PI = 3.141593
80      A   =   N1 * SIN (I * PI / 180) / N2
90      B   =   ATN (A / SQR (1 - A * A))
100     I2 =   B * 180 / PI
110     SCREEN 1
120     LINE (160,70) - (160,170),   1
130     LINE (80,120) - (240,120),,, &H2727
140     PSET (160,120)
150     TEMP =   I + 180
160     DRAW "TA =   TEMP; R 85
170     PSET  (160,120)
180     PRINT "Angle of Refraction ="; I2; "degrees"
190     DRAW "TA  =  I2; R 85
200     END
```

SNELLA,* is the angle of incidence I, the refractive index N1 to the left of the surface, and the index N2 to the right of the surface (lines 40 through 60). Snell's law is used in the form shown in Equation [2.4-7], with the angle of incidence, I, converted to radians (line 80).

To find the angle of refraction we use Equation [2.4-9]; it becomes line 90. To convert the angle of refraction to degrees we use Equation [2.4-10], that is line 100. The result, I′, is shown both numerically (line 180), and graphically. The graphical display is produced by lines 110–170 and 190; an example is shown by Figure 2.4-2.

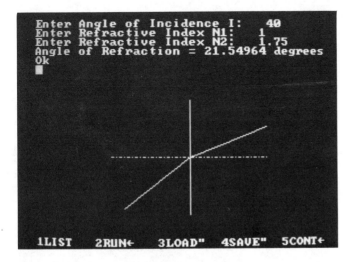

Figure 2.4-2 Computer display of refraction at a plane surface. I = 40°; N1 = 1; N2 = 1.75.

---

*Program SNELLA was originally written by Karen Meyer-Arendt.

TRACING A RAY THROUGH A SPHERICAL SURFACE

Next we let the ray pass through a spherical surface. The surface has a radius of curvature R. We assume that the incident ray is parallel to the axis and proceeding at a height H, and that the refractive index to the left of the surface is unity, $N1 = 1$. The input, in short, is the radius R, the height H, and the index to the right of the surface, $N2$. These are lines 40 through 60 in Program 2.4-5, SURTAC.

PROGRAM 2.4-5   TRACING A RAY THROUGH A SPHERICAL SURFACE

```
10      REM   Tracing a Ray Through a Spherical Surface
20      REM   Program SURTAC
30      PRINT
40      INPUT "Enter Radius of Surface, R:   ", R
50      INPUT "Enter Height of Ray, H:       ", H
60      INPUT "Enter Refractive Index N2:    ", N2
70      PRINT
80      A   =  H / R
90      B   =  ATN (A / SQR (1 - A * A))
100     PI  =  3.141593
110     I   =  B * 180 / PI
120     PRINT "Angle of Incidence ="; I; "degrees"
130     K   =  SIN(B) / N2
140     M   =  ATN (K / SQR (1 - K * K))
150     I2  =  M * 180 / PI
160     PRINT "Angle of Refraction ="; I2; "degrees"
170     W   =  M - B
180     Q2  =  R * SIN(M) - R * SIN(W)
190     L2  =  -Q2 / SIN(W)
200     PRINT "Axial Intercept = "; L2
210     PRINT
220     END
```

As we have seen in Chapter 2.1, we begin the ray trace with the *opening equation:*

$$QC = \sin I - \sin U \qquad\qquad [2.4\text{-}11]$$

The curvature C is the reciprocal of the length of the radius, $C = 1/R$. For a ray parallel to the axis, $Q = H$ and $U = 0$; thus, the last term in Equation [2.4-11] drops out and the opening equation, in accordance with Figure 2.4-3, becomes

$$\frac{H}{R} = \sin I$$

Since we know H and R, we can determine the angle of incidence:

$$I = \arcsin \frac{H}{R} \qquad\qquad [2.4\text{-}12]$$

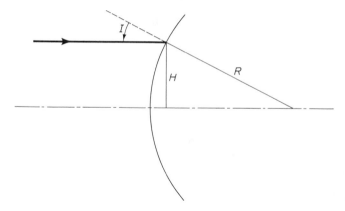

Figure 2.4-3   Ray incident on a spherical surface.

Now we proceed in the same way as earlier when tracing a ray through a plane surface. We replace H/R by a variable, A (line 80). The angle as such, arcsin I, is found from line 90:

$$B = ATN (A / SQR (1 - A * A)) \qquad [2.4\text{-}13]$$

To convert B into degrees, we multiply it by 180 / PI, which is line 110. If desired, we could display on the screen ("PRINT") the angle of incidence (line 120).

Next come the various *refraction equations*. For $n = 1$, Snell's law is simply

$$\sin I' = \frac{\sin I}{n'} \qquad [2.4\text{-}14]$$

But in BASIC the angles must be given in radians. So we replace I by B and substitute for sin I' the variable K; that changes Equation [2.4-14] to

$$K = SIN(B) / N2 \qquad [2.4\text{-}15]$$

which is line 130.

Again, we convert the sine of the angle of refraction into the angle itself,

$$I' = \arcsin(\sin I/n') \qquad [2.4\text{-}16]$$

which in BASIC becomes

$$M = ATN (K / SQR (1 - K * K)) \qquad [2.4\text{-}17]$$

where M is yet another variable, the angle of refraction in radians (line 140). Again, we may convert the angle of refraction into degrees and PRINT it (lines 150–160).

The second refraction equation is

$$U - I = U' - I' \qquad [2.4\text{-}18]$$

With I, I', and U known ($U = 0$), we can solve for U' (or U2 as we call it in BASIC):

$$U2 = I2 - I \qquad [2.4\text{-}19]$$

which holds true no matter whether the angles are given in degrees or radians. Equation [2.4-19], with all three angles in radians, becomes

$$W = M - B \qquad\qquad [2.4\text{-}20]$$

which is line 170.

The third refraction equation is

$$Q'C = \sin I' - \sin U' \qquad\qquad [2.4\text{-}21]$$

We divide both sides by C and again change $1/C \rightarrow R$. That becomes

$$Q2 = R * SIN(M) - R * SIN(W) \qquad\qquad [2.4\text{-}22]$$

With all the variables to the right of the equal sign known, we can now determine Q2 (line 180).

Finally, we come to the *closing equation,*

$$L' = -\frac{Q'}{\sin U'} \qquad\qquad [2.4\text{-}23]$$

or

$$L2 = -Q2/SIN(W) \qquad\qquad [2.4\text{-}24]$$

which is line 190. Since we know both Q2 and W (which is U2 in radians), we can now find L2, which is the distance from the surface to the point where the refracted ray intersects the axis (Figure 2.4-4).

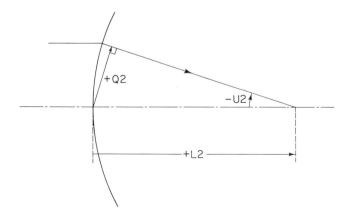

Figure 2.4-4 Ray refracted by a single spherical surface.

TRACING A RAY THROUGH A SYSTEM

There are many computer programs available that help in tracing a ray through a complete system.* Some of these programs can handle up to 100 surfaces, including cylindrical and aspheric surfaces. Most of the elements are lined up along the axis (making the system *coaxial*), but several can also be tilted or decentered.

---

*Some examples of commercially available ray-tracing programs are the following: Optical Research Associates' *Code V* program; Sinclair Optics' optical design programs *OSLO* and *Super-Oslo;* Don Small Optics' Small Optical Design & Analysis program *SODA;* Kidger Optics Ltd.'s *SIGMA;* Stellar Software's *BEAM2* and *BEAM3.*

First, we enter the characteristics of the system as far as they are known, the approximate curvatures of the surfaces, their spacing, and the types of glass. Based on these data, the program will draw a graphical layout.

Next comes a first-order ray trace, using at least several rays. If the result looks promising, we follow that with an exact, trigonometric ray trace including, perhaps, a number of skew rays. This will tell us which third-order aberrations are present, and how much each accounts for.

We then choose some of the curvatures, thicknesses, and types of glass and use them as variables. Depending on the system's purpose, we also attach weighting factors to the individual aberrations. That will permit *optimization* which tells us which curvatures, and which spacings between surfaces, are best. The most common types of glass are usually listed in the program, some times as many as 100 types with room to add more in case new developments come along.

For illustration I show a simple example, Program 2.4-6, tracing several rays through a combination of three lenses.* The input to the program involves two parts, an *optics table*, TRIPLE.OPT, that lists the index preceding the respective surface, the vertex position of the surface along the z-axis, and the curvature of the surface, the reciprocal of the length of the radius. The *ray table*, TRIPLE.RAY, lists the direction cosines X and Y of five rays. The graphics output produced by the program is shown in Figure 2.4-5.

PROGRAM 2.4-6   TRACING RAYS THROUGH A TRIPLET

```
   8 surfaces      TRIPLE.OPT
  index     Zvx      curv
---.----:----.----:----.---:------------------------------
  1.00  :   0.00  :   0.4  : lens   first element, convex front
  1.62  :   0.40  :  -0.1  : lens   first element, convex rear
  1.00  :   0.70  :  -0.3  : lens   second element, concave front
  1.59  :   0.88  :   0.4  : lens   second element, concave rear
  1.00  :   1.25  :   1.0  : aperture stop
  1.00  :   1.50  :   0.1  : lens   third element, convex front
  1.62  :   1.90  :  -0.2  : lens   third element, convex rear
  1.00  :  10.25  :        : image plane
        :         :        :
```

```
  5 rays       TRIPLE.RAY
     X0         Y0         U0
----.----:----.----:----.----:
    1.0  :   0.0  :   0.1  :
    0.7  :   0.7  :   0.1  :
    0.0  :  -0.7  :   0.1  :
   -0.7  :   0.7  :   0.1  :
   -1.0  :   0.0  :   0.1  :
         :        :        :
```

---

*For this demonstration I have used the BEAM2 lens design program.

Figure 2.4-5   Computer graphics showing rays passing through a combination of four lenses.

CONCLUSIONS

There can be no doubt that computers greatly facilitate any lens design. Just imagine how Petzval or Conrady, in bewilderment perhaps but also with appreciation, would look at a modern computer churning out ray traces in rapid succession. Still, in spite of all the publicity, the role of the computer is probably overrated. After all, it is the lens designer who designs lenses, and who comes up with the basic concept and tells the computer what to do.

Even optimization is not the answer to our problems. Indeed, there are cases where by making some changes in the basic layout (even at the price of a temporary deterioration) we may find a new avenue that leads to a superior design. Perhaps we may even find a completely different design, far removed from the original "optimized" sequence.

The role of the computer, in short, is to help the designer make decisions, and to take the repetitiveness out of the work. *Lens design as such is still an art.*

SUGGESTIONS FOR FURTHER READING

V. C. HARE, JR., *Introduction to Programming: A BASIC Approach* (New York: Harcourt Brace & World, Inc., 1970).

D. COOPER and M. CLANCY, *Oh! Pascal!*, 2nd edition (New York: W. W. Norton & Company, Inc., 1985).

PROBLEMS

2.4-1. Write a program that will produce the sines of the angles from 0 to 90°, in steps of 5°.

2.4-2. Print a table of the trigonometric functions SIN(X), COS(X), and TAN(X) of the angles from 10° to 80°, in steps of 5°.

2.4-3. Find the distance between two points, one with the coordinates X1, Y1 and the other with the coordinates X2, Y2, considering the distance between the two points is the hypotenuse of a right-angled triangle.

2.4-4. Using the CIRCLE statement and the coordinates of five properly placed points as centers, draw the five-ring symbol of the Olympic Games.

2.4-5. A ray of light in water ($n = 1.333$) is incident on a plate of glass ($n' = 1.523$) at an angle of $-65°$. What is the angle of refraction?

2.4-6. How far from true vertical does the setting sun appear to an observer under water ($n' = 1.333$)? Compare with Problem 1.2-4.

2.4-7. Trace a ray through a spherical surface that has a radius of curvature of $+48$ mm. How high above the axis must the ray be to produce an angle of incidence of 30°?

2.4-8. Trace a ray through a spherical surface, using Program SURTAC and assuming the radius of curvature is $+20$ mm, the height of the ray 10 mm, and the index $n' = 1.5$. What is:
   (a) The angle of incidence?
   (b) The angle of refraction?
   (c) The axial intercept?

2.4-9. Trace several rays through a surface that has a radius of curvature of $+40$ mm and separates a medium of index 1 on the left from a medium of index 1.6 on the right. Consider rays that are parallel to the axis, their heights increasing from 5 mm to 30 mm in steps of 5 mm, and determine how far from the vertex they intersect the axis.

2.4-10. If the radius of curvature of a single surface is $+50$ mm, the height of the ray 40 mm, the index to the left of the surface unity, and the axial intercept 100 mm, what must be the index to the right of the surface?

2.5

Optical Systems

THERE EXIST SO MANY TYPES of optical systems that it would be futile to even try to list them. Thus, instead of trying, I will limit myself to a discussion of a few representative groups of systems.

TELESCOPES

Astronomical telescope. Kepler's *astronomical telescope** has two converging lenses or groups of lenses, an *objective* in front and an *eyepiece* next to the eye. With the object at infinity, the light comes in parallel and the objective forms in its right-hand focal plane a real image, the *air image* (Figure 2.5-1). If this focal plane coincides with the left-hand focal plane of the eyepiece, rays forming the air image are made again parallel by the eyepiece and are still parallel when they enter the eye. Without accommodation, such rays form a real image on the retina which the observer then sees projected out to infinity. In the

---

*Johannes Kepler (1571–1630), German astronomer and physicist. Kepler's most important contributions were to astronomy. Analyzing Tycho Brahe's data he formulated his three laws of planetary motion; recognized the need for better optics in astronomy; investigated both theoretically and experimentally the formation of images by lenses, mirrors, and the eye; explained the rainbow; and was the first, using a pinhole camera, to observe sunspots. In one of his books, *Ad Vitellionem Paralipomena . . . Astronomiae Pars Optica,* 1604, he laid the groundwork for much of today's geometrical optics. In another, *Mathematici Dioptrice,* 1611, he described his telescope. Kepler believed in mysticism, cast horoscopes to finance his studies, once tried to find the sounds that belong to each planet producing the "music of the spheres."

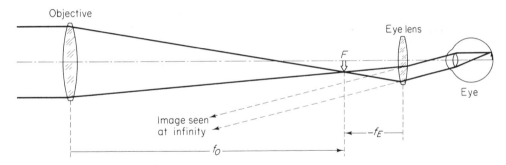

Figure 2.5-1 Astronomical telescope focused for infinity on both object side and image side. Air image located at *F*.

afocal mode, therefore, the objective and the eyepiece are separated by a distance, *d*, that is equal to the sum of their focal lengths:

$$d = f_O + f_E \qquad\qquad [2.5\text{-}1]$$

But experience shows that often, when a person looks into an optical instrument, he or she involuntarily accommodates, a phenomenon called *instrument myopia*.* If this is so, the light entering the eye is slightly divergent, as when reading, the air image formed by the objective must have moved *inside* the focal length of the eyepiece, and the image is seen at the distance of most distinct vision, 25 cm in front of the eye (Figure 2.5-2).

Magnification of a telescope.   The magnification produced by a telescope is a matter of *angular* magnification; it is the ratio of the angular size of the image seen through the telescope to the angular size of the object, seen without the telescope:

$$M_{\text{Telescope}} = \frac{\tan \theta'}{\tan \theta} \qquad\qquad [2.5\text{-}2]$$

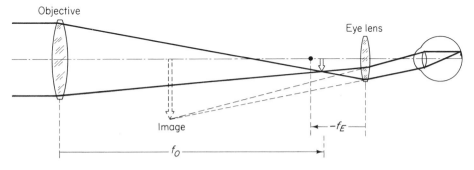

Figure 2.5-2 Astronomical telescope focused for object at infinity and image at the near-point.

---

*Instrument myopia can be avoided by first moving the eyepiece *out* of the telescope beyond the point of best focus, and then moving it in and *stopping* when the image is sharp. Viewing while accommodating is tiring; it also makes the lines of sight converge, which serves no useful purpose.

When focusing at a distant object, and looking into the telescope without accommodation, both object and image are seen at infinity.

From Figure 2.5-3 it follows that

$$\tan \theta = \frac{-y'}{f_O} \quad \text{and} \quad \tan \theta' = \frac{-y'}{-f_E}$$

Substituting in Equation [2.5-2] and canceling $-y'$ gives

$$M_{\text{Telescope}} = -\frac{f_O}{f_E} \qquad [2.5\text{-}3]$$

where the minus sign means that the image is inverted.*

Now consider a ray, parallel to the optic axis, that enters the objective next to its upper rim (Figure 2.5-4, top left). The ray goes through F and, with the telescope in the afocal mode, emerges from the eyepiece again parallel to the axis (bottom right). This ray, following our earlier notation (page 40), is a *parallel ray* with respect to the objective and a *focal ray* with respect to the eye lens.

Next consider a *chief ray* (the dashed ray in Figure 2.5-4). The ray comes from the upper rim of the objective, goes through the center of the eye lens (without refraction), and at some distance to the right intersects the focal ray. The point of intersection locates the image (of the objective). If, as usually is the case, the aperture of the objective is the *entrance pupil*, EP, then the image is the *exit pupil*, XP. The diameters of these two pupils are related through similar triangles:

$$\frac{\text{EP}}{f_O} = \frac{\text{XP}}{-f_E}$$

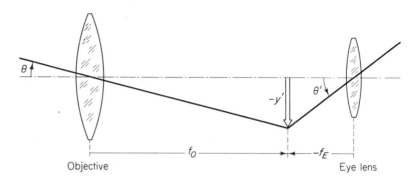

Figure 2.5-3   Deriving magnification of telescope.

---

*This inversion is immaterial with many astronomical objects such as fixed stars. But why do we see *more* stars with a larger telescope? Not because of higher magnification; fixed stars appear as points no matter how much they are magnified. Instead of "magnifying power," it is the *light-gathering power* of the telescope that counts.

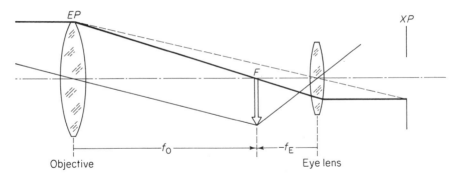

Figure 2.5-4   Entrance pupil and exit pupil of an astronomical telescope.

Rearranging and substituting in Equation [2.5-3] shows that the magnification of a telescope is also given by

$$M_{\text{Telescope}} = \frac{\text{diameter of entrance pupil}}{\text{diameter of exit pupil}}$$ [2.5-4]

Based on this definition, there is a simple way for determining the magnification. Place a square aperture of known size in front of the objective. Aim the telescope at the sky or some other diffuse target. Hold a sheet of paper a short distance behind the eyepiece and move it back and forth until the aperture is in focus. Measure the size of the image. Divide the size of the aperture by the size of its image; this gives the telescope's magnification.

If a telescope has the specifications 6×30, this means that it has $6 \times$ angular magnification and an objective 30 mm in diameter. The inverse ratio, 30/6, is the size of the exit pupil (in millimeters), and the square of that, 25, is the light-gathering power.

Eyepieces.   The field of view of a two-lens telescope is unnecessarily restricted because of vignetting. As shown in Figure 2.5-5, top, some of the light (dashed ray) does not reach the eye at all. Correction is possible by placing an additional lens, called a *field lens*, at, or close to, the air image. This lens has no effect on magnification; it merely directs all rays passing through into the last lens, the *eye lens*. Consequently, since the field lens is by necessity closer to the entrance pupil than the eye lens, the field of view is larger, without vignetting, than with the eye lens alone.

Such considerations have led to the development of *eyepieces*. A *Huygens eyepiece* consists of two plano-convex lenses, their convex sides facing the object. The field lens has about twice the focal length of the eye lens, with a separation of about one-half the sum of both. A *Ramsden eyepiece* has two equal, or nearly equal, plano-convex lenses, their convex sides facing each other. Their separation is nearly the same as their focal length. A *Kellner eyepiece* is essentially a Ramsden eyepiece with the eye lens made achromatic (Figure 2.5-6, left).

An *orthoscopic eyepiece* is like a Ramsden eyepiece but has a triplet as the field lens; it has a wide field and good color correction. The *Erfle eyepiece* (Figure 2.5-6, right) has two or

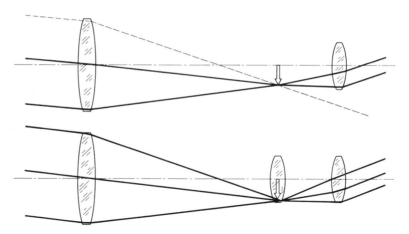

Figure 2.5-5 Astronomical telescope containing only two lenses (*top*) and with an additional field lens (*bottom*).

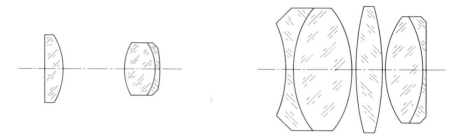

Figure 2.5-6 Kellner eyepiece (*left*) and Erfle eyepiece (*right*). To scale. (From *Optical Design,* MIL-HDBK-141.)

even three achromatic doublets; it is often used in high-quality astronomical telescopes.* A *Barlow lens* is a negative lens placed just ahead of the focal plane of the objective, doubling and sometimes tripling the magnification. A *Gauss eyepiece* contains a reticle, a beamsplitter, and a light bulb to illuminate the reticle (which helps in aiming at a target at night). In a *filar micrometer eyepiece,* a hair line can be moved across the field and the distance of motion read on a drum (which is useful for measuring the image size or the spacing of image details).

Terrestrial telescopes.   The drawback of the astronomical telescope is that the image is inverted. That would be awkward for terrestrial observations. Correction is possible by adding a *relay* or *erector lens* (as we will discuss in detail in the example that follows), by using a *prism,* or by designing a *Galilean telescope.*

Prism telescopes, if used for binocular vision, are called *prism binoculars.* Older types of prism binoculars have two sets of Porro prisms. Newer models contain Pechan prisms, also known as Malmros, Schmidt, Thompson, or Z prisms (Figure 2.5-7).

---

*H. Erfle, *Ocular,* U.S. Patent 1,478,704, Dec. 25, 1923.

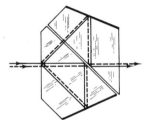

Figure 2.5-7   Pechan prism.

Galileo's telescope.   In the *Galilean telescope,** as in the astronomical tele-scope, the objective is positive, but in contrast to it, the eyepiece is negative. Still, the two focal points coincide. In Figure 2.5-8, top, *F* is both the second (right-hand) focal point of the first lens and the first (left-hand) focal point of the second lens. In Galileo's type (bottom), *F* is still the second focal point of the first lens and again the first (but now the *right*-hand) focal point of the (negative) second lens. Parallel light incident on the system is refracted toward *F* but before it reaches the focus, the light is intercepted by the negative lens

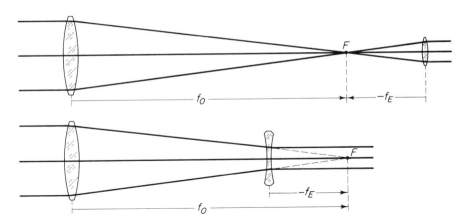

Figure 2.5-8   Focal points coincide in both the astronomical telescope (*top*) and Galileo's telescope (*bottom*).

*Named after Galileo Galilei (1564–1642), Italian scientist. While, as a medical student, attending mass in a cathedral in Pisa, Galileo is said to have watched a bronze chandelier swinging in the breeze, discovered that the frequency of oscillation, no matter what the amplitude, was constant. When in the early 1600s Dutch opticians had found that a combination of two lenses showed distinct objects upright as well as magnified. Galileo quickly figured out the lenses necessary, stating in his *Il Saggiatore* (Roma: Appresso Giacomo Mascardi, MDCXXIII) that he had "discovered the same instrument, not by chance, but by the way of pure reasoning." In 1609 he demonstrated his "cannocchiale" in public, looking from the Campanile of San Marco in Venice at the Campanile San Giustinio in Padua (where he taught mathematics), 33 km away. The word *telescope* was coined later by a Greek poet and theologian, Ioannis Demisiani of Cephalonia, combining τῆλη = far and σκοπὸς = viewer. The second of more than 100 telescopes that Galileo made was built into a lead tube 70 cm long and had 3× angular magnification; the fifth, called *Discoverer,* had 30×. With it he resolved the Milky Way into a myriad of stars and discovered the moons of Jupiter and the phases of Venus—which proved its rotation around the sun.

and the emergent light is parallel again. The image, therefore, is seen at infinity. *With ac-commodation, the image is seen at the distance of most distinct vision (not shown).*

The distance between the lenses is the same as with Kepler's telescope,

$$d = f_O + f_E \qquad\qquad\qquad \text{[2.5-1A]}$$

where the numerical value substituted for f_E must now be negative. The magnification, like-wise, is the same,

$$M_{\text{Telescope}} = -\frac{f_O}{f_E} \qquad\qquad\qquad \text{[2.5-3A]}$$

but this time it is positive and the image upright. Telescopes of the Galilean type do not give much magnification (because the exit pupil lies to the left of the last lens and the eye cannot be brought there) but they are short, and hence are often used as opera glasses.

Example

Design a telescopic sight, to be used on a hunting rifle.

Solution.   With no more information given than this, consider first the diameters of the various pupils and the magnification desired. The diameter of the pupil of the eye varies from 2 to 8 mm, depending on the ambient light. In daylight, the pupil is about 3 mm in size and not much is gained by making the exit pupil of the telescope larger, except that aligning the eye with the eyepiece becomes easier. Let us begin with an exit pupil 3.75 mm in diameter (Figure 2.5-9a).

Since a telescope's magnification is equal to the ratio of the diameter of the objective to the diameter of the exit pupil, a higher magnification, for a given exit pupil, requires a larger-diameter objective. For daylight use, reasonable length, light weight, and moderate cost, we may settle for $M_T = 8\times$. The objective lens, then, should have a diameter of

$$D = (8)(3.75) = 30 \text{ mm}$$

Certainly, we must have an upright image. But for reasons explained earlier, an $8\times$ telescope cannot be built as a Galilean type. Instead, we use an astronomical telescope together with an *erector* (Figure 2.5-9b). We choose, somewhat arbitrarily, 12 cm as the focal length, f_O, of the objective. Since $M = f_O/f_E$, this requires an eyepiece of focal length

$$f_E = \frac{f_O}{M} = \frac{120 \text{ mm}}{8} = 15 \text{ mm}$$

With this focal length, the exit pupil is

$$s' = \frac{sf}{s+f} = \frac{-(120 + 15)(15)}{-120 - 15 + 15} = +16.9 \text{ mm}$$

to the right of the eyepiece. But this does not give enough *eye relief* ($=$ distance from the last lens to the exit pupil which should coincide with the pupil of the eye). The pupil of the eye is about 3.6 mm behind the cornea and the lens is held in a mount; therefore, the actual clearance between the telescope and the eye might be less than 10 mm, not enough for comfort and to avoid the danger of recoil, and certainly not enough for people who wear glasses.

The solution lies in making the erector provide some magnification too. For example, if the eyepiece is given a more reasonable focal length, perhaps 30 mm, the erector can make up

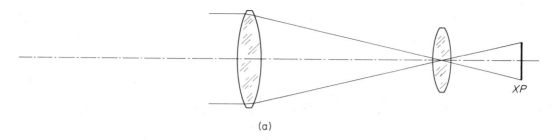

(a)

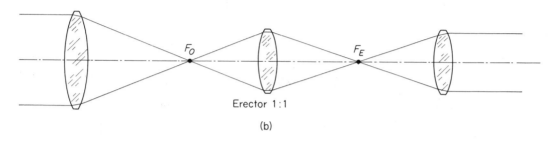

Erector 1:1

(b)

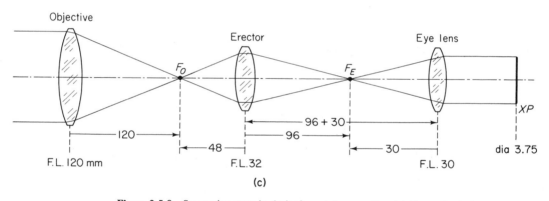

Figure 2.5-9   Successive steps in designing a telescope. For details see the text.

for the deficit and provide $2\times$. Assume that the erector has a focal length of 32 mm. Solving $s'/s = -2$ for s' and inserting this in $s' = sf/(s + f)$ gives

$$(-2)(s) = \frac{(s)(32)}{s + 32}$$

and therefore,

$$s = -48 \text{ mm}$$

and

$$s' = 96 \text{ mm}$$

The design so far looks as shown in Figure 2.5-9c. The total length is now about 30 cm, acceptable for a riflescope.

The system still has too narrow a *field of view*. Draw a chief ray (*A* in Figure 2.5-10) through the center of the objective and just caught by the erector. Ray *B*, parallel to *A*, also goes through the erector; but *C*, also parallel to *A*, will not. In other words, the top half of the objective aperture is not used; light entering there is lost, and vignetting is 50 percent. The angle at which this occurs is α, subtended by *A* and the axis.

At which angle does *all* the light go through, without vignetting? That happens when *D* is caught by the erector ($\beta < \alpha$). We could provide a larger field, and reduce vignetting, by using a larger erector. But that would mean more severe aberrations, and correcting these is expensive. Instead, we place a *field lens* at F_O. Its focal length is chosen so that the objective is imaged into the erector:

$$f_{\text{field lens}} = \frac{ss'}{s - s'} = \frac{(-120)(48)}{-120 - 48} = 34.3 \text{ mm}$$

This is an important step; it shows how by adding a simple lens (which does not even contribute to image formation) we get a wide field, without vignetting.

The diameter of the erector is determined by rays that come from the rim of the objective and pass through F_O (chief rays with respect to the field lens). In our example,

$$D_{\text{erector}} = (30)\left(\frac{48}{120}\right) = 12 \text{ mm}$$

The same result is obtained by making the erector 96/30 times larger than the exit pupil: $(96/30)(3.75) = 12$ mm. (All diameters refer to free apertures; the actual lenses may be 2 mm larger.) A 12-mm aperture would, *without* a field lens, give at 50 percent vignetting a (total) field of

$$2 \arctan\left(\frac{6}{120 + 48}\right) = 4.1°$$

But *with* a field lens at F_O (assuming its diameter is the same as that of the erector), the field of view is

$$2 \arctan\left(\frac{6}{120}\right) = 5.7°$$

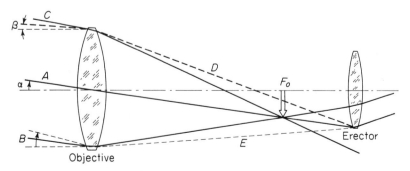

Figure 2.5-10   Various degrees of vignetting.

There is another benefit besides the larger field. The lens placed at F_O limits the angular field for *all* rays, which means that up to this angle there is no vignetting: The boundary between no vignetting and complete vignetting is sharp.

Finally, we place another field lens at F_E, designing in effect an eyepiece. This lens should have a focal length of

$$f = \frac{(-96)(30)}{-96 - 30} = 22.9 \text{ mm}$$

Its diameter, and the diameter of the eye lens, however, should be relatively large. The field of 5.7°, just mentioned, is the field of the telescope; when looking into the eyepiece, the field actually seen is $8 \times 5.7 = 45°$. This is a nice wide field, which should not be restricted by making the field lens (at F_E) and the eye lens too small.

We have now arrived at the *thin-lens solution.* Next we change to realistic, "thick" lenses, move the two field lenses slightly away from the foci (so that dust accumulating on them will not show), achromatize the system, trace a series of rays including skew rays through it, and correct for other aberrations. Finally, we place cross-hairs at either F_O or F_E. This completes our design.

THE MICROSCOPE

A microscope is an instrument for viewing small objects. A *simple microscope,* first used by van Leeuwenhoek,* is merely a single lens of very short focal length. A *compound microscope* has two lenses, an objective and an eyepiece. The objective forms a real, inverted, magnified image. This image is further magnified by the eyepiece.

So far, that looks the same as with a telescope. What, then, is the difference between the two? Consider a telescope that is first focused at infinity and used without accommodation. The inner foci coincide (as shown earlier in Figure 2.5-1). As the object is slowly brought closer, the air image moves away, through a distance that in essence is Newton's image distance with respect to the objective. It is this distance, from the right-hand focus of the objective, F_O in Figure 2.5-11, to the left-hand focus of the eyepiece, F_E, that is characteristic of a *micro*scope, in contrast to a *tele*scope. This distance, called the *optical tube length, T,* is usually set at 160 mm.

From the shaded triangles in Figure 2.5-11 we see that the height of the object, y, and the height of the air image, $-y'$, are related as

$$\frac{y}{-y'} = \frac{f_O}{-T}$$

---

*Antoni van Leeuwenhoek (1632–1723), Dutch clerk and biologist. In 401 publications, including 375 letters to the Royal Society of London and a book, *Arcana Naturae per Microscopum Detecta* (Secrets of Nature Discovered through the Microscope), van Leeuwenhoek described bacteria and yeast cells, the circulation of blood, red blood cells (which had been misinterpreted by Malpighi as fat cells), spermatozoa (though he fancied seeing tiny human heads inside), and the cross-striation of muscle fibers. He also explained that certain insects do not develop from mud, but from eggs laid there.

Low-power compound microscopes exist even longer. In 1590 Hans and Zacharias Janssen of Middelburg, Germany, combined a biconvex objective with a biconcave eye lens. The word *micro · scope* derives from the Greek μικρός = small and σκοπεῖν = to view. Descartes called these low-power instruments "perspicilia pulicaria," Latin for "viewing glasses for fleas."

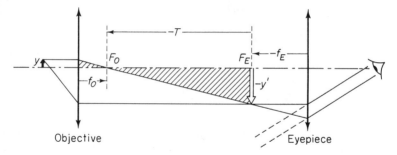

Figure 2.5-11 Ray trace through a compound microscope.

The inverse ratio, $-y'/y$, by definition, is the magnification provided by the objective,

$$M_O = \frac{-T}{f_O} \qquad [2.5\text{-}5]$$

The eyepiece, used without accommodation and following Equation [1.4-19], gives a magnification

$$M_E = \frac{25}{f_E} \qquad (f \text{ in cm}) \qquad [2.5\text{-}6]$$

The total magnification of the microscope is the product of both,

$$\boxed{M_{\text{Microscope}} = -\frac{T}{f_O}\frac{25}{f_E}} \qquad [2.5\text{-}7]$$

Emphasizing again the similarity between the two, we can in fact derive the magnification of a microscope from that of a telescope. Consider a telescope aimed at an object at infinity. The light, therefore, is parallel and the objective, L in Figure 2.5-12, brings it to a focus at F. Then place a small object, B, in front of L. The size of B and its distance from L are chosen such that angle θ, and therefore θ', stay the same. In order to focus B into S', another lens, A, must be added. *Its* focal length, f_A, must equal the distance A-B, because only then will rays originating at B be parallel when entering L.

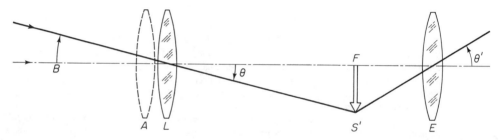

Figure 2.5-12 Changing a telescope into a microscope by adding a lens (dashed outline on left). Heavy line is chief ray.

Now consider the two magnifications. For the telescope we have

$$M_{\text{Telescope}} = \frac{\tan \theta'}{\tan \theta}$$

[2.5-2A]

and for the microscope

$$M_{\text{Microscope}} = \frac{\tan \theta'}{B/25}$$

[2.5-8]

Dividing Equation [2.5-8] by [2.5-2A] gives

$$\frac{M_M}{M_T} = \frac{\tan \theta}{B/25} = \frac{-S'/f_L}{B/25} = \frac{(-S')(25)}{(B)(f_L)}$$

[2.5-9]

From similar triangles we find that

$$\frac{B}{-S'} = \frac{-f_A}{f_L}$$

and by substituting in Equation [2.5-9] that

$$\frac{M_M}{M_T} = \frac{25}{-f_A}$$

We substitute for M_T the ratio of the two focal lengths, i.e., the magnification of the telescope,

$$M_{\text{Telescope}} = -\frac{f_L}{f_E}$$

[2.5-3A]

That gives for the magnification of the microscope

$$M_{\text{Microscope}} = -\frac{(f_L)(25)}{(f_E)(-f_A)}$$

[2.5-10]

But, rather than express the microscope's magnification as a function of f_A, we add the powers of the two lenses A and L,

$$\frac{1}{f_A} + \frac{1}{f_L} = \frac{1}{\mathbf{f}}$$

where $\mathbf{f}$ is the equivalent focal length of the complete (two-lens) objective. We solve for $1/f_A$ and substitute in Equation [2.5-10]; that gives

$$M_{\text{Microscope}} = \frac{(f_L)(25)}{f_E} \left(\frac{1}{\mathbf{f}} - \frac{1}{f_L} \right) = \frac{(25)(f_L - \mathbf{f})}{(\mathbf{f})(f_E)}$$

[2.5-11]

But $f_L - \mathbf{f}$ is the (negative) optical tube length, $-T$, and, hence, Equation [2.5-11] is precisely the same as [2.5-7].

Example

The light follows a different path depending on whether the microscope is used for projection and microphotography or for viewing. For instance, if the objective bears the designation "16 mm" and the eyepiece "12.5×," by how much must the tube of the microscope be raised or lowered when changing from visual observation (without accommodation) to photography, assuming that the photographic film is 60 mm away from the eyepiece?

Solution. With the microscope focused for projection, the air image is formed *outside,* ahead of, the first focal length of the eyepiece. The eyepiece relays the image to the screen or film (Figure 2.5-13, left). With the microscope used for viewing, the objective is brought closer to the specimen. This causes the air image to move farther away, to a position *inside* the focal length of the eyepiece, which now acts as a magnifier (right).

From the data given, we determine first the focal length of the eyepiece,

$$f_E = \frac{25 \text{ cm}}{12.5} = 20 \text{ mm}$$

If the film, and therefore the image, is 60 mm away (from the eyepiece), the object distance (for the eyepiece) is

$$s_E = \frac{s'f}{f - s'} = \frac{(60)(20)}{20 - 60} = -30 \text{ mm}$$

or **10** mm more than for visual observation (-20 mm).

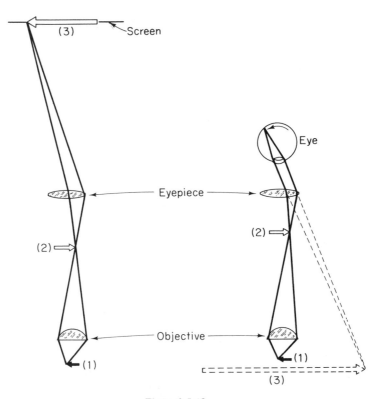

Figure 2.5-13

Next, we use Gauss' thin-lens equation to find the two object distances (for the objective). For visual observation,

$$s_v = \frac{s'f}{f - s'} = \frac{(16 + 160)(16)}{16 - (16 + 160)} = -17.6 \text{ mm}$$

while for photography,

$$s_p = \frac{(16 + 160 - \mathbf{10})(16)}{16 - (16 + 160 - \mathbf{10})} = -17.7 \text{ mm}$$

Thus the tube must be *raised*

$$17.7 - 17.6 = \boxed{0.1 \text{ mm}}$$

The *numerical aperture* of a microscope objective is defined the same as that of an optical fiber,

$$\text{NA} = n_0 \sin I \qquad\qquad [2.5\text{-}12]$$

The term n_0 is the index of the medium between the cover glass placed on top of the specimen and the front lens of the objective, and I is the angle of total internal reflection at the glass–air boundary of the cover (Figure 2.5-14, left).

With air ($n' = 1.00$) filling the space between cover glass and front lens, the numerical aperture cannot exceed unity. But, if the air is replaced by oil ($n = 1.515$, right), the angle of total reflection becomes larger and the cone of light wider, producing a better and brighter image.

As a rule of thumb, the *useful magnification* of a microscope is about 600 times the numerical aperture. Although it is easy to go beyond this limit by using a higher-power eyepiece or by projecting the image on a distant screen, the result is merely a larger image but not the disclosure of more detail. Any magnification beyond the useful limit is an "empty magnification."

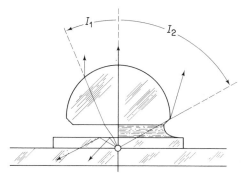

Figure 2.5-14 Rays limiting the cone of light entering a "dry" microscope objective (*left*) and an oil-immersion objective (*right*).

THE EYE

On first sight, the human eye seems to be a simple optical system containing a lens that forms an image on a light-sensitive screen. But it is much more than that. The eye's lens is of the GRIN type, and in addition capable of changing its shape.

On entering the eye, the light passes first through the *cornea* and then through the *anterior chamber* (Figure 2.5-15). The anterior chamber, filled with the *aqueous humor,* is bounded by an aperture stop, the *iris,* with a central hole in it, the *pupil.* The diameter of the pupil varies as a function of the amount (and wavelength) of the light.

Next comes the *crystalline lens.* An elastic, jellylike composite of long cells, the crystalline lens is particularly rich in proteins (33%, more than any other tissue in the human body) and potassium (25 times the concentration in the surrounding liquids), factors that help maintain its transparency and refractive index. The index is highest in the center (1.41), lowest at the equator (1.37), and intermediate (around 1.39) at the vertices where the lens borders the aqueous in front, and the *vitreous humor* in the rear. The curvatures of the surfaces of the lens are controlled by the circumferential *ciliary muscle.* After passing through the vitreous, the light reaches the *retina,* which through the *optic nerve* connects to the brain.

The optical properties of the eye can be represented either by the simpler *reduced eye* or by the more realistic *schematic eye.* The reduced eye (see page 34) is assumed to have only one surface, with a radius of curvature of $R = +5.7$ mm. The two focal lengths are $f_1 = -16.8$ mm and $f_2 = +22.5$ mm, and the power is $+60$ diopters.

Gullstrand's *schematic eye* has three surfaces, the front surface of the cornea and the front and rear surfaces of the lens. Their dimensions are listed in Table 2.5-1. The total power is again 60 diopters. Of this, about 43 diopters is contributed by the cornea, the remainder by the lens.

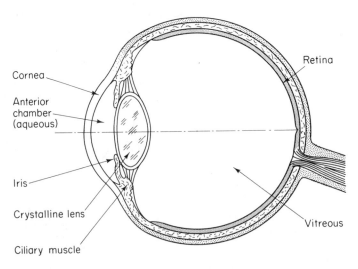

Figure 2.5-15   Cross section through the human eye.

TABLE 2.5-1 SCHEMATIC EYE

Radii of curvature	
Cornea	+7.80 mm
Lens	
Anterior surface	+10.00 mm
Posterior surface	−6.00 mm
Refractive indices	
Aqueous	1.336
Lens (cumulative)	1.413
Vitreous	1.336

Focusing on a nearby object is brought about by *accommodation.* Accommodation is accomplished by a change of power of the lens, resulting from a change in shape. In its relaxed state, the lens is nearly spherical. But the lens is suspended by ligaments and these ligaments are attached to the circumferential ciliary muscle; thus, with the muscle relaxed, the lens is stretched into a flatter shape. During accommodation, the ciliary muscle contracts, allowing the lens to return to its more spherical shape. This change affects mainly the front surface, whose radius of curvature shrinks, from +10.0 mm to +5.3 mm.

Refractive anomalies.   The two most common refractive errors of the eye are *myopia* (nearsightedness) and *hyperopia* (hypermetropia, farsightedness). In myopia, the image of a distant object is formed in front of the retina and thus is seen out of focus (Figure 2.5-16a). Only objects that are not farther away than the *far point* will focus on the retina (b).

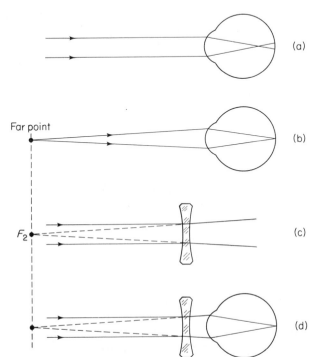

Figure 2.5-16 Myopia and its correction.

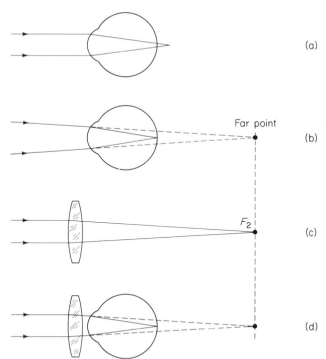

(a)

(b)

(c)

(d)

Figure 2.5-17 Hyperopia and its correction.

Correction for myopia requires a minus lens. Its second (left-hand) focal point should coincide with the far point (c). Then, parallel light incident on the eye becomes divergent, so that it appears to come from the far point and *will* focus on the retina (d).

In *hyperopia,* parallel light comes to a focus behind the eye (Figure 2.5-17a). The far point is to the right of the retina. It is defined as the point to which rays from an object must converge for a sharp image to be formed on the retina (b). For correction, again the second focal point of the correcting lens should coincide with the far point (c). Then rays entering the eye are already converging and will focus on the retina (d).

CAMERA LENSES

Camera lenses are a good example for a study of evolution. They have gradually developed from very simple to highly complex systems, their progress punctuated from time to time by fundamentally new concepts.

The simplest type of a camera lens, known since the late sixteenth century, is the *biconvex lens.* It suffers from every type of aberration, and is all but useless for any except the most rudimentary form of photography.

Some 250 years later it was found that a *meniscus,* placed with its concave side facing the object and with an aperture stop in front of it, has not much astigmatism or coma. However, spherical and chromatic aberration, distortion, and curvature of field are still severe and limit its use. Its aperture can be no larger than $f/16$. A slight improvement is possible by using an *achromatic meniscus,* the so-called "landscape lens."

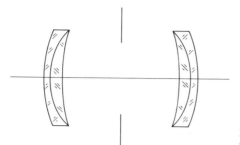

Figure 2.5-18 *Rapid rectilinear* camera lens.

A combination of *two achromatic menisci,* however, their concave sides facing each other as in Figure 2.5-18, is much better. An example is the *Rapid Rectilinear,* developed in 1866. It has very little coma, distortion, and lateral color but still has considerable spherical aberration, and either astigmatism or curvature of field. Its aperture can be as large as $f/8$.

Several other symmetrical lenses followed, many of them of the *double Gauss* type. A particularly flat field, for example, is obtained with the *Planar,* which, in addition to the two menisci, has two plano-convex lenses, as illustrated in Figure 2.5-19.

The design of camera lenses then took a dramatically different turn, due to the invention of the *Taylor–Cooke triplet.** This is the three-lens system which I have used repeatedly as an illustration to introduce various chapters.

The Taylor triplet consists of two positive lenses, in front and back, both made of crown, and a negative lens, in between, made of flint. The negative lens is the crucial element. Its distance from the first (positive) lens is relatively large. Thus the light, as it reaches the minus lens, has fairly high vergence. Hence, the power of the minus lens can also be kept high to correct for spherical and chromatic aberration, and to reduce coma, astigmatism, curvature of field, and distortion, and still leave sufficient power in the system. Because of its good correction, the Taylor triplet could be made with much higher speeds ($f/6.3$ and better) and a wider field than any other camera lens known at that time.

From this concept of the Taylor–Cooke triplet there derived numerous other camera lenses. The most successful of these is the *Zeiss Tessar.*† As shown in Figure 2.5-20, the rear

---

*Designed in 1893 by Harold Dennis Taylor (1862–1943), optical manager of the T. Cooke & Sons Optical Company, Bishophill, York, England. It seems that early camera lenses were often "corrected" for aberrations by placing stops to eliminate unwanted marginal rays (and making long exposure times a necessity). Taylor broke with tradition. In his invention disclosure he describes his "idea of eliminating the diaphragm corrections and throwing the whole burden of flattening the final image and correcting marginal astigmatism upon a negative lens." H. D. Taylor, *A Simplified Form and Improved Type of Photographic Lens,* Brit. Patent 22,607, 6 Oct. 1894, and *Lens,* U.S. Patent 540,122, May 28, 1895. In 1906, Taylor (who had started out to become an architect) published a book, *A System of Applied Optics,* where he set forth the principles of lens design he had used, an extension of Coddington's earlier work.

†Invented by Paul Rudolph (1858–1935), lens designer at the Carl Zeiss optical company in Jena, Germany. In 1895 Rudolph designed the Planar, in 1902 the Tessar. In his invention disclosures, *Sphärisch, chromatisch und astigmatisch korrigiertes Objektiv aus vier, durch die Blende in zwei Gruppen geteilte Linsen,* Dtsch. Reichspatent 142 294, 25 Apr. 1902, and *Photographic Objective,* U.S. Patent 721,240, Feb. 24, 1903, Rudolph describes how with a "comparatively small number of components" he developed a "spherically, chromatically, and astigmatically corrected objective" that is "fruitful in an extraordinary degree." For details on modern Tessars, see R. Kingslake, *Lens Design Fundamentals,* pp. 277–86 (New York: Academic Press, Inc., 1978).

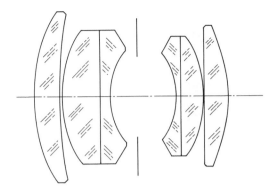

Figure 2.5-19 *Planar* lens.

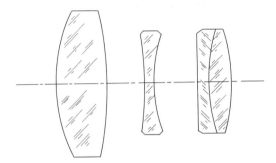

Figure 2.5-20 *Tessar* camera lens.

element became an achromatic cemented doublet. The Tessar has excellent correction for spherical aberration, good achromatism, sufficiently high speed ($f/3.5$ and even $f/2.8$), very little astigmatism, and a field free of curvature and distortion out to about 60°. Today the Tessar and its derivatives are the most widely used high-quality camera lenses. Table 2.5-2 lists some typical parameters.

TABLE 2.5-2 TESSAR CAMERA LENS*

Axial thickness, d (mm) Refractive index, n	Radius of curvature (mm)	Air space (mm)
Lens 1		
$d = 3.57$	$R_1 = +16.28$	
$n = 1.6116$	$R_2 = -275.7$	
		1.89
Lens 2		
$d = 0.81$	$R_3 = -34.57$	
$n = 1.6053$	$R_4 = +15.82$	
		3.25
Lens 3		
$d = 2.17$	$R_5 = \infty$	
$n = 1.5123$	$R_6 = +19.20$	
Lens 4		
$d = 3.96$	$R_7 = +19.20$	
$n = 1.6116$	$R_8 = -24.00$	

*Focal length: 50.82 mm.

A *telephoto lens* has a focal length longer than the diagonal across the film negative.

A *wide-angle lens* has a focal length shorter than the diagonal. A lens that covers a 180° field (a hemisphere) is called a *fisheye* or *sky lens.* An example is shown in Figure 2.5-21.

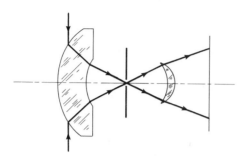

Figure 2.5-21   Sky lens.

A *zoom lens* is a system whose focal length can be varied without moving the image out of focus; a zoom lens, in other words, provides variable magnification. Many zoom lenses are derived from the Taylor triplet, plus–minus–plus. The various lenses, or groups of lenses, move in a complex, often nonlinear way. The front lens in Figure 2.5-22, for example, is stationary, to provide a tight seal. The second lens and the third lens both move, the rates and directions of motion being governed by cams, or slots cut into a rotatable cylinder.

With the minus lens in midposition, the image has a certain size. Moving the minus lens forward makes the image smaller, as with a wide-angle lens; moving it backward makes the image larger, as with a telephoto lens.

But why is this so? Look at the angles subtended by peripheral rays at the image and consider the *Lagrange invariant, nyu*, the product of refractive index, image size, and slope angle (page 43). Since the product *yu* must be necessity be constant, a larger angle (the

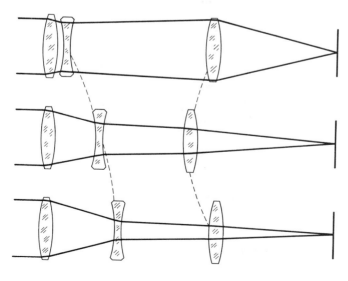

Figure 2.5-22   Zoom lens showing wide-angle position (*top*), telephoto position (*bottom*).

upper example in Figure 2.5-22) produces a smaller image, as with a wide-angle lens; and a smaller angle (bottom) produces a larger image, as with a telephoto lens.

From simple landscape lenses to highly complex 18-element zoom systems,* camera lenses can be built to remarkable specifications. For example, cameras in high-flying aircraft can from a height of more than 12 km (40,000 ft) resolve details as small as the numbers on an automobile license plate.

Example

Telephoto lenses for cameras are often built in the form of a Galilean telescope. Assume that the front lens has +50 mm focal length and the second lens −25 mm focal length. The distance between both is 30 mm. Determine:

(a) The focal length.
(b) The actual physical length of the system.

Solution.   (a) The term *focal length* refers, more precisely, to *equivalent* focal length. With the dimensions given, the equivalent focal length of the system, from Table 1.4-1, is

$$\mathbf{f} = \frac{f_1 f_2}{f_1 + f_2 - d} = \frac{(+50)(-25)}{+50 - 25 - 30} = \boxed{+250 \text{ mm}}$$

This focal length, by definition, is the distance from the second principal plane, $\mathbf{H}_2$, of the system to the film in the camera (Figure 2.5-23).

(b) In order to find the physical length of the system, we determine first the back vertex power (of the system), using Equation [1.4-14]. We convert the focal lengths given into powers, express the distance d between the lenses in meters, and set $n = 1$, because the medium between the lenses is air. Then

$$\mathbf{P}_2 = \frac{+20}{1 - (+20)(0.03)} + (-40) = 10 \text{ m}^{-1}$$

The reciprocal of $\mathbf{P}_2$ is the back vertex focal length, $\mathbf{v}_2$, the distance from the second lens to the plane of the film,

$$\mathbf{v}_2 = \frac{1}{10} = 0.1 \text{ m} = 100 \text{ mm}$$

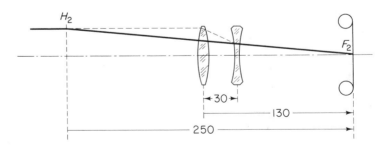

Figure 2.5-23   Example of telephoto lens. To scale.

*H. Basista and E. Glatzel, *Photographic Lens of Continuously Variable Focal Length,* U.S. Patent 4,451,124, May 29, 1984.

The total length of the system, from the first lens to the film, therefore, is

$$30 + 100 = \boxed{130 \text{ mm}}$$

considerably less than its focal length, 250 mm. This is very convenient when toting around such gear.

REFLECTING SYSTEMS

Conventional telescopes and microscopes are combinations of *lenses;* they are *dioptric systems.* But telescopes and microscopes can also be built as combinations of *mirrors;* these are *catoptric systems.* Combinations of lenses and mirrors are *catadioptric systems.* *

Catoptric systems.   Mirrors have no chromatic aberration and, in addition to visible light, they also work in the UV, the IR, and even in the X-ray region where no refractive materials are known. Reflecting telescopes are widely used in astronomy. In principle, an astronomical reflector is similar to a refractor, except that the objective *lens* is replaced by a *mirror.* In early reflectors, a small plane mirror or a prism was used to deflect the beam out of the tube for viewing; that is the essence of *Newton's telescope* (Figure 2.5-24).

Figure 2.5-24 Newton's telescope. [Reproduced from I. Newton, "An Account of a New Catadioptrical Telescope Invented by Mr. Newton," *Philos. Trans. Roy. Soc. London* **7** (1672), 4004–10.]

---

*The term *dioptric* comes from the Greek $\delta\iota\grave{\alpha}$ = [light passing] through, and catoptric comes from $\kappa\acute{\alpha}\tau o\pi\tau\rho o\nu$ = mirror; catadioptric is a combination of both.

*Gregory's telescope** is the equivalent of the astronomical telescope; it has two curved mirrors, a concave primary (mirror) with a central hole, and a concave secondary mounted as shown in Figure 2.5-25, left. The distance between the two mirrors is equal to, or slightly larger than, the sum of their focal lengths.

Cassegrain's telescope† is the equivalent of Galileo's type, with the secondary being convex (Figure 2.5-25, center). The secondary contributes most to the overall magnification because its distance from the primary's focus is much less (about one-fourth) than that from the secondary to the image. Cassegrainian telescopes are widely used; they are much shorter than a Gregorian of the same focal length.

The *Schwarzschild type* (right), in principle similar to the Cassegrainian, is used in microscopy.‡ Its focal length is easily derived from the equivalent-power equation,

$$\mathbf{P} = P_1 + P_2 - P_1 P_2 d$$

If we substitute $-2/R$ for the powers of the two mirrors and, because the light changes direction, change the sign in the R_2 term, then

$$\frac{1}{f} = \frac{-2}{R_1} + \frac{2}{R_2} + \frac{4d}{R_1 R_2} \qquad [2.5\text{-}13]$$

where d is the distance between the mirrors.

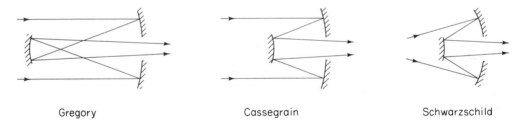

Gregory   Cassegrain   Schwarzschild

Figure 2.5-25   (*Left to right*) Gregory's telescope, Cassegrain's telescope, Schwarzschild's microscope.

---

*James Gregory (1638–1675), Scottish mathematician and astronomer, suggested the system named after him in 1663, but opticians at that time could not grind the mirrors required.

†Guillaume Cassegrain, French. Little is known about Cassegrain's life; according to some, he was a professor of physics at the College of Chartres, according to others, a sculptor at the court of Louis XIV. In 1672, nine years after Gregory, Cassegrain combined a concave and a *convex* mirror, intercepting the rays before they came to a focus (an idea quickly dismissed by Newton as only a minor modification of Gregory's precedent). The real virtue of Cassegrain's design is that the spherical aberration introduced by one mirror is partially canceled by the other, a fact established by Ramsden a century later. Cassegrain's original system had a paraboloid primary and a hyperboloid secondary; since then the term "Cassegrainian" has broadened to mean any combination of a concave primary and a convex secondary. In the *Ritchey–Chrétien* type, for example, both mirrors are hyperboloid, completely eliminating spherical aberration and coma.

‡D. S. Grey, "A New Series of Microscope Objectives: II. Preliminary Investigation of Catadioptric Schwarzschild Systems," *J. Opt. Soc. Am.* **39** (1949), 723–28.

Example

A microscope objective of the Schwarzschild type may have a primary (mirror) of 50 mm and a secondary of 10 mm radius of curvature. The distance between the two mirrors is 40 mm. Where must the object be placed to produce an image 18 cm from the primary?

Solution.   While the light is reflected back and forth between the mirrors, it is much easier to work such a problem with the system *unfolded,* as shown in Figure 2.5-26. First determine the powers of the two mirrors:

$$P_1 = \frac{-2}{R_1} = \frac{-2}{-0.05} = +40 \text{ m}^{-1}$$

$$P_2 = \frac{-2}{+0.01} = -200 \text{ m}^{-1}$$

With the image 18 cm from the primary, or $18 + 4 = 22$ cm from the secondary, the exit vergence at the secondary is

$$\mathbf{V}_2' = \frac{1}{+0.22} = +4.545 \text{ m}^{-1}$$

and the entrance vergence (at the secondary) is

$$\mathbf{V}_2 = \mathbf{V}_2' - P_2 = +4.545 - (-200) = +204.545 \text{ m}^{-1}$$

Without the secondary in place, wavefronts of such vergence would come to a focus $1/204.545 = 4.89$ mm to the right of the secondary (considering the *unfolded* system), or 44.9 mm to the right of the primary. The exit vergence at the primary, therefore, is

$$\mathbf{V}_1' = \frac{1}{+0.0449} = +22.27 \text{ m}^{-1}$$

and the entrance vergence is

$$\mathbf{V}_1 = 22.27 - (+40) = -17.73 \text{ m}^{-1}$$

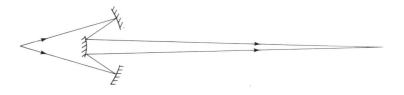

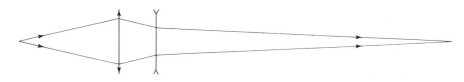

Figure 2.5-26 Schwarzschild-type microscope objective in its actual configuration (*top*) and unfolded (*bottom*).

This means that the object distance is

$$s = \frac{1}{-17.73} = -56.4 \text{ mm}$$

and the object must be placed

$$-56.5 + 40 = \boxed{-16.4 \text{ mm}}$$

in front of the secondary.

Catadioptric systems. The two-mirror systems described so far have serious off-axis aberrations and their fields are small. A significant improvement is possible by adding a highly aspherical glass plate, called a *Schmidt corrector,* shown in Figure 2.5-27.*

Schmidt's concept of a corrector plate revolutionized the design of large-aperture, wide-field reflectors. But these correctors are difficult to design and produce. This difficulty was overcome by *Maksutov*† who used *only spherical surfaces.* The two radii of curvature of the corrector, R_a and R_b in Figure 2.5-28, are used as variables to eliminate spherical aberration, coma, and axial color. Minimum axial color results when

$$\frac{R_a - R_b}{t} = \frac{n^2 - 1}{n^2} \qquad [2.5\text{-}14]$$

where t is the thickness of the glass and n its refractive index for yellow. The secondary is either applied directly to the corrector, or a *Mangin mirror*‡ is added.

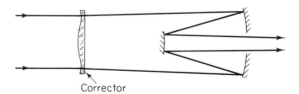

Corrector

Figure 2.5-27 Cassegrain–Schmidt reflector.

---

*Bernhard Voldemar Schmidt (1879–1935), Estonian optician and instrument maker. A rather independent, self-taught, highly skilled craftsman, despite the loss of his right forearm, Schmidt set up shop in Mittweida near Jena, Germany, later joined the Bergedorf astronomical observatory near Hamburg, where in 1930 he designed and built the reflectors named after him. His first reflector had a speed of $f/1.75$, unheard of at that time, and gave photographs of superb quality.

†Dmitry Dmitrievich Maksutov (1896–1964), Russian physicist and astronomer. After building his first reflector at age 13, Maksutov at 15 was elected a member of the Russian Astronomical Society, in 1914 graduated from a military engineering school, in 1944 became professor and head of Instrument Construction at the Pulkovo Observatory. Maksutov wrote several books, among them *Astronomical Optics,* Moscow 1946, and *Preparation and Testing of Astronomical Optics,* Moscow 1948. His most significant contribution is "Novye katadioptricheskie meniskovye sistemy," *Dokl. Akad. Nauk SSSR* **37**, 147–52 (1942), and *Zh. Tekh. Fiz.* **13**, 87–108 (1943), translated as "New Catadioptric Meniscus Systems," *J. Opt. Soc. Am.* **34** (1944), 270–84. In his original design, Maksutov used the following parameters (in millimeters): primary R_1 −823.2; meniscus diameter 100, thickness 10.0, radius R_a −152.8, R_b −158.6, $n = 1.5163$, $\nu = 64.1$; meniscus–mirror distance 539.1; f-stop number $f/4$.

‡A Mangin mirror is a thick negative meniscus, the rear surface made reflective. The negative lens introduces negative spherical aberration (overcorrection) which compensates for the positive aberration (undercorrection) of the mirror component. Compared with a paraboloid of the same focal length, a Mangin mirror has about half the coma.

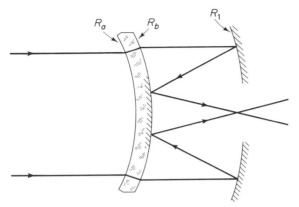

Figure 2.5-28   Cassegrain–Maksutov system showing meniscus corrector (*left*) and primary (*right*).

The Cassegrain–Maksutov telescope is short, sturdy, and of light weight; it has a large aperture and a wide field and is completely sealed, making it impervious to air turbulence and contamination. It is a very popular design.

The trend in the design of astronomical telescopes is toward *deformable* mirrors. Such mirrors consist of a thin, flexible face plate, often made of beryllium and supported on a rigid backplate by a series of thick piezoelectric disks. Telescopes of this kind include a sensor that monitors distortions of the incoming wavefronts (distorted because of air turbulence) and an elaborate real-time servo and feedback system to compensate for these distortions, eliminating the poor "seeing" (image quality) that so often afflicts astronomical observations. All of this goes by the name of *adaptive optics*.*

SUGGESTIONS FOR FURTHER READING

A. G. INGALLS, editor, *Amateur Telescope Making,* Vols. 1–3 (New York: Scientific American, Inc., 1955).

R. B. McLaughlin, *Special Methods in Light Microscopy* (London: Microscope Publications Ltd., 1977).

S. Inoué, *Video Microscopy* (New York: Plenum Press, 1986).

A. Cox, *Photographic Optics,* 15th edition (New York: Focal Press, Inc., 1974).

R. Kingslake, "The Development of the Zoom Lens," *J. Soc. Motion Pict. Telev. Eng.* **69** (1960), 534–44.

J. Maxwell, *Catadioptric Imaging Systems* (New York: American Elsevier Publishing Company, Inc., 1972).

D. F. Horne, *Optical Instruments and Their Applications* (Bristol, England: Adam Hilger Ltd., 1980).

---

*See R. H. Freeman and J. E. Pearson, "Deformable Mirrors for All Seasons and Reasons," *Appl. Opt.* **21** (1982), 580–88.

PROBLEMS

2.5-1. An astronomical telescope consists of two lenses separated by a distance of 80 cm. If the angular magnification, when focusing at infinity, is $5\times$, what are the powers of the two lenses?

2.5-2. A telescope has an objective of $+60$ cm focal length and an eyepiece of $+5$ cm focal length. With the telescope focused at a target 3 m away and used without accommodation, what is:
(a) The distance between the two lenses?
(b) The magnification?

2.5-3. If the objective of an astronomical telescope is 50 mm in diameter and its focal length 40 cm, and if the angular magnification is $8\times$, what is:
(a) The diameter of the exit pupil?
(b) The power of the eyepiece?

2.5-4. The objective of a telescope has a diameter of 40 mm and a focal length of 32 cm. When focused at infinity, the exit pupil is 2.5 mm in diameter. What is:
(a) The magnification of the telescope?
(b) The focal length of the eyepiece?

2.5-5. A $10\times$ telescope is pointed at the moon and focused to give an image at infinity. If the focal length of the objective is 60 cm, how far and in which direction must the eyepiece be moved to project an image of the moon on a screen 30 cm away from the eyepiece?

2.5-6. An astronomical telescope is focused for an object at infinity, without accommodation. How must the distance between the telescope objective and the eyepiece be changed if the telescope is used:
(a) For taking photographs with a camera focused for infinity?
(b) For taking photographs using only the camera body, without the camera lens?
(c) By an observer who does accommodate?

2.5-7. What is the magnification of a telescope that has an objective of 30 cm focal length and an eyepiece of 3.5 cm focal length and that is used for looking, *with* accommodation, at an object 13.12 m away?

2.5-8. A telescope has a $+2.00$-diopter objective and gives $10\times$ angular magnification. In which direction and how much must a 5.00-diopter myope, not wearing corrective glasses, move the eyepiece in order to see the image in focus?

2.5-9. A $2.5\times$ Galilean telescope has a -25.00-diopter eye lens. When focused for infinity and used without accommodation, what is the distance between the lenses?

2.5-10. A Galilean telescope, containing a $+50$-cm F.L. objective and a -8-cm F.L. eyepiece, is focused at an object 6.25 m away. What must be the distance between the two lenses if the image is to be at infinity? Solve by the vergence method.

2.5-11. A low-vision aid can be realized by wearing $+20.00$-diopter spectacle lenses over -32.00-diopter contact lenses. To focus at infinity, what is:
(a) The angular magnification?
(b) The axial distance between the lenses?

2.5-12. A low-vision aid has an objective of $+12$ cm focal length, a distance between lenses of 10 cm, and is designed for a working distance (from the eye) of 40 cm.
(a) What is the power of the eye lens?
(b) What is the magnification?

2.5-13. Van Leeuwenhoek's early *brass microscope* gave a magnification of $125\times$. What was the focal length of the lens he used?

2.5-14. If a microscope has an objective of 16 mm focal length, an optical tube length of 16 cm, and an eyepiece marked $5\times$, what is the total magnification?

2.5-15. A microscope is made from two equal lenses of $+50.00$ diopters power each, separated by a distance of 14 cm. If used without accommodation, what is:
(a) The object distance?
(b) The magnification?

2.5-16. A pocket microscope used for viewing a gem has an objective of 5 mm focal length and an eyepiece of 2 cm focal length. When adjusted for minimum eyestrain (no accommodation), the lenses are 5 cm apart. What is:
(a) The object distance?
(b) The magnification?

2.5-17. A $6\times$ telescope is converted into a microscope by adding a 50-mm-focal-length camera lens in front of the objective. When used without accommodation, what is:
(a) The object distance?
(b) The magnification?

2.5-18. An oil-immersion microscope objective is marked 1.6 mm, 1.25 N.A. It is used, at 16 cm optical tube length, together with an eyepiece marked $10\times$. If the oil has an index of 1.515, is the microscope's total magnification within the useful limit?

2.5-19. A lighted candle is placed in front of the eye in order to view the various *Purkinje–Sanson images* caused by reflection at the refracting surfaces of the eye. Show:
(a) The orientation and the relative sizes of the three images.
(b) The change that occurs in accommodation.

2.5-20. Assume that an *aphakic* eye (an eye from which the crystalline lens has been removed) has the radius of curvature of the cornea and the refractive index of a *schematic* eye and the length of a *reduced* eye. Find the location of the far point.

2.5-21. If a nearsighted person cannot see objects clearly that are farther away from the eye than 40 cm, what power lenses are needed to see distant objects in focus?

2.5-22. A person wants to look at the image of his or her own eyes, without accommodation, using a concave mirror of 60 cm radius of curvature. How far must the mirror be from the eyes?

2.5-23. A telephoto lens of the Galilean type has two elements, of focal lengths $+150$ mm and -60 mm, respectively, separated by a distance of 120 mm.
(a) What is the focal length of the system?
(b) If the object is at infinity, how far from the second lens is the image?

2.5-24. A telephoto lens can also be used in reverse, with the minus lens in front. If $f_1 = -30$ mm, $d = 60$ mm, and $f_2 = +45$ mm:
(a) What is the focal length of the combination?
(b) What is its purpose?

2.5-25. A Cassegrainian telescope of $R_1 = -40$ cm, $R_2 = -13\frac{1}{3}$ cm, and $d = 15$ cm is focused at the moon. Where is the image located? Solve by the vergence method.

2.5-26. What should be the radius of curvature of the back surface of a Maksutov corrector if its front surface has a radius of -74 mm, its thickness is 5 mm, and the material spectacle crown ($n = 1.523$)?

2.6

Systems Evaluation

With the design of the system completed, we now turn to testing and evaluation. The modern way of evaluating a system is by the use of *transfer functions,* a concept that is easy to define: *Transfer functions are a measure of performance.* They let us predict theoretically, as well as confirm or disprove experimentally, how the individual lenses perform that constitute the system, and how the system as a whole measures up to specifications and expectations. Transfer functions can be used also to evaluate peripheral components such as photographic film, video equipment, the eye, and even the atmosphere through which the light passes on its way to the system.

CONTRAST

In earlier times, optical systems were often tested by using as the object a target like that shown in Figure 2.6-1. Such targets have *high* contrast: the lines are a deep black on a white background.

But realistic objects are rarely ever black and white. Most often they come in shades of gray. A gray object in a gray fog has *low* contrast and, surely, two lines of low contrast cannot be resolved as well as two lines of the same spacing and high contrast. Any evaluation of an optical system, therefore, must take into account the *contrast.*

The contrast of a test pattern, object or image, depends on the amount of light, L, that is returned by the details of the pattern. If the pattern is of repetitive periodicity, such as a series of dark bars on a bright background, the term used is *contrast modulation,* γ. Con-

186

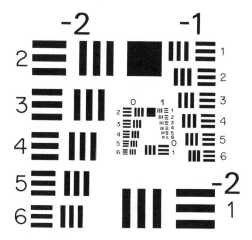

Figure 2.6-1 USAF test chart.

trast modulation is defined as the ratio of the difference between the amount of light, L_{max}, returned by the bright intervals, and the amount of light, L_{min}, returned by the dark bars, to the sum of such amounts,

$$\gamma = \frac{L_{max} - L_{min}}{L_{max} + L_{min}} \qquad [2.6\text{-}1]$$

For instance, L_{max} may have any value but if L_{min} is zero, the contrast modulation is 100%. If $L_{max} = L_{min}$, the modulation is zero. A value of 80% means high contrast; such objects are well visible. A value of 20% means low contrast; such objects are barely visible.

If the test object is nonrepetitive, containing dark letters on a gray background, the contrast may be defined as

$$C = \frac{L_B - L_O}{L_B} = 1 - \frac{L_O}{L_B} \qquad [2.6\text{-}2]$$

where the subscript O refers to the object and B to the background. If the object is darker than the background, the contrast is positive and less than 100%. If the object is lighter than the background, C is negative and can have any value.

Next, consider an object of repetitive, sinusoidal light distribution. Scanning across the object, we find an alternating sequence of maxima and minima (Figure 2.6-2). But,

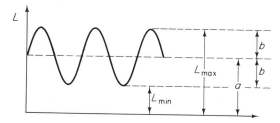

Figure 2.6-2 Sinusoidal light distribution as a function of space (or time).

instead of recording these variations, it is often more practical to introduce two new parameters, the mean light level, $a = (L_{max} - L_{min})/2 + L_{min}$, and the variation around the mean, $b = L_{max} - a$. The ratio of both is called the *normalized modulation, M*,

$$M = \frac{b}{a}$$ [2.6-3]

which is independent of the absolute amount of light received.

TRANSFER FUNCTIONS

Modulation transfer.   Actually, whenever we test for performance, we are not as interested in the modulation contrast as such as in how much modulation can be transferred from object to image. It is the *modulation transfer* that counts. The ratio of modulation in the image to modulation in the object is called the *modulation transfer factor, **T***,

$$\mathbf{T} = \frac{M_{image}}{M_{object}}$$ [2.6-4]

In Figure 2.6-3, for instance, the image modulation is one-half the object modulation, and hence $\mathbf{T} = \frac{1}{2}$.

Is the image contrast always less than the object contrast? Not necessarily. For example, a portrait photograph taken on high-contrast film used for technical reproductions would show harsh, unnatural shadows, of a contrast *higher* than that of the object.

Contrary to what Equation [2.6-4] suggests, the transfer factor **T** of a given lens does not have a single, unique value. Instead, *it varies as a function of spatial frequency, R*:

$$\mathbf{T} = f(R)$$ [2.6-5]

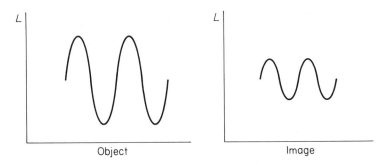

Figure 2.6-3   Scanning across lines shows high contrast in the object (*left*) and low contrast in the image (*right*).

The term *spatial frequency* refers to the number of lines, or other detail, within a given length. A technical line drawing, for example, may have four lines, and four intervals adjoining these lines, per centimeter. The spatial frequency, then, is 4 lines/cm, but since spatial frequency is customarily written in units of inverse *millimeters,* we have $R = 0.4$ mm^{-1}. If 1000 lines, and the intervals between them, fit into 1 mm, $R = 1000$ mm^{-1}. If one bar alone is 1 cm wide, $R = 0.05$ mm^{-1}. The higher the spatial frequency that a given optical system can resolve, the better the system.

In the propagation of light, in wave motion in general, and in electric circuits, frequency usually means *time frequency.* In transfer functions and in optical data processing, it means *space frequency.*

According to Equation [2.6-5] the transfer factor is a function of spatial frequency. Therefore, instead of a single-number transfer *factor,* **T**, we have a transfer *function,* **T**(R), called *modulation transfer function,* **MTF**.

Spread functions.   Assume that an object point (a pixel) is focused by a lens into the image plane. But because of aberrations and other limitations, the conjugate pixel is not a point any longer; instead, the pixel spreads out into a more diffuse patch of light. When we scan this patch using a pinhole aperture, we find a light distribution across the patch that is typical of the lens used, called the *point spread function, S(y, z)*, of the lens.

Since it is difficult to scan precisely across a point, it is better to use a line object and a slit aperture. The scan trace then represents the *line spread function, S(z)*, shown in Figure 2.6-4. The line spread function is the point spread function integrated over the length of the slit, y,

$$S(z) = \int_{-\infty}^{+\infty} S(y, z)dy \qquad [2.6\text{-}6]$$

Next consider an extended object, composed of many pixels. With the light being incoherent, each pixel forms its own patch, the point spread function of the lens, and the

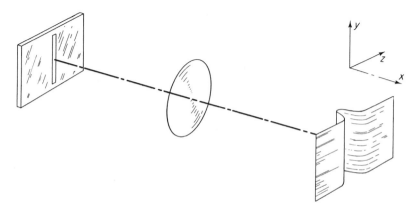

Figure 2.6-4   Line spread function. Object left, image right. The distribution of light in the image is shown as a cylindrical mound.

image is the sum of the individual spread functions. For simplicity, we integrate over one dimension only, calling E the final light distribution, I the initial distribution, and S the spread function of the lens,

$$E(z) = \int_{-\infty}^{+\infty} I(z)S(z)\, dz \qquad\qquad [2.6\text{-}7]$$

The converse holds true as well: the spread function is the derivative of the light distribution,

$$\frac{dE(z)}{dz} = I(z)S(z) \qquad\qquad [2.6\text{-}8]$$

This means that the slope of the scanning trace, $dE(z)/dz$, at a given point, z_0, in the image is proportional to the line spread function at that point. Therefore, if the trace is found experimentally, a plot of its slope versus position in the image is the spread function of the lens.

Now, since the input, the system's response, and the output are all functions of the same variable, R, we can write the spread function in the form of a *Fourier transformation*, a concept to be explored more fully in Chapter 4.4. At this point, we merely note that the *modulation transfer function of a lens is the Fourier transform of the spread function of that lens*,

$$\boxed{\mathbf{MTF} = \int_{-\infty}^{+\infty} S(z)e^{-2\pi iRz}\, dz} \qquad\qquad [2.6\text{-}9]$$

Phase transfer.   Due to aberrations, in particular coma and distortion, some image pixels may become dislodged with respect to their nominal ("correct") positions; such dislocation is shown in Figure 2.6-5.

The extent of the dislocation is measured by the *phase* (shift) ϕ. The same as with T, the phase ϕ likewise is a function of spatial frequency, $\phi = f(R)$. Thus, in addition to the

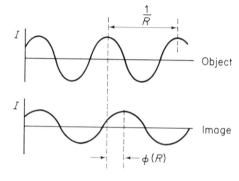

Figure 2.6-5   Dislocation of image (*bottom*) with respect to object (*top*). R, spatial frequency; ϕ, phase shift. Also note lower contrast in image compared to object.

modulation transfer function, we need to consider also a *phase transfer function*. Both together, written in exponential notation, form the *optical transfer function*, **OTF**:

$$\boxed{\text{OTF} = \text{MTF} \cdot e^{i\phi(R)}} \qquad [2.6\text{-}10]$$

The optical transfer function is a space-frequency-dependent complex quantity whose modulus, appropriately, is the modulation transfer function and whose phase is the phase transfer function. The optical transfer function, therefore, is the *ratio of the Fourier transform of the light distribution in the image to the Fourier transform of the light distribution in the object:**

$$\text{OTF} = \frac{\text{Fourier transform of light distribution in image}}{\text{Fourier transform of light distribution in object}} \qquad [2.6\text{-}11]$$

In the most general terms, the optical transfer function describes the degradation of an image, at different space frequencies. A perfect lens should have a modulation transfer factor **T** of unity, and a phase transfer factor ϕ of zero, *at all space frequencies*. In reality, this can rarely, if ever, be so, because of diffraction and aberrations. At low space frequencies, the transfer factor of a good lens may be close to unity. At higher frequencies, the $T(R)$ curve gradually declines, approaching $T = 0$ somewhere between 100 and 1000 mm^{-1}. The $\phi(R)$ curve, at low frequencies, will be close to zero. At higher frequencies, it increases. In a way, the $\phi(R)$ curve is a mirror image of the $T(R)$ curve, as shown in Figure 2.6-6.

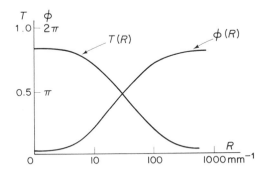

Figure 2.6-6 Modulation transfer factor, $T(R)$, and phase transfer factor, $\phi(R)$, both as functions of spatial frequency.

---

*This important relationship was first formulated by Pierre-Michel Duffieux (1891–1976), French physicist, erstwhile assistant to Charles Fabry, and later professor at the universities of Rennes and Besançon. Duffieux, like Fourier before him, started out by working on the seasonal transfer of heat through the earth's soil, went on to Fourier transform spectroscopy, and wrote a highly original book, *L'Intégrale de Fourier et ses Applications à l'Optique* (Paris: Masson et Cie, 1970), in which he discussed the notion of spatial frequency and the existence of a cutoff frequency, laying the foundation for transfer functions as we use them today for image evaluation.

THE EXPERIMENTAL DETERMINATION OF TRANSFER FUNCTIONS

The essential point in determining the optical transfer function of a lens or lens system is to form and scan an image of a test grid. Often used for this purpose are *Foucault grids* (also called Sayce targets); these contain a series of black-and-white, square-wave or sinusoidal, parallel bars of varying space frequency, widely spaced at one end and close together at the other. The grid can be used as the object, with a scanning slit in the conjugate image plane; or the slit can be the object, with the grid placed in the image plane. Sometimes the grid is made into a drum, as shown in Figure 2.6-7.

At the low space frequencies (widely spaced slots), the photocell faithfully registers all bars and intervals. At the high frequencies, depending on the quality of the lens, the individual signals fuse together, the lines are no longer "resolved," and the envelope across the trace goes *down*.

If a lens *system* is tested, the total **MTF** is the *product* of the **MTF**s of the individual lenses. (The individual transfer curves are multiplied, ordinate by ordinate, at each point of the abscissa; see, for example, Problem 2.6-11.) But that holds only if any aberrations introduced by one lens are not counteracted by aberrations of another.

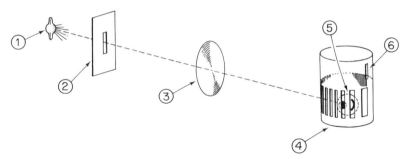

Figure 2.6-7   Experimental arrangement for measuring transfer functions. 1, Light source; 2, slit; 3, lens under test; 4, rotating drum; 5, photodetector; 6, reference mark for determining phase transfer.

Example 1

Photographs are taken from a high-altitude aircraft of a cruise ship, painted white. Assume that the **MTF** of a typical camera lens is that shown in Figure 2.6-6. What focal length is necessary to obtain enough detail to identify the ship?

Solution.   The ship may have a brightness of 5 arbitrary units, and the surrounding ocean may have 2 units. Then its contrast is

$$\gamma = \frac{5 - 2}{5 + 2} = 43\%$$

Initially, the focal length of the system may be chosen so that the ship's image is 0.5 mm wide. This means a spatial frequency of $R = 1$ mm^{-1}. From Figure 2.6-6 we find that at this frequency the contrast transfer factor **T** is 0.8. At $R = 10$, **T** is 0.7, and at $R = 100$, it is 0.2.

Knowing that the object is of limited contrast (0.43), we now find that for different space frequencies the contrast, transferred to the image, changes as follows:

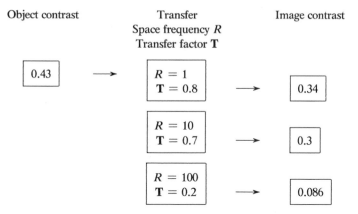

Object contrast   Transfer   Image contrast
Space frequency R
Transfer factor T

0.43 $\longrightarrow$ $R = 1$ / $T = 0.8$ $\longrightarrow$ 0.34

$R = 10$ / $T = 0.7$ $\longrightarrow$ 0.3

$R = 100$ / $T = 0.2$ $\longrightarrow$ 0.086

A contrast of 34% is satisfactory for visual observation. That means the ship is easy to see. But fine details, at $R = 100$, cannot be seen. At this frequency the image contrast is so low, less than 9%, that a lens of longer focal length is needed to obtain a larger image. The larger image contains *lower* frequencies, which in turn permit more contrast to be transferred.

Example 2

What effect does atmospheric turbulence have on aerial reconnaissance?

Solution.   From Figure 2.6-8 we conclude that if the details on the ground have high contrast, details of a given space frequency can be resolved well both through calm air and through turbulent air. Turbulence merely causes the contrast to become less (arrow A) but since the details have high contrast, they remain well visible. On the other hand, if the details have low contrast, for example if the target is camouflaged, only large details can be identified (arrow B), a result that surely we have anticipated by intuition.

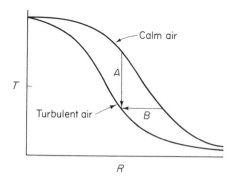

Figure 2.6-8   Modulation transfer through calm air and turbulent air.

Example 3

Visual acuity is often determined using a *Snellen acuity chart*, a series of black letters of different size printed on a white background. No provision is made for an evaluation of *contrast*. But, short of printing numerous such charts, what can we do?

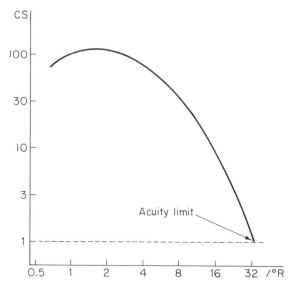

Figure 2.6-9   Plot of contrast sensitivity versus spatial frequency.

Solution.   We use a sinusoidal test pattern generated on a computer display. Beginning at zero, the pattern's contrast is gradually increased until the pattern becomes visible and the patient notes its orientation; that is repeated at different space frequencies.

Then we plot the reciprocal of the contrast, from a minimum of 1 (the reciprocal of 100%) to a practical limit of 1000 (the reciprocal of 0.1%), versus the spatial frequency; that is called the *contrast sensitivity function* (Figure 2.6-9).

Toward the higher space frequencies, the function falls off rapidly toward a contrast sensitivity of unity. The frequency at which it reaches that limit is the *acuity limit*. Beyond that, the pattern cannot be resolved.

The interesting point is that good acuity at one space frequency does not necessarily mean good acuity at another. In fact, some patients can read Snellen letters reasonably well but complain that objects as they occur in nature appear foggy and "washed out." Clearly, more than one space frequency is needed to evaluate any transfer of contrast.

PATENT CONSIDERATIONS

Perhaps the system we have designed and tested looks promising, and we consider whether it might be worth trying to obtain a *patent*. Patents are granted for developments in applied technology; they are not granted for ideas, scientific principles, or for newly discovered laws of nature. Only when the principles are applied to the invention of a "new and useful process, machine, manufacture, or composition of matter, or any new and useful improvements thereof," to quote the United States Patent Act of 1952, may a patent be issued. To be patentable, in short, an invention must be new and useful, it must not result from merely exercising mechanical skill or substituting materials: It must be different from "prior art."

If you think some of your work may lead to an invention, you should keep a written record of the work, best in a notebook that is bound, has numbered pages, and has no pages torn out or left blank. Entries in such a notebook need not be an elaborate literary effort; simply state how the elements are arranged and how they work. Include a drawing and date and sign what you have written.

Have two witnesses add a statement, "I have read and understood this disclosure," which they also should date and sign. Note that simple reading is not enough; the witnesses should have understood how the invention works, which means they must have some familiarity with the subject matter.

Next, things are made easier if you consult a *patent attorney.* Take along the written description. The attorney will quickly understand what the invention is all about and will give further advice as to its merits and chances of patentability.

If you decide to go ahead, the patent attorney will initiate a *patent search.* That search will usually turn up several patents that come close. Carefully read these patents and discuss with the attorney their merits and differences, compared to your invention.

If your disclosure seems clearly superior, a formal *Patent Application,* couched in legal terms, is filed, with drawings made to the Patent Office's specifications. All of this takes time; in general, start the application process no later than six months after the initial notebook entry. When the Patent Office receives the application, the patent is "pending."

More often than not, the patent examiner will reject some of the claims made in the application. That will necessitate a further exchange of letters, handled through the patent attorney. If things go well, the patent will issue within one to a few years. Legal protection, however, begins with the patent "pending." (This advantage is sometimes used by makers of certain gadgets, who quickly tool up, manufacture the product, and saturate the market. Such items often become obsolete soon; thus, when the patent is finally denied, the product is no longer marketable anyway.) Needless to say, a worthwhile *optical design* does not suffer such an ignominious fate.

SUGGESTIONS FOR FURTHER READING

F. H. PERRIN, "Methods of Appraising Photographic Systems: I. Historical Review," *J. Soc. Motion Pict. Telev. Eng.* **69** (1960), 151–56; II. "Manipulation and Significance of the Sine-Wave Response Function," *ibid.* **69** (1960), 239–49.

K. R. BARNES, *The Optical Transfer Function* (New York: American Elsevier Publishing Company, Inc., 1971).

W. B. WETHERELL, "The Calculation of Image Quality," in R. R. Shannon and J. C. Wyant, editors, *Applied Optics and Optical Engineering,* Vol. 8, Chap. 6, pp. 171–315 (New York: Academic Press, Inc., 1980).

P. M. DUFFIEUX, *The Fourier Transform and Its Applications to Optics,* 2nd edition (New York: Wiley-Interscience, 1983).

General Information Concerning Patents (Washington, DC: U.S. Department of Commerce, Patent and Trademark Office, 1986).

PROBLEMS

2.6-1. A target contains a series of black lines, each line 8 mm wide and separated from the next line by an interval also 8 mm wide. What is the spatial frequency?

2.6-2. What is the spatial frequency, in mm^{-1}, of the bar pattern in the USAF test chart shown in Figure 2.6-1:
(a) Near the left-hand lower corner, number 5?
(b) At the right-hand margin, number 4?

2.6-3. A camera lens of 28.6 mm focal length resolves details that in the image plane have a spatial frequency of 10 mm^{-1}. What is the angular resolution, in degrees?

2.6-4. What is the power of a lens that has an angular resolution of 0.05° and resolves details in the focal plane which have a spatial frequency of 7 mm^{-1}?

2.6-5. If in a periodic test pattern the minima receive two-thirds the amount of light that the maxima receive, what is the contrast?

2.6-6. If the contrast in a modulation test pattern is 60%, and if the maxima receive 20 units of light, how much do the minima receive?

2.6-7. If a target has bright bars emitting 5 arbitrary units of light, and dark bars emitting 3 units, how much contrast does the target have assuming:
(a) The target is repetitive (periodic)?
(b) The target has letters that are nonrepetitive?

2.6-8. At which light levels will both types of contrast (see Problem 2.6-7) yield the same result?

2.6-9. If in a periodic test pattern the maxima receive 10 units of light and the minima 6 units, does the contrast modulation differ from the *normalized* modulation?

2.6-10. Continue with Problem 2.6-9. Double the amount of light incident on the test pattern.
(a) How much light is now returned by the maxima and by the minima?
(b) Again determine the ordinary and the normalized modulation.

2.6-11. A test object of 76% modulation is photographed through a camera lens that for 4 lines mm^{-1} has a transfer factor $\mathbf{T} = 0.85$. The image is formed on film whose transfer factor, at the same frequency, is 0.62. What is the total modulation?

2.6-12. Assume that the primary mirror of a reflecting telescope is:
(a) Of perfect paraboloidal shape.
(b) Distorted and not truly paraboloidal.
(c) Of the correct shape but merely ground to a silky finish, rather than polished.
Plot the **MTF**s that correspond to these three conditions.

2.6-13. An extended object is projected by a converging lens into a real image, at about unit magnification. Assume that a sheet of distorted window glass is placed:
(a) Directly over the object.
(b) Immediately in front of, or closely behind, the lens.
(c) Directly in front of the image.
Neglecting the need for refocusing, how will the image change?

2.6-14. Continue with Problem 2.6-13 and assume now that the sheet of glass is precisely plane-parallel but only finely ground, rather than polished. How does this change the image?

3.1

Light as a Wave
Phenomenon

MANY OF THE LIMITATIONS of geometrical optics can be explained by assuming that light is made up of *waves*. Light waves behave much like other waves. They bend around obstacles, they combine with like waves to reinforce or weaken each other, and, in the most general terms, they deviate from rectilinear propagation in ways that cannot be predicted from geometrical optics.

We begin our discussion of wave optics with a study of *wave motion*. A motion that repeats itself periodically, in equal intervals of time, is called a *periodic motion*. If the motion follows a sine, or a cosine, function and the waveform therefore is sinusoidal or cosinusoidal, it is called a *simple harmonic motion*.

SIMPLE HARMONIC MOTION

A good way of describing simple harmonic motion is by considering it a projection of uniform circular motion. Look at the *reference circle* (Figure 3.1-1) where a point P_0 travels in a circular path, counterclockwise, at an angular velocity ω. After a given interval of time, the point has reached P_1. The projection of P_1 on the y axis gives point Q. Thus as P moves around the circle, Q will move up and down along the y axis.

It is customary to plot on the x axis the independent variable, *time,* and on the y axis the dependent variable, the *displacement*. Let point P begin its motion at the initial position P_0, where $t = 0$ and $y = 0$. When P has advanced through $\phi = 45°$, it has reached P_1. Since the total angle around a point is $360°$ or 2π radians, point P has moved through $2\pi(45°/360°) = \frac{1}{4}\pi$ rad.

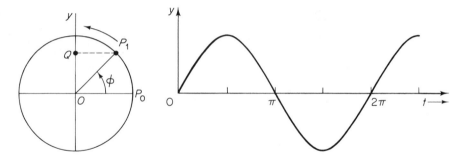

Figure 3.1-1   Reference circle (*left*) and simple harmonic motion (*right*).

After another equal interval of time, P has moved through a total of $\phi = 90°$, or $\frac{1}{2}\pi$ rad. The projection, y, has reached a maximum. After turning through $\phi = 180° = \pi$ rad, P is at a position where again $y = 0$. After $\phi = 270° = \frac{3}{2}\pi$ rad, P is at the lower maximum, and after $\phi = 360° = 2\pi$ rad, it is back at its initial position. Connecting all points gives the sinusoidal curve shown in Figure 3.1-1, right.

We call the radius of the reference circle **A**, for *amplitude,* and the angle ϕ the *phase.* Then

$$\sin \phi = \frac{y}{\mathbf{A}}$$

where y is the displacement of Q above O. Substituting the definition of angular velocity, $\omega = \phi/t$, solved for ϕ, gives

$$\boxed{y = \mathbf{A} \sin \omega t}$$
[3.1-1]

which describes the height of the displacement, measured along the y axis. Along the x axis, we have

$$x = \mathbf{A} \cos \omega t$$
[3.1-2]

In a more general sense, point P may begin its motion, not at P_0, but at a certain initial phase angle ϕ. Then, instead of Equations [3.1-1] and [3.1-2], we use

$$y = \mathbf{A} \sin(\omega t + \phi)$$
[3.1-3]

or

$$x = \mathbf{A} \cos(\omega t + \phi)$$
[3.1-4]

Note that point P moves around the reference circle at constant velocity but the velocity of the projected point Q changes. At $y = 0$, its velocity, u, is highest. Toward the maximum value of the displacement, both $+y$ and $-y$, point Q slows down. At the maxima themselves, $u = 0$. Then u changes direction, increasing again toward maximum velocity at $y = 0$, and so on.

So far, I have always referred to the maximum displacement of a point on a wave as the *amplitude.* But in practice the amount of light generated, propagated, or received is a matter of *energy* or, more precisely, it is a matter of energy per unit of time. The energy of light is commonly called *intensity, I*. In principle similar to the energy contained in a mechanical wave, the *intensity of light is proportional to the square of the amplitude:*

$$I \propto \mathbf{A}^2 \qquad [3.1\text{-}5]$$

Complex notation.   Simple harmonic motion can be represented by either a sine function or a cosine function:

$$y = \mathbf{A} \sin(\omega t + \phi)$$
$$y = \mathbf{A} \cos(\omega t + \phi) \qquad [3.1\text{-}6]$$

Such functions can be combined into *complex numbers*. The concept of a complex number often instills unnecessary fears. In reality, it is quite simple. A complex number is a number that has both a real part and an imaginary part. The two parts can be drawn along orthogonal axes, a real axis, x, and an imaginary axis, y (Figure 3.1-2). Point P, which represents the "complex" number, then has the *rectangular* coordinates (x, y). But the amplitude $\mathbf{A}$ can also be represented by *polar* coordinates (A, ϕ):

$$\sin \phi = \frac{y}{A} \qquad \text{and} \qquad \cos \phi = \frac{x}{A} \qquad [3.1\text{-}7]$$

Furthermore, we see from Figure 3.1-2 that

$$\cos(90° - \phi) = \frac{y}{A} \qquad [3.1\text{-}8]$$

In short, the sine function and the cosine function (of a wave vector) are essentially the same except that they *differ in phase by 90°*.

Assume now that we have two vectors, $\mathbf{a}$ and $\mathbf{b}$, that differ by 90° (they are *orthogonal* to one another). In complex notation, we preface $\mathbf{b}$ by a symbol, i. This symbol means that $\mathbf{b}$, as compared with $\mathbf{a}$, has made a *counterclockwise rotation through 90°*. The resultant of $\mathbf{a}$ and $i\mathbf{b}$ is then

$$\mathbf{A} = \mathbf{a} + i\mathbf{b} = A \cos \phi + iA \sin \phi \qquad [3.1\text{-}9]$$

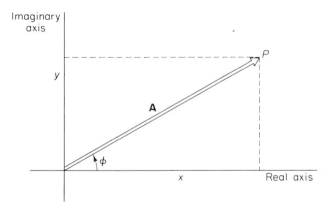

Figure 3.1-2 *Argand diagram* shows vector, **A**, extending from zero to (complex) point P.

Using Euler's formula,*

$$\cos \phi + i \sin \phi = e^{i\phi} \qquad\qquad [3.1\text{-}10]$$

we can now write Equation [3.1-9] in the form of

$$\mathbf{A} = Ae^{i\phi} \qquad\qquad [3.1\text{-}11]$$

where A is the absolute value of the (complex) *amplitude* $\mathbf{A}$ of the wave:

$$A = |\mathbf{a} + i\mathbf{b}| = \sqrt{a^2 + b^2} \qquad\qquad [3.1\text{-}12]$$

and ϕ is its *phase:*

$$\phi = \tan^{-1}\left(\frac{b}{a}\right) \qquad\qquad [3.1\text{-}13]$$

Term e is the base of natural logarithms, 2.71828. . . .

Whenever we are concerned with the *energy* of light, we need the *square* of the amplitude. Using complex notation, this is easy to do. We merely multiply the complex number by its *complex conjugate*. The complex conjugate of $\mathbf{A}$, for example, is written $\mathbf{A}*$. Conjugate terms, whenever they contain i, change signs; thus the product of $\mathbf{A}$ and $\mathbf{A}*$ becomes

$$\mathbf{A}\mathbf{A}* = Ae^{i\phi}Ae^{-i\phi} = A^2 \qquad\qquad [3.1\text{-}14]$$

which is (proportional to) the energy.

Finally, since the presence of i implies a 90° rotation, the $iA \sin \phi$ term in Equation [3.1-9] becomes redundant and we may write the *real part* $y = A \cos(\omega t)$ as

$$\boxed{y = Ae^{i\omega t}} \qquad\qquad [3.1\text{-}15]$$

This is a very useful expression.

MOVING WAVES

So far we have discussed the simple harmonic motion of a *single* point. Now we proceed to a *series* of points, which together make up a wave, and see how this wave propagates through space.

There are two ways of looking at moving waves. For example, consider the waves on the surface of a lake at a given instant of time. If we take a snapshot of the waves, their motion is frozen at time $t = 0$. That means that we consider the shape of the wave *as a function of space,*

$$y = f(x) \qquad\qquad [3.1\text{-}16]$$

---

*Leonhard Euler (1707–1783), Swiss mathematician. Euler worked on a wide range of topics, from astronomy to perturbation theory to harmonic wave motion, laying the groundwork for Thomas Young, who some 50 years later showed that light can be represented by waves. Euler's formula is most remarkable in that it connects geometry (trigonometric functions) with algebra (exponential functions).

But we can also look at the wave as a *function of time*. For example, if we look through a narrow slot, as in Figure 3.1-3, we will see a given point on the water surface merely move up and down (the x position remaining constant). In contrast to Equation [3.1-16], we now have

$$y = f(t) \qquad [3.1\text{-}17]$$

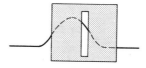

Figure 3.1-3   Moving wave as seen through narrow slot.

Now assume that the wave configuration moves forward, to the right. If v is the velocity of the wave, then after time t the wave has moved through distance $x = vt$. Therefore, in order to retain the equality in $y = f(x)$, vt must be *subtracted* from x_0:

$$y = f(x_0 - vt) \qquad [3.1\text{-}18]$$

Since angular velocity $\omega = \phi/t$, we have for *one* oscillation,

$$\omega = \frac{2\pi}{T} \qquad [3.1\text{-}19]$$

where T is the period of the wave, the reciprocal of the frequency, $1/T = \nu$. Substituting this expression in Equation [3.1-20] gives

$$\omega = 2\pi\nu \qquad [3.1\text{-}20]$$

and substituting this in Equation [3.1-1], $y = \mathbf{A}\sin(\omega t)$ leads to

$$y = \mathbf{A}\sin(2\pi\nu t) \qquad [3.1\text{-}21]$$

This describes the displacement in terms of *time*.

Then take the relationship

$$v = \lambda\nu = \frac{x}{t} \qquad [3.1\text{-}22]$$

Retain the two right-hand terms and solve for ν.

$$\nu = \frac{x}{\lambda t} \qquad [3.1\text{-}23]$$

Substitute Equation [3.1-23] in [3.1-21] and cancel t:

$$y = \mathbf{A}\sin\left(2\pi\,\frac{x}{\lambda}\right) \qquad [3.1\text{-}24]$$

This describes the displacement in terms of *space*.

We are now ready to represent the displacement in *space and time,* that is, we make the wave *move.* We combine Equations [3.1-18] and [3.1-24]. This gives

$$y = \mathbf{A} \sin\left[\frac{2\pi}{\lambda}(x - vt)\right]$$

[3.1-25]

which is an important equation. It describes the displacement of any point on a (sinusoidal) wave in both space and time.

THE GENERAL WAVE EQUATION

Returning to Equation [3.1-25], how does the displacement, y, vary as a function of space, x, and time, t? This requires partial differentiation of y with respect to x and t:

$$\frac{\partial y}{\partial x} = \frac{2\pi}{\lambda}\mathbf{A}\cos\left[\frac{2\pi}{\lambda}(x - vt)\right]$$

[3.1-26]

and

$$\frac{\partial y}{\partial t} = \frac{2\pi v}{\lambda}\mathbf{A}\cos\left[\frac{2\pi}{\lambda}(x - vt)\right]$$

[3.1-27]

Combining both equations and eliminating equal factors gives

$$\frac{\partial y}{\partial x} = -\frac{1}{v}\frac{\partial y}{\partial t}$$

[3.1-28]

which is the differential equation describing the wave.

The *second* derivative,

$$\frac{\partial^2 y}{\partial x^2} = \frac{1}{v^2}\frac{\partial^2 y}{\partial t^2}$$

[3.1-29]

holds for any (sinusoidal) wave, independent of the direction of travel, either $-x$ or $+x$.

There are two ramifications that follow. First, I have referred to a displacement, y, which might give the impression that something material really *moves.* This, however, need not be so. We replace y by the more general term ξ which stands for *any disturbance,* without prejudice as to its nature, and instead of using Equation [3.1-29] write

$$\frac{\partial^2 \xi}{\partial t^2} = v^2\frac{\partial^2 \xi}{\partial x^2}$$

[3.1-30]

This is the *one-dimensional wave equation.* It connects variations in time and space to the velocity of propagation of the wave.

Second, we do not want to restrict ourselves to waves that propagate along the x axis. But if we are to include waves propagating in *any* direction, we need to extend the right-hand term in Equation [3.1-30] to the y and z axes, and replace it by

$$\frac{\partial^2 \xi}{\partial x^2} + \frac{\partial^2 \xi}{\partial y^2} + \frac{\partial^2 \xi}{\partial z^2} \tag{3.1-31}$$

Such an expression can be written, much shorter, using the *vector differential operator* ∇, read "del." The operator ∇ stands for

$$\nabla = \frac{\partial}{\partial x} + \frac{\partial}{\partial y} + \frac{\partial}{\partial z} \tag{3.1-32}$$

If we multiply ∇ by itself, we obtain a *scalar* (operator), ∇^2:

$$\nabla \cdot \nabla = \nabla^2 \tag{3.1-33}$$

This is the *Laplacian operator;* it represents the sum of three second-order partial derivatives:

$$\nabla^2 = \frac{\partial^2}{\partial x^2} + \frac{\partial^2}{\partial y^2} + \frac{\partial^2}{\partial z^2} \tag{3.1-34}$$

For a wave propagating in any direction in three-dimensional space, therefore, Equation [3.1-30] becomes

$$\boxed{\frac{\partial^2 \xi}{\partial t^2} = v^2 \nabla^2 \xi} \tag{3.1-35}$$

which is the *general three-dimensional wave equation.*

At this point we will temporarily leave the theory of wave motion and defer discussion of what actually constitutes ξ to Chapter 4.1, when we come to the electromagnetic nature of light. We turn now to some more tangible aspects of waves, interference (Chapters 3.2 through 3.4) and diffraction (Chapters 3.5 and 3.6).

SUGGESTIONS FOR FURTHER READING

J. R. PIERCE, *Almost All about Waves* (Cambridge, MA: The MIT Press, 1974).

W. C. ELMORE and M. A. HEALD, *Physics of Waves* (New York: Dover Publications, Inc., 1985).

I. G. MAIN, *Vibrations and Waves in Physics,* 2nd edition (New York: Cambridge University Press, 1984).

For an introduction to optics' close cousin, acoustics, see

J. BACKUS, *The Acoustical Foundations of Music,* 2nd edition (New York: W. W. Norton & Company, Inc., 1977).

PROBLEMS

3.1-1.   If a young person can hear sound over a frequency range from 22 Hz to 15 kHz and if the velocity of sound is $v = 331 \text{ m s}^{-1}$, what are the wavelengths corresponding to these limits?

3.1-2.   If sound waves of 6.85 kHz frequency propagate at a velocity of 344 m/s, what is the phase angle difference (in radians) of two points on the wave 4 cm apart?

3.1-3. Assume that point P on the reference circle moves at a constant angular velocity of 4 rad/s. How long does it take the projection, Q, to move from $y = 0$ to a position halfway up to the maximum amplitude?

3.1-4. The sinusoidal curve on the right in Figure 3.1-1 may represent a natural-size snapshot of a 50-Hz wave (vertical dimension exaggerated). How fast does the wave move forward?

3.1-5. Starting at $y = 0$, a point P on a sinusoidal wave has moved to a position $\frac{2}{3}$ of the height of the amplitude, $y = \frac{2}{3}\mathbf{A}$. If $\lambda = 10$ cm, how much has P advanced along the x axis?

3.1-6. What is the wavelength and the velocity of propagation of the wave that can be represented by $y = 4 \sin(10x - 20t)$?

3.1-7. If the amplitude of a certain wave is 2 cm, the period 0.02 s, and the velocity 40 cm s^{-1}, what is:
(a) The frequency?
(b) The wavelength?
(c) The equation describing the displacement as a function of space and time?

3.1-8. If surface waves on a lake can be described by the equation $y = 0.4 \sin(2t)$, what is:
(a) The maximum displacement?
(b) The velocity, at time $t = 0$, of a cork floating on the water?

3.1-9. A wave has a period of 0.2 s and an amplitude of 4 cm. What is the displacement and velocity of an element of the wave at the time the phase angle has increased from zero to 60°?

3.1-10. A point moves counterclockwise in a circle at a constant speed of 5 cm s^{-1}. If the period is 6 s and if at time $t = 0$ the point is 30° away from the $+x$ direction, what is:
(a) Its displacement, y?
(b) The displacement at $t = 4$ s?
(c) The equation of motion in terms of A, ω, and ϵ?

Interference

The term interference refers to the phenomenon that some parts of light, under certain conditions, temporarily intensify or weaken each other. Expressions such as "constructive" and "destructive" interference should not be used because they seem to imply that somehow there could be a "destruction" of light. Certainly, this is not so. If less light reaches a given point, more light reaches some other point: interference merely causes a *redistribution* of the light.

YOUNG'S DOUBLE-SLIT EXPERIMENT

The classical interference experiment is *Thomas Young's double-slit experiment.* Young let light pass through two close-by openings in an opaque screen.* At first, he tried two pinholes, but quickly realized that the fringes were much brighter when he used two parallel *slits* (S_1 and S_2 in Figure 3.2-1).

---

*Thomas Young (1773–1829), British physician and physicist. The son of a Quaker family, Young could read at age 2, at 6 began studying Latin, and at 13 had also mastered Greek, Hebrew, Italian, and French. At 19 he entered medical school, correctly explained the accommodation of the eye, and was elected Fellow of the Royal Society. In 1796, Young graduated from the University of Göttingen Medical School, opened a practice in London, and five years later became Professor of Natural Philosophy (physics) at the Royal Institution of London. That same year, 1801, he read before the Royal Society the first of several papers presenting the wave theory of light and the principle of interference, much to the opposition of Newton's followers. Young made noteworthy contributions also to acoustics, atmospheric refraction, elasticity, fluid dynamics, and color vision; later in his life, he helped decipher the Egyptian Rosetta Stone hieroglyphics. Th. Young, "On the Theory of Light and Colours," *Philos. Trans. Roy. Soc. London* **92** (1802), 12–48; and "Experiments and calculations relative to Physical Optics," *ibid.* **94** (1804), 1–16.

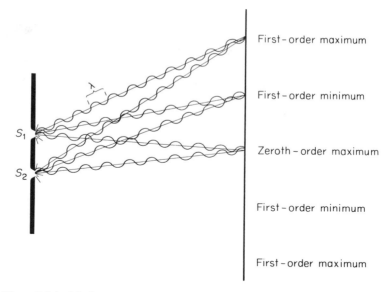

Figure 3.2-1 Maxima and minima in Young's double-slit experiment. Vertical dimensions exaggerated.

In the center of the field, where the contributions from the two slits have traveled through equal distances and where the path difference is zero:

$$\Gamma = 0$$

we have the *zeroth-order maximum*. But maxima will also occur whenever the path difference is one wavelength, λ, or an integral multiple of a wavelength, $m\lambda$:

$$\Gamma = m\lambda \qquad [3.2\text{-}1]$$

The integer m is called the *order* of interference.

In order to calculate the positions of the maxima, we call d the distance between the centers of the two slits, θ the change in direction of the light, and Γ the path difference between the two contributions. Then from Figure 3.2-2,

$$\sin \theta = \frac{\Gamma}{d} \qquad [3.2\text{-}2]$$

Combining Equations [3.2-1] and [3.2-2] gives

$$\boxed{d \sin \theta = m\lambda} \qquad m = 0, 1, 2, \ldots \qquad [3.2\text{-}3]$$

which is *Young's double-slit equation for maxima.* Interference *minima* occur whenever one of the contributions has shifted in phase by $\frac{1}{2}\lambda$, that is, when

$$\boxed{d \sin \theta = (m - \tfrac{1}{2})\lambda} \qquad m = 1, 2, 3, \ldots \qquad [3.2\text{-}4]$$

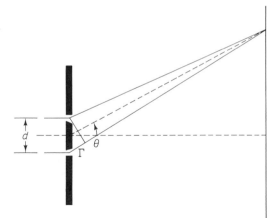

Figure 3.2-2 Deriving the double-slit equation.

(Note that now $m = 1$ is the lowest possible order; in double-slit interference there is no zeroth-order minimum.)

In a typical double-slit experiment, the angle θ is small and, since θ and d are easy to measure, Young's experiment can be used to determine the wavelength.

Example

Light passes through two slits separated by a distance $d = 0.8$ mm. On a screen 1.6 m away, the distance between the two second-order maxima is 5 mm. What is the wavelength of the light?

Solution.   We call x the distance from the double slits to the screen and y the distance of a given maximum, or minimum, from the optic axis. For small angles $\sin \theta$ can be set equal to y/x; thus, when Equation [3.2-3] is solved for λ,

$$\lambda = \frac{dy}{xm} \qquad [3.2\text{-}5]$$

Since the 5-mm distance given is the distance *between* the two maxima, rather than the distance of one maximum from the axis, we use one-half that distance, $y = 2.5$ mm, so that

$$\lambda = \frac{(0.8 \text{ mm})(2.5 \text{ mm})}{(1600 \text{ mm})(2)} = \boxed{625 \text{ nm}}$$

SUPERPOSITION OF WAVES

We now turn to the more general aspects of interference and discuss the *superposition of waves*. We consider four cases.

1. Superposition of waves of equal phase and frequency. Assume that two sinusoidal waves of the same frequency are traveling side by side in the same direction. The waves are of *equal phase*, without any phase angle difference between them. We plot the amplitudes of

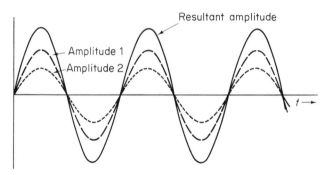

Figure 3.2-3 Superposition of waves of equal phase and frequency.

the waves, as in Figure 3.2-3, and add them, point by point. The resultant amplitude is the sum of the individual amplitudes,

$$\mathbf{A} = \mathbf{A}_1 + \mathbf{A}_2 + \mathbf{A}_3 + \cdots + \mathbf{A}_N \qquad [3.2\text{-}6]$$

and the resultant intensity, from Equation [3.1-5], is proportional to the square of the sum of the amplitudes,

$$I \propto (\mathbf{A}_1 + \mathbf{A}_2 + \mathbf{A}_3 + \cdots + \mathbf{A}_N)^2 \qquad [3.2\text{-}7]$$

2. Superposition of waves of constant phase difference. Now consider two waves that have the same frequency but that have a certain *constant phase angle difference* between them. The two waves have certain *different initial phase angles,* ϕ. The resultant wave can be found by several methods: by graphical construction, by trigonometry, by vector addition, and by the use of complex numbers.

(a) *Graphical construction.* As before, we draw the two wave amplitudes on top of each other and add corresponding points. This gives the resultant wave.

(b) *Trigonometry.* Again, the two waves have the same frequency (and therefore the same angular velocity ω) but different phase angles, ϕ_1 and ϕ_2:

$$y_1 = \mathbf{A}_1 \sin(\omega t + \phi_1)$$

and $\hspace{10cm}$ [3.2-8]

$$y_2 = \mathbf{A}_2 \sin(\omega t + \phi_2)$$

Since

$$\sin(\alpha + \beta) = \sin \alpha \cos \beta + \cos \alpha \sin \beta \qquad [3.2\text{-}9]$$

Equations [3.2-8] can be written

$$y_1 = \mathbf{A}_1(\sin \omega t \cos \phi_1 + \cos \omega t \sin \phi_1)$$

and $\hspace{10cm}$ [3.2-10]

$$y_2 = \mathbf{A}_2(\sin \omega t \cos \phi_2 + \cos \omega t \sin \phi_2)$$

The resultant displacement y is the algebraic sum of the two individual displacements:

$$y = y_1 + y_2$$

$$= \mathbf{A}_1 \sin \omega t \cos \phi_1 + \mathbf{A}_1 \cos \omega t \sin \phi_1 + \mathbf{A}_2 \sin \omega t \cos \phi_2 + \mathbf{A}_2 \cos \omega t \sin \phi_2$$

$$= (\mathbf{A}_1 \cos \phi_1 + \mathbf{A}_2 \cos \phi_2)\sin \omega t + (\mathbf{A}_1 \sin \phi_1 + \mathbf{A}_2 \sin \phi_2)\cos \omega t \qquad [3.2\text{-}11]$$

The terms in parentheses are constant in time. Thus we can set

$$\mathbf{A}_1 \cos \phi_1 + \mathbf{A}_2 \cos \phi_2 = \mathbf{A} \cos \epsilon \qquad [3.2\text{-}12]$$

and

$$\mathbf{A}_1 \sin \phi_1 + \mathbf{A}_2 \sin \phi_2 = \mathbf{A} \sin \epsilon \qquad [3.2\text{-}13]$$

where **A** is the amplitude of the resultant wave and ϵ the new initial phase angle. In order to solve for **A** and ϵ, we square Equations [3.2-12] and [3.2-13] and add them. This gives

$$\mathbf{A}^2 \cos^2 \epsilon + \mathbf{A}^2 \sin^2 \epsilon = \mathbf{A}_1^2(\cos^2 \phi_1 + \sin^2 \phi_1)$$

$$+ 2\mathbf{A}_1\mathbf{A}_2(\cos \phi_1 \cos \phi_2 + \sin \phi_1 \sin \phi_2)$$

$$+ \mathbf{A}_2^2(\cos^2 \phi_2 + \sin^2 \phi_2) \qquad [3.2\text{-}14]$$

Since

$$\sin^2 \alpha + \cos^2 \alpha = 1$$

and

$$\cos(\alpha - \beta) = \cos \alpha \cos \beta + \sin \alpha \sin \beta$$

this simplifies to

$$\mathbf{A}^2 = \mathbf{A}_1^2 + \mathbf{A}_2^2 + 2\mathbf{A}_1\mathbf{A}_2 \cos(\phi_1 - \phi_2) \qquad [3.2\text{-}15]$$

We can now substitute Equations [3.2-12] and [3.2-13] in [3.2-11]. This gives

$$y = \mathbf{A} \cos \epsilon \sin \omega t + \mathbf{A} \sin \epsilon \cos \omega t \qquad [3.2\text{-}16]$$

and, using again Equation [3.2-9],

$$\boxed{y = \mathbf{A} \sin(\omega t + \epsilon)} \qquad [3.2\text{-}17]$$

which shows that *the resultant of two sinusoidal waves is again a sinusoidal wave,* of the same frequency but with a new amplitude, **A**, and a new phase angle, ϵ.

(c) *Vector addition.* Assume that one of the two waves follows the equation

$$y_1 = \mathbf{A}_1 \sin(\omega t)$$

which is actually

$$y_1 = \mathbf{A}_1 \sin(\omega t + \phi_1) \qquad [3.2\text{-}18]$$

The rotating vector, or *phasor,* representing the wave at time t_0 is shown in Figure 3.2-4, left. As before, the phasor rotates counterclockwise with angular velocity ω. The other wave has

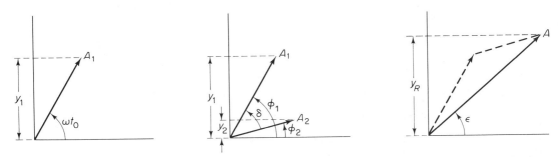

Figure 3.2-4 (*Left*) Phasor representing wave amplitude $\mathbf{A}_1$; (*center*) two phasors representing waves of different amplitudes and with constant phase difference δ; (*right*) vector triangle showing resultant phasor.

the same angular velocity but a different amplitude, $\mathbf{A}_2$; it also lags behind the first wave by a (constant) phase difference δ,

$$y_2 = \mathbf{A}_2 \sin(\omega t - \delta) = \mathbf{A}_2 \sin(\omega t + \phi_2) \qquad [3.2\text{-}19]$$

The resultant of the two phasors can be found by the parallelogram method of vector addition or, more easily, by a *vector triangle*. Note that both phasors, and the vector triangle formed by them, *rotate as a unit*.

If we then plot the displacements, y, versus time, t, for the two component waves and for the resultant, we arrive at Figure 3.2-5. As before, *the sum of two, or more, sinusoidal waves of the same frequency is a sinusoidal wave*.

(d) *Complex numbers.* The advantage of using complex numbers is that the (vector) addition of real amplitudes can be written more easily in the form of an (algebraic) addition of complex amplitudes. Assume that the two waves follow the equations

$$\mathbf{A}_1 = A_1 e^{i(\omega t + \phi_1)}$$

and $\qquad\qquad\qquad\qquad\qquad\qquad\qquad\qquad\qquad\qquad\qquad [3.2\text{-}20]$

$$\mathbf{A}_2 = A_2 e^{i(\omega t + \phi_2)}$$

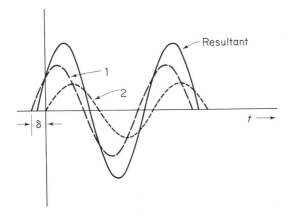

Figure 3.2-5   Two sine waves, 1 and 2, of constant phase difference, δ, and the *resultant*, also a sine wave.

where **A** is the complex amplitude, ω the angular frequency, t the time, and ϕ_1 and ϕ_2 are the two initial phase angles. Adding these two equations gives

$$\mathbf{A} = \mathbf{A}_1 + \mathbf{A}_2 = A_1 e^{i(\omega t + \phi_1)} + A_2 e^{i(\omega t + \phi_2)}$$

The square of the resultant, $\mathbf{A}^2$, is found by multiplying the complex terms by their complex conjugates:

$$\mathbf{A}^2 = (A_1 e^{i\phi_1} + A_2 e^{i\phi_2})(A_1 e^{-i\phi_1} + A_2 e^{-i\phi_2})$$
$$= \mathbf{A}_1^2 + \mathbf{A}_2^2 + A_1 A_2 [e^{i(\phi_1 - \phi_2)} + e^{-i(\phi_1 - \phi_2)}] \qquad [3.2\text{-}21]$$

From Euler's formula (page 200),

$$e^{i\phi} + e^{-i\phi} = \cos\phi + i\sin\phi + \cos\phi - i\sin\phi = 2\cos\phi \qquad [3.2\text{-}22]$$

and therefore, Equation [3.2-21] becomes

$$\mathbf{A}^2 = \mathbf{A}_1^2 + \mathbf{A}_2^2 + 2\mathbf{A}_1 \mathbf{A}_2 \cos(\phi_1 - \phi_2) \qquad [3.2\text{-}23]$$

the same as Equation [3.2-15].

Interference fringes.   Since the intensity of a wave is proportional to the square of the amplitude, $I \propto \mathbf{A}^2$, and since $\phi_1 - \phi_2$ is the phase angle difference, δ, Equation [3.2-23] may be written

$$\mathbf{I} = I_1 + I_2 + 2\sqrt{I_1 I_2} \cos\delta \qquad [3.2\text{-}24]$$

Whenever the phase difference is zero, $\delta = 0$, we have a *maximum* amount of light,

$$\mathbf{I}_{max} = I_1 + I_2 + 2\sqrt{I_1 I_2}$$

which, when $I_1 = I_2$, becomes

$$\mathbf{I}_{max} = 4I_1 \qquad [3.2\text{-}25]$$

On the other hand, whenever the phase difference $\delta = 180°$, $\cos 180° = -1$ and we have a *minimum* amount of light:

$$\mathbf{I}_{min} = I_1 + I_2 - 2\sqrt{I_1 I_2}$$

which, when $I_1 = I_2$, becomes

$$\mathbf{I}_{min} = 0 \qquad [3.2\text{-}26]$$

At points that lie between the maxima and minima, we find, following Equation [3.2-24] and assuming that $I_1 = I_2$, that

$$\mathbf{I} = I_1 + I_1 + 2I_1 \cos\delta$$
$$= 2I_1(1 + \cos\delta)$$
$$= 4I_1 \cos^2 \tfrac{1}{2}\delta \qquad [3.2\text{-}27]$$

Consequently, the light distribution resulting from a superposition of waves consists of *interference fringes*.

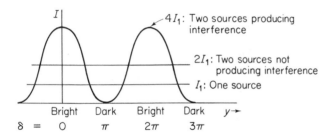

Figure 3.2-6 Intensity distribution as produced by one source, two sources not causing interference, and two sources causing interference.

Figure 3.2-6 shows the intensity distribution across a few fringes projected on a screen. If only one beam of light illuminates the screen, the distribution is uniform throughout (lower horizontal line marked I_1). If we have two beams of light of equal intensity, but lacking the property necessary to produce interference, the distribution is again uniform, but twice as much light will reach the screen (upper horizontal line marked $2I_1$). But if the two beams *are capable of producing interference,** they form alternate maxima and minima. If the initial contributions have equal amplitudes, and therefore equal intensities, then the maxima, from Equation [3.2-25], contain *four times* the intensity of the individual contribution; the minima, from Equation [3.2-26], contain nothing. The result is the $\cos^2$ curve labeled $4I_1$. Note that the *integrated* values, that is, the areas under the curves labeled $2I_1$ and $4I_1$, *are the same:* The light, indeed, is neither created nor destroyed; interference merely causes the energy to be *redistributed.*

3. Superposition of waves of different frequencies. So far we have assumed that the waves that come to interference have the same frequency. But light is never truly "monochromatic" and never has but one frequency; instead, it always has a certain range of frequencies (*quasimonochromatic* light). When such waves superimpose, the result is much more complicated. This situation will be discussed in Chapter 4.4.

4. Superposition of waves of random phase difference. Finally, we consider waves that have random phase differences between them, or that have widely different frequencies. If such waves superimpose, they do not produce any discernible interference. In this case the resultant intensity is found from adding the intensities,

$$\sum^{N} \mathbf{A}^2 = \mathbf{A}_1^2 + \mathbf{A}_2^2 + \mathbf{A}_3^2 + \cdots + \mathbf{A}_N^2 \qquad [3.2\text{-}28]$$

rather than adding the amplitudes and squaring their sum.

We return now to the more practical applications of interference.

---

*The property of light necessary to produce interference is called *coherence,* to be discussed in Chapter 3.4.

THE MICHELSON INTERFEROMETER

General construction.   There is no interferometer that is as versatile as the *Michelson interferometer.** It makes it possible to bring two optical planes into coincidence and to move them virtually through one another. In addition, Michelson's instrument can be used with an extended source; therefore, compared with a set of double slits, it gives much brighter fringes.

In its most elementary form, the Michelson interferometer consists of a beamsplitter, *A*, and of two mirrors, *C* and *D* (Figure 3.2-7). The beamsplitter divides the light into two parts, one part being transmitted toward mirror *C*, the other part being reflected toward mirror *D*. The two mirrors, *C* and *D*, return the light to *A*. There the beams recombine and proceed toward *E* where interference is observed.

One of the mirrors is mounted so that it can be moved along the axis. But note that, if reflection at *A* occurs at the rear surface as shown, the light reflected at *D* will pass through *A* three times while the light reflected at *C* will pass through only once. For this reason, a compensating plate, *B*, of the same thickness and inclination as *A*, is inserted into the *A*–*C* path. If we look into the instrument from *E*, we see mirror *D*, and in addition we see a virtual image, *C'*, of mirror *C*. Depending on the positions of the mirrors, image *C'* may be in front of, or behind, or exactly coincident with mirror *D*.

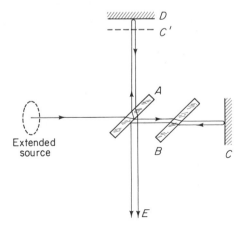

Figure 3.2-7   The Michelson interferometer.

*Albert Abraham Michelson (1852–1931). Born in Strelno, Prussia (now Strzelno, Poland), Michelson was two years old when his parents brought him to the United States. He graduated from, and taught at, the U.S. Naval Academy and later worked at the Case School of Applied Science in Cleveland, at Clark University in Worcester, Massachusetts, and at the University of Chicago. In 1907 he was awarded the Nobel Prize in physics, the first American scientist to be so honored. Michelson is best known for his precise determination of the velocity of light, for inventing the interferometer that bears his name, for the Michelson–Morley aether drift experiment, and for establishing the length of the meter in terms of wavelength of light. He also made noteworthy contributions to astronomy, spectroscopy, and geophysics, was proficient in tennis and other sports, played the violin, and liked to paint landscapes. The interferometer was first described in A. A. Michelson, "Interference Phenomena in a new form of Refractometer," *Am. J. Sci.* (3) **23** (1882), 395–400, and *Philos. Mag.* (5) **13** (1882), 236–42.

Fringe formation.   If indeed image C' coincides with mirror D, the two arms of the interferometer are equal in length. If C' and D do not coincide, the distance between them is finite, $C'D = d$. In addition, the observer may be looking into the system subtending an angle I with the optic axis (Figure 3.2-8).

Assume that the light comes from a point S and is reflected by both C' and D. The observer will see two virtual images: S', which is due to reflection at C', and S'', which is due to reflection at D. Sighting along the axis, the two images S' and S'' are a distance $2d$ apart, which is the *path difference* Γ,

$$\Gamma = 2d$$

But to the observer looking into the system at an angle I, the apparent distance between S' and S'' is *less* by the cosine of I,

$$\Gamma = 2d \cos I$$

Replacing Γ by $m\lambda$, following Equation [3.2-1], results in

$$\boxed{2d \cos I = m\lambda} \qquad [3.2\text{-}29]$$

which is the *Michelson interferometer equation.*

For a given mirror separation d, and a given order m and wavelength λ, angle I is constant. This means that the fringes are of rotational symmetry: They are *circles* concentric around the axis. They are *fringes of equal inclination.*

At the coincidence position where $d = 0$, we expect the two waves to reinforce each other and to form a maximum. But this is not so. The reason is a 180° or π *phase change*

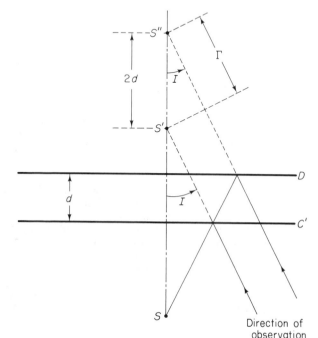

Direction of observation

Figure 3.2-8 Looking off-axis into the Michelson interferometer.

π phase change   No phase change

Figure 3.2-9 A π phase change occurs on rare-to-dense reflection (*left*) but not on dense-to-rare reflection (*right*).

that occurs on external (air-to-glass) reflection (Figure 3.2-9). It does not occur on internal (glass-to-air) reflection, and it does not occur on transmission or refraction either.

Look again at Figure 3.2-7 and note: It is only the light that comes from C and goes to E that undergoes rare-to-dense reflection and therefore a π phase change. Because of this disparity between the two beams, there will at the coincidence position be a *minimum:* The center of the field will be *dark.*

If we now *move* one of the mirrors through a distance $\frac{1}{4}\lambda$, the path length changes by $\frac{1}{2}$ λ, the two contributions get out of phase by 180°, the phase change compensates for that, and we have a *maximum.* Moving the mirror by another $\frac{1}{4}\lambda$ gives another minimum, moving it another $\frac{1}{4}\lambda$ again a maximum, and so on.

As d is gradually made larger, a given ring—which has a certain order m—increases in size because the product $2d\cos I$ in Equation [3.2-29] must by necessity remain constant. Consequently, the cosine of I must become smaller, and the angle I and the rings larger, a new ring appearing in the center of the field each time one of the mirrors has moved through $\frac{1}{2}\lambda$. Moreover, as d increases, new rings appear in the center faster than rings already present disappear in the periphery; thus the field will become more crowded with thinner rings (Figure 3.2-10). Conversely, as d is made smaller, the rings contract and disappear in the center.

As the mirrors are *tilted,* and are no longer perpendicular to each other, the fringes appear to be straight. Actually, they are sections of large hyperbolas. If now distance d is changed, that is, if one of the mirrors is moved, the fringes move across the field, and can easily be counted as they pass a reference mark. For each fringe that passes, the mirror has moved through one-half of a wavelength; hence, if m fringes are counted as the mirror is moved through d, the wavelength of the light is

$$\lambda = \frac{2d}{m} \qquad\qquad [3.2\text{-}30]$$

With care, $\frac{1}{100}$ of the width of a fringe can be measured.

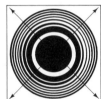

Figure 3.2-10 Appearance of fringes in the Michelson interferometer as the mirrors are moved away from each other. Arrows on the far right indicate motion of the fringes.

A   *B*   *C*   *D*   *E*   *F*

Figure 3.2-11   Aligning the Michelson interferometer.

Aligning the Michelson Interferometer. Tape a black mark to a ground glass placed between light source and beamsplitter. Make the two paths equal to within 1 mm. Look into the interferometer from *E*. Three reflected images of the mark will be seen (Figure 3.2-11A). A slight turn of the mirror tilt controls will show which of these can be moved. Superimpose the movable image exactly on the right-hand image of the stationary pair of images, aligning the movable image first in *height* and then *sideways* (B). At this point, interference fringes should become visible (C).

Using the tilt controls, bring the center of the fringes into the center of the field of view (D). Move the other mirror along the axis so that the fringes move *inward* (E). Near the zeroth order, alignment becomes critical. Even placing your hand over and close to the paths—which will raise the temperature of the air and lower its refractive index—will make the fringes wiggle and move in and out of the field. At the zeroth order, the center of the field will be dark (F).

APPLICATION AND EVALUATION

The most important application of interferometry is to *precision measurements*. In principle almost any interferometer will do, but for particular purposes some types are better suited than others. We distinguish two groups, depending on whether the instrument is used for *reflecting* or for *transparent* objects.

The first category includes the *examination of surfaces, metrology,* and *alignment.* Here we use the Michelson interferometer, or one of its many variations. For example, if we want to know whether there are scratch marks or other defects in a surface, we use that surface in place of one of the mirrors. The other mirror serves as a reference. Any defect present in the surface shows as a characteristic distortion of the fringe pattern, as we see from the following example.

Example

When examining the surface of a polished workpiece in thallium light (535 nm), some fringes are seen to be distorted by $\frac{4}{10}$ the distance between them (Figure 3.2-12). How deep is the defect?

Solution. As always in interferometry, consecutive fringes indicate a path difference of one wavelength, $\Gamma = 1\lambda$. But the light is reflected, and goes back and forth through the depth of the defect; therefore, a Γ path difference is produced by a defect only one-half that deep. In our example, the distortion is $\frac{4}{10}$ of the fringe separation and hence the depth of the defect, from Equation [3.2-29], is

$$d = (0.4)(\tfrac{1}{2})(535 \times 10^{-9}) = \boxed{0.1 \ \mu\text{m}}$$

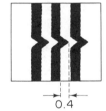

0.4   **Figure 3.2-12**

For transparent objects the *Mach-Zehnder interferometer** is preferred, in part because the light passes through the sampling field only once and evaluation is easier. The
Mach–Zehnder interferometer, illustrated in Figure 3.2-13, consists of two mirrors and two
beamsplitters. The first beamsplitter divides the incoming light, the mirrors reflect the
beams as shown, and the second beamsplitter brings the beams together again. The two
paths may be widely separated, which permits testing large objects, as in a windtunnel.

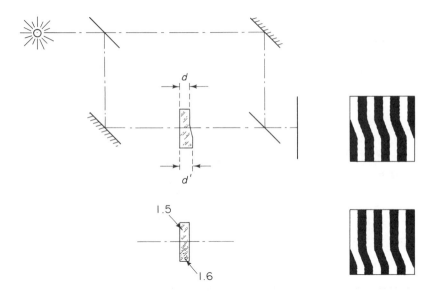

Figure 3.2-13 Fringe shifts in an interferogram (*right*) will result if the sample inserted in one arm of the interferometer has local variations of thickness (*top*) or variations of refractive index (*bottom*). Interferometer shown in Mach–Zehnder type.

---

*Ernst Waldfried Joseph Wenzel Mach (1838–1916), Austrian physicist, best known for his investigation of
shock waves caused by a projectile as it reaches the speed of sound, the ratio of the speed of the projectile to the speed
of sound called the *Mach number*. An ardent experimentalist, Mach did not believe in then unobservable and, to
him, rather mystical ideas such as the atomic structure of matter and the theory of relativity. The interferometer is
described in L. Mach, "Modifikation und Anwendung des Jamin Interferenz-Refraktometers," *Anz. Akad. Wiss.
Wien math. naturwiss.-Klasse* **28** (1891), 223–24, and L. Zehnder, "Ein neuer Interferenzrefraktor," *Z. Instrumen-
tenkd.* **11** (1891), 275–85.

Interferometers respond to differences in *optical path length*. From Equation [1.1-4], we recall that optical path length is the product of thickness and refractive index:

$$S = Ln \qquad [3.2\text{-}31]$$

The *path difference,* Γ, is simply the difference between two such path lengths:

$$\Gamma = S_2 - S_1 \qquad [3.2\text{-}32]$$

Note that any fringe shift seen in an interferometer may be due to either a change in thickness or a change in refractive index. In the upper example in Figure 3.2-13 we assume that the refractive index within the sample is constant but the thickness is not; thus from Equations [3.2-31] and [3.2-32], we have an optical path difference of

$$\Gamma = n(L_2 - L_1) \qquad [3.2\text{-}33]$$

In the lower example, the thickness is constant but the index is not; thus we have

$$\Gamma = L(n_2 - n_1) \qquad [3.2\text{-}34]$$

Since from Equation [3.2-1],

$$\Gamma = m\lambda$$

a given path difference will cause either

$$m = \frac{1}{\lambda} n(L_2 - L_1) \qquad [3.2\text{-}35]$$

or

$$m = \frac{1}{\lambda} L(n_2 - n_1) \qquad [3.2\text{-}36]$$

additional wavelengths to be present within the thicker, or denser, medium and a fringe shift by m fringes will result.

This is a distinction that the two samples shown in Figure 3.2-13 and the two Equations [3.2-35] and [3.2-36] illustrate very well.

SUGGESTIONS FOR FURTHER READING

W. H. STEEL, *Interferometry,* 2nd edition (New York: Cambridge University Press, 1985).

S. TOLANSKY, *An Introduction to Interferometry* (New York: John Wiley & Sons, Inc., 1973).

J. DYSON, *Interferometry as a Measuring Tool* (Brighton, England: The Machinery Publishing Co. Ltd., 1970).

PROBLEMS

3.2-1. Monochromatic light passes through a double slit, producing interference. The distance between the slit centers is 1.2 mm and the distance between consecutive fringes on a screen 5 m away is 0.3 cm. What is:

(a) The wavelength?

(b) The color of the light?

3.2-2. Light passes through two narrow slits 0.8 mm apart. If on a screen 80 cm away the distance between the two second-order maxima is 2 mm, what is the wavelength of the light?

3.2-3. Light of 600 nm wavelength passes through a double slit and forms interference fringes on a screen 1.2 m away. If the slits are 0.2 mm apart, what is the distance between the zeroth and:

(a) A third-order maximum?

(b) A third-order minimum?

3.2-4. When two narrow slits are 0.375 mm apart, then on a screen 1.2 m away the distance between the two first-order maxima is found to be 4 mm. What is the wavelength of the light?

3.2-5. When one of the slits in Young's experiment is covered by a film of transparent material, the zeroth order is seen to shift by 2.2 fringes. If the refractive index of the material is 1.4 and the wavelength of the light 500 nm, how thick is the film?

3.2-6. A double slit is illuminated by light containing two wavelengths, 450 nm and 600 nm. What is the least order at which a maximum of one wavelength will fall exactly on a minimum of the other?

3.2-7. A series of double slits are cut into an opaque screen such that the separation, d, between slits is constant but the spacing between different double slits varies at random (Figure 3.2-14). Describe the resulting interference pattern.

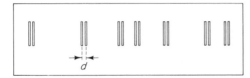

Figure 3.2-14

3.2-8. A satellite circling Earth is transmitting microwaves of 15 cm wavelength. When the satellite is above a ground station which has two antennas 100 m apart and located in the plane of the orbit, a signal is received that fluctuates with a period of $\frac{1}{10}$ s. If the satellite is known to be at an altitude of 400 km and if we neglect the curvature of Earth, what is the satellite's velocity?

3.2-9. Two sinusoidal waves of equal amplitude are $\frac{1}{4}$ of a wavelength out of phase. What is the amplitude of the resultant wave?

3.2-10. Two harmonic waves, given by $y_1 = 7\sin(\omega t + \pi/2)$ and $y_2 = 5\sin(\omega t + \pi/3)$, superimpose on each other. Find the resultant amplitude.

3.2-11. Two waves, $y_1 = \sin x$ and $y_2 = \frac{1}{2}\sin 2x$, are added to one another. Find the equation of the resultant wave and draw plots of the three waves.

3.2-12. Two waves are given by $y_1 = 2\sin x$ and $y_2 = \sin(x + \pi/3)$, respectively. Determine the amplitude, the new phase angle, and the equation of the resultant wave.

3.2-13. Using red cadmium light, $\lambda = 643.8$ nm, Michelson in his original experiment could still see interference fringes after he had moved one of the mirrors 25 cm away from the coincidence position. How many fringes did he count?

3.2-14. How far must one of the mirrors of a Michelson interferometer be moved for 100 fringes of light of 630 nm wavelength to cross the center of the field of view?

3.2-15. As one of the mirrors in a Michelson interferometer is moved, 500 fringes are counted crossing the field of view. If the micrometer reads 5.070 mm at the beginning of the count and 5.245 mm at the end, what is the wavelength of the light?

3.2-16. A Michelson interferometer is used with blue light of 475 nm wavelength and set up for a path difference of 5 μm. What is the angular radius of:
(a) The tenth-order ring?
(b) The twentieth-order ring?

3.2-17. If one arm of a Michelson interferometer contains a tube 2.5 cm long which is first evacuated and then slowly filled with air ($n = 1.0003$), how many fringes will cross the center? Assume light of 600 nm wavelength.

3.2-18. The crystalline lens of the human eye has in its center a refractive index of 1.41 while near the periphery the index is 1.38. If a slice $\frac{1}{10}$ mm thick is cut out of the lens and examined in 600 nm light in a Mach–Zehnder interferometer, what is the distortion of the fringes?

3.2-19. The *Rayleigh interferometer* is derived from the double-slit design. It has two test chambers placed side by side, parallel to each other, that can be evacuated (Figure 3.2-15). Assume that the chambers are 30 cm long and that one of them is left evacuated while the other is gradually filled with a certain gas. If, with light of 500 nm wavelength, 240 fringes are seen to cross the field of view, what is the refractive index of the gas?

Figure 3.2-15

3.2-20. The two tubes of a Rayleigh interferometer are first filled with air ($n = 1.0003$). The air in one tube is then gradually replaced by a gas of unknown refractive index. If the tubes are 10 cm long, the wavelength is 500 nm, and 70 fringes cross the center of the field, what is the refractive index of the gas?

Thin Films

SOME OF THE MOST LUSTROUS COLORS in nature are the iridescent colors of insects. These colors are due to interference, unlike the colors caused by pigments in water colors or oil paints which are due to absorption. Colors in thin layers of materials, from soap bubbles to oil slicks, are also caused by interference. Of greater significance, theoretical as well as practical, are interference filters, antireflection coatings, and especially the high-resolution Fabry–Perot interferometer, the prototype of the laser cavity. What these systems have in common is that they are based on multiple-beam interference in *thin films*.

PLANE-PARALLEL PLATES

Consider two plane surfaces, parallel to one another and a short distance apart. Light incident on the two surfaces is reflected as shown in Figure 3.3-1, left. This kind of reflection reminds us of the reflection of light at the two surfaces, C' and D, in a Michelson interferometer. Indeed, the same equation that we had derived there also applies here; we only need to include the index of refraction of the medium between the surfaces, replacing the distance d by the optical path length nd:

$$\boxed{2nd \cos I = m\lambda} \qquad m = 1, 2, 3 \qquad [3.3\text{-}1]$$

(Again, I use $m = 1$ as the lowest order because $m = 0$ would imply that either n or d is zero or that $I = 90°$, none of which is possible.)

But the light need not be *reflected*. It could as well be *transmitted* (Figure 3.3-1,

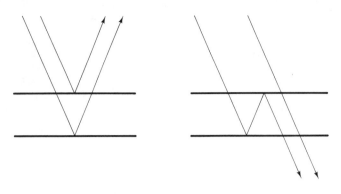

Figure 3.3-1 Interference in thin films occurs both on reflection (*left*) and transmission (*right*). Refraction at the boundaries has been neglected.

right). Furthermore, the space between the two surfaces may either be filled by an actual thin film, or it may be an empty gap, an "air film" (between plates). But the film may also be a layer of oil floating on water, or it may be a coating applied to a lens. In either case, we need to consider again whether or not there is a *phase change*.

A π phase change occurs only on external (rare-to-dense) reflection. On internal (dense-to-rare) reflection and on transmission and refraction there is no phase change. For example, if the light is *transmitted* through a film in air, both reflections are dense-to-rare and neither will cause a π change. In that case, Equation [3.3-1] describes the condition for *maxima*. Conversely, the same as in double-slit interference, *minima* will occur whenever

$$2nd \cos I = (m - \tfrac{1}{2})\lambda \qquad m = 1, 2, 3 \qquad\qquad [3.3\text{-}2]$$

But if the light is *reflected* by a film in air, the light reflected at the upper surface *will* undergo a π phase change (because that is now a rare-to-dense reflection), but the light reflected at the lower surface *will not*. Then the two equations change places: Equation [3.3-1] refers to minima and Equation [3.3-2] to maxima.

Example

A soap bubble, made with soapy water of $n = 1.38$ and known to be 114 nm thick, is seen at normal incidence in reflected white light. In what color will it appear?

Solution. While the soapy film, both inside and outside the bubble, borders on air, a π phase change occurs only at the outside front surface. Hence, to find a reflection *maximum* we use Equation [3.3-2],

$$2nd \cos I = (m - \tfrac{1}{2})\lambda$$

For light at normal incidence, $\cos I = 1$; therefore, the wavelength of the reflected light is

$$\lambda = \frac{(2)(1.38)(114 \times 10^{-9})}{1 - 0.5} = 629 \text{ nm}$$

which means that the bubble's reflection is $\boxed{\text{red}}$.

THE FABRY—PEROT INTERFEROMETER

The *Fabry–Perot interferometer** consists of two plane, parallel glass plates. The plates are separated by a distance d, usually several millimeters (Figure 3.3-2). The fringes formed are due to multiple-beam interference. They are exceedingly narrow, much more so than in any other type of interferometer. The resolvance, therefore, is very high, which makes the Fabry–Perot instrument a very powerful *spectrometer*.

The medium between the two surfaces is air. Thus π phase changes occur on both of these (air-to-glass) surfaces and

$$2d \cos I = m\lambda \qquad\qquad [3.3\text{-}3]$$

is the equation for *maxima*. Lens L may be a separate lens, as shown, or it may simply be the eye of the observer, in which case the retina serves as the screen. As in Michelson's interferometer, the maxima and minima, for given values of I and m, appear in the form of rings, concentric around the axis of the lens. The maximum nearest the center has the highest value of m. Maxima farther away have lower m's. If d is large, the rings are close together and a telescope may be needed to see them.

In the earlier Fabry–Perot interferometers, the distance between the plates could be varied, much as in the Michelson instrument. But keeping the plates exactly parallel during movement is very difficult. Today these systems have become obsolete and have been replaced by plates kept at *fixed distances* from each other. They are called *etalons*.

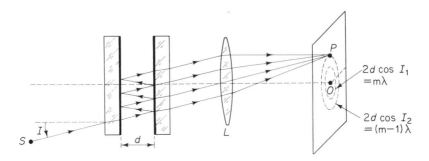

Figure 3.3-2   Fabry-Perot interferometer. Point S is part of an extended light source.

---

*Marie Paul Auguste Charles Fabry (1867–1945), French physicist. While at the University of Marseilles, Fabry taught medical students, later became director of the French Institut d'Optique and professor of physics at the École Polytechnique (Sorbonne) in Paris, and is remembered as a brilliant lecturer of sometimes caustic wit. Besides his work on electricity, spectroscopy, atmospheric ozone, and astrophysics, Fabry today is best known for the instrument he developed together with Jean Baptiste Gaspard Gustave Alfred Perot (1863–1925), which he referred to as "this interferometer whose name I have the honor to bear myself." Ch. Fabry and A. Perot, "Théorie et applications d'une nouvelle méthode de spectroscopie interférentielle," *Ann. Chim. Phys.* (7) **16** (1899), 115–44.

Variable spacing, though, is still used in the *scanning Fabry–Perot interferometer*. Here the plates are mounted on a spacer made out of barium titanate or some other piezoelectric material, driven by an alternating voltage of about 1 kV. Such spectrometers are generally used in the "central-spot scanning" mode.

Resolvance of the Fabry–Perot instrument. We call $\Delta\lambda$ the difference between two wavelengths and ΔI the angular separation between maxima formed by these wavelengths. The problem then becomes one of expressing $\Delta\lambda$ in terms of ΔI. To do so, we solve Equation [3.3-3] for λ and differentiate with respect to I:

$$\frac{d\lambda}{dI} = -\frac{2d}{m}\sin I \qquad [3.3\text{-}4]$$

For paraxial rays, and for small changes in wavelength and angle,

$$\frac{d\lambda}{dI} = \frac{\Delta\lambda}{\Delta I}$$

so that

$$\Delta\lambda = -\frac{2d}{m} I\,\Delta I \qquad [3.3\text{-}5]$$

(The minus sign is of no consequence because it is immaterial whether $\Delta\lambda$ is measured upward or downward from the mean wavelength.)

For paraxial rays I is small, $\cos I \approx 1$, and Equation [3.3-3] reduces to

$$2d = m\lambda \qquad [3.3\text{-}6]$$

Now, if for a given ring

$$2d\cos I = m\lambda$$

then for the next larger ring,

$$2d\cos I_2 = (m' - 1)\lambda$$

Substituting for $\cos I$ the approximation $\cos I \approx 1 - I^2/2$ yields

$$2d\left(1 - \frac{I^2}{2}\right) = (m' - 1)\lambda \qquad [3.3\text{-}7]$$

Subtracting Equation [3.3-6] from [3.3-7] gives

$$I^2 d = \lambda$$

and dividing this by Equation [3.3-5] yields

$$\frac{\lambda}{\Delta\lambda} = \frac{m}{2}\frac{I}{\Delta I} \qquad [3.3\text{-}8]$$

which relates chromatic resolvance to *angular* resolvance.

Next we consider the *phase difference* between adjacent rays. First, we realize that the *path* difference Γ, combining Equations [3.2-1] and [3.3-3], is

$$\Gamma = 2d\cos I$$

Since a wave of length λ corresponds to a full 2π on the reference circle, path difference Γ and phase difference δ are connected by

$$\frac{\Gamma}{\lambda} = \frac{\delta}{2\pi} \qquad [3.3\text{-}9]$$

Thus, the phase difference between adjacent rays is

$$\delta = 2\pi \frac{\Gamma}{\lambda} = \frac{2\pi}{\lambda} 2d \cos I = \frac{4\pi d}{\lambda} \cos I \qquad [3.3\text{-}10]$$

Differentiating with respect to I gives

$$\frac{d\delta}{dI} = -\frac{4\pi d}{\lambda} \sin I$$

and, again assuming small angles,

$$\Delta\delta = -\frac{4\pi d}{\lambda} I \,\Delta I \qquad [3.3\text{-}11]$$

To determine how much light is reflected and transmitted at the two surfaces we use complex notation. We call $\mathbf{A}_0$ the amplitude of the light incident on the first surface. A certain fraction of this light is reflected, $\mathbf{A}_0\rho$, and another fraction, $\mathbf{A}_0\tau$, is transmitted. The factors ρ and τ are called the *amplitude reflection coefficient* and the *amplitude transmission coefficient*. If there is no loss, the sum of the reflected and transmitted energies (that is, the squared values of the amplitudes) must equal the incident energy,

$$(\mathbf{A}_0\rho)^2 + (\mathbf{A}_0\tau)^2 = \mathbf{A}_0^2$$

and thus

$$\rho^2 + \tau^2 = 1 \qquad [3.3\text{-}12]$$

Again, at the second surface part of the light is reflected (with an amplitude $\mathbf{A}_0\rho^2$) and part is transmitted (with an amplitude $\mathbf{A}_0\rho\tau$). The next ray is transmitted with an amplitude $\mathbf{A}_0\rho^2\tau^2$, the next one after that with $\mathbf{A}_0\rho^4\tau^2$, and so on (Figure 3.3-3). In general,

$$\mathbf{A}_N = \mathbf{A}_0\rho^{2(N-1)}\tau^2 \qquad [3.3\text{-}13]$$

In complex notation, the incident amplitude

$$\mathbf{A}_0 \quad \text{becomes} \quad A_0 e^{i\omega t}$$

and

$$\mathbf{A}_1 \rightarrow A_0\tau^2 e^{i\omega t}$$
$$\mathbf{A}_2 \rightarrow A_0\rho^2\tau^2 e^{i(\omega t - \delta)}, \ldots$$
$$\mathbf{A}_N \rightarrow A_0\rho^{2(N-1)}\tau^2 e^{i[\omega t - (N-1)\delta]} \qquad [3.3\text{-}14]$$

If the number of terms approaches infinity, the series converges, and the transmitted amplitude reduces to

$$\mathbf{A}_T = A_0 e^{i\omega t}\left(\frac{\tau^2}{1 - \rho^2 e^{-i\delta}}\right) \qquad [3.3\text{-}15]$$

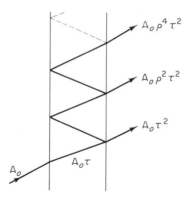

Figure 3.3-3 Amplitudes of successive rays passing through Fabry–Perot etalon.

Multiplying this equation by its complex conjugate yields the (transmitted) *energy*,

$$I_T = I_0 \frac{\tau^2}{1 - 2\rho^2 \cos \delta + \rho^4} \qquad [3.3\text{-}16]$$

Using the identity $\cos \delta = 1 - 2 \sin^2(\delta/2)$, Equation [3.3-16] can be transformed into

$$I_T = \frac{I_0}{1 + [4r \sin^2(\delta/2)]/(1 - r)^2} \qquad [3.3\text{-}17]$$

Now we plot the angular position of the maxima (Figure 3.3-4). The minimum angular separation, ΔI, necessary to see *two* maxima must be such that the curves cross at the half-intensity point, $I_T = \frac{1}{2} I_0$. If they do, the denominator in Equation [3.3-17] must be 2.0 for each. This condition is satisfied when the second term in the denominator is unity,

$$\frac{4r \sin^2(\delta/2)}{(1 - r)^2} = 1$$

or

$$\sin^2\left(\frac{\delta}{2}\right) = \frac{(1 - r)^2}{4r} \qquad [3.3\text{-}18]$$

For the fringes to be narrow, $\delta/2$ must assume values of either near zero or near a multiple of π; any change of $\delta/2$ from these values is small. Then the sine may be set equal to the angle.

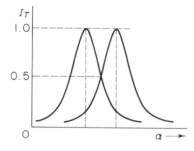

Figure 3.3-4 Plot of intensity versus angular position of two fringes just resolved.

Furthermore, if $\Delta\delta$ is the phase angle difference between adjacent maxima, Equation [3.3-18] becomes

$$\sin \tfrac{1}{2}\left(\frac{\Delta\delta}{2}\right) \approx \frac{\Delta\delta}{4} = \frac{1-r}{2\sqrt{r}} \qquad [3.3\text{-}19]$$

Solving Equation [3.3-11] for $\Delta\delta/4$ and combining it with Equation [3.3-19] gives

$$-\left(\frac{\pi d}{\lambda}\right) I \, \Delta I = \frac{1-r}{2\sqrt{r}} \qquad [3.3\text{-}20]$$

and solving Equation [3.3-5] for $I\,\Delta I$, inserting it into Equation [3.3-20], and eliminating $2d$, we have

$$\boxed{\frac{\lambda}{\Delta\lambda} = \frac{m\pi\sqrt{r}}{1-r}} \qquad [3.3\text{-}21]$$

The quantity $\lambda/\Delta\lambda$ is the *chromatic resolvance* (resolving power). It depends on the wavelength λ, on the order m (and indirectly, therefore, on d), and on the reflectance, r, of the two surfaces. If the reflectance is high (close to unity), the resolvance is high, as illustrated in Figure 3.3-5. For adjacent fringes, that is, if $m = 1$, the right-hand term in Equation [3.3-21] is called the *finesse* of the interferometer, which may reach values between 30 and 1000.

As the two wavelengths, λ_1 and λ_2, become more and more different, the rings formed by them separate. Of course, rings formed by any wavelength will satisfy Equation [3.3-5]. But since we have two wavelengths, λ_1 and λ_2 (where $\lambda_2 > \lambda_1$), light of λ_2 forms rings *smaller* than those formed by λ_1, and, at some wavelength difference, an mth-order λ_2 ring coincides with an $(m + 1)$th-order λ_1 ring,

$$m\lambda_2 = (m + 1)\lambda_1$$

The wavelength difference at which this occurs is

$$\delta\lambda = \frac{\lambda_1}{m}$$

or, substituting Equation [3.3-3] solved for m and assuming normal incidence, we obtain

$$\delta\lambda = \frac{\lambda^2}{2d} \qquad [3.3\text{-}22]$$

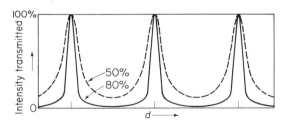

Figure 3.3-5 Plot of transmission versus plate separation in a Fabry–Perot interferometer. Percentages refer to reflectance of surfaces.

The interval $\delta\lambda$ is called the *free spectral range;* it is the change in wavelength necessary to shift the fringe system by one fringe.

Example

Determine the least separation of two spectrum lines, near 500 nm, that can be resolved by an etalon that is 10 mm long (deep) and has a reflectance of 90%.

Solution. From Equation [3.3-3] we find that

$$m = \frac{2d}{\lambda} = \frac{(2)(0.01)}{5 \times 10^{-7}} = 40{,}000$$

Inserting this in Equation [3.3-21],

$$\frac{\lambda}{\Delta\lambda} = \frac{m\,\pi\,\sqrt{r}}{1-r}$$

shows that

$$\Delta\lambda = \frac{\lambda(1-r)}{m\,\pi\,\sqrt{r}}$$

$$= \frac{(500 \times 10^{-9})(1-0.9)}{(40{,}000)(\pi)\sqrt{0.9}} = \boxed{0.0042 \text{ Å}}$$

Hence, two lines only 0.0042 Å apart can be resolved—a much better result than from any other type of interferometer/spectrometer.

NEWTON'S RINGS

If we try to show them, they are hard to see. If we do not want them, they stand out pretty well. They are familiar to anyone who has looked at oil spilled on wet pavement or has mounted photographic slides between glass plates and is annoyed by the amoeboid fringes that appear between surfaces in contact. They are *Newton's rings.*

Newton's rings result when a fairly flat convex surface is placed in contact with a plane surface. That produces a system of circular fringes around the point of contact. As illus-

---

*Sir Isaac Newton (1642–1727), British scientist, professor of mathematics at Cambridge University, long-time president of the Royal Society. After graduating from Trinity College, the bubonic plague then raging forced Newton into seclusion for two years. During that time he invented differential and integral calculus, discovered the composition of white light, formulated his three laws of motion, and conceived the idea of universal gravitation and the dynamics of the solar system. Around 1670 at Cambridge he gave a series of lectures on optics, five years later sent a copy of his notes to the secretary of the Royal Society, and in 1704 published them under the title *Opticks: or, a Treatise of the Reflexions, Refractions, Inflexions and Colours of Light.* A Latin version came out two years later. The English edition begins with these words: "My Design in this Book is not to explain the Properties of Light by Hypotheses, but to propose and prove them by Reason and Experiments." There is no question of the extraordinary significance of Newton's work. He often found the solution to problems that had escaped others, adding the mathematical reasoning later. Opinions of Newton the man vary. Some consider him the last of the magicians; others adore him unqualifyingly as English Poet Alexander Pope did when he wrote: "Nature and Nature's laws lay hid in night: God said, *Let Newton be!* and all was light."

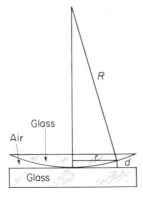

Figure 3.3-6   Notation as used with Newton's rings.

trated in Figure 3.3-6, we call R the radius of curvature of the surface, r the radius of a given Newton ring, and d the width of the gap between the plates at this ring. Then we take the vertex depth formula,

$$d \approx \frac{r^2}{2R} \qquad [3.3\text{-}23]$$

and insert it into the expression for minima in reflected light, Equation [3.3-1]. That gives

$$r^2 \approx m\lambda R \qquad [3.3\text{-}24]$$

Newton's rings are an example of *fringes of equal thickness*. They are widely used in the testing of surfaces and other types of quality control. But quantitative measurements are easily possible too. For example, following Equation [3.3-24], if we know the wavelength, we can determine the radius of curvature of a surface by measuring the diameter of just one Newton ring.

Example

Assume that Newton's rings are formed between a plano-convex lens of 50 cm radius of curvature and a plano-concave lens, placed as shown in Figure 3.3-7. If the twentieth dark ring, seen in reflected light of 550 nm wavelength, has a diameter of 12 mm, what is the radius of curvature of the concave surface?

Figure 3.3-7

Solution.   First, we *assume* that the lower surface is *plane*. Then from Equation [3.3-24] the radius of curvature (of a different lens on top) would have to be

$$R = \frac{r^2}{m\lambda} = \frac{(6 \times 10^{-3})^2}{(20)(550 \times 10^{-9})} = 3.273 \text{ m}$$

At the twentieth ring, the air gap between the actual lens and a plane surface, from Equation [3.3-23], is

$$d = \frac{r^2}{2R} = \frac{(6 \times 10^{-3})^2}{(2)(0.5)} = 0.036 \text{ mm}$$

But the air gap *needed* to produce the Newton rings, from the same equation, is

$$d = \frac{(6 \times 10^{-3})^2}{(2)(3.273)} = 0.0055 \text{ mm}$$

Thus, at the twentieth ring the air gap must be *less,* and the lower surface by *raised,* by

$$0.036 - 0.0055 = 0.03 \text{ mm}$$

Finally, we again use Equation [3.3-23] and find the radius needed:

$$R = \frac{r^2}{2d} = \frac{(6 \times 10^{-3})^2}{(2)(0.03 \times 10^{-3})} = \boxed{60 \text{ cm}}$$

INTERFERENCE FILTERS

Color filters are based either on absorption or on interference. Both types are characterized by three parameters: the *peak wavelength,* λ_{max}, in nm; the peak transmittance or *efficiency,* T_{max}, as a percentage of the total transmittance possible at this wavelength; and the *halfwidth,* HW, the width of the passband at the level of one-half the peak transmittance, in nm (Figure 3.3-8).

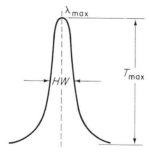

Figure 3.3-8   Characteristics of a color filter: peak wavelength, λ_{max}; peak transmittance, T_{max}; and halfwidth, HW.

Transmission-type *interference filters** consist of a spacer layer, for example a thin transparent layer of cryolite, Na_3AlF_6, sandwiched between two reflective coatings. (In older bandpass filters these coatings were metallic; in modern filters they are *dielectric,* that is, nonmetallic; they do not conduct electricity. Dielectric coatings are more efficient.) The coatings and the spacer are vacuum-deposited on a plate of glass, and then another plate is cemented on top for protection (Figure 3.3-9). The separation between the two coatings, usually one-half of a wavelength or a multiple thereof, determines the wavelength. Interference filters can be made for any wavelength desired and to specifications much more rigid than those met by other filters. Table 3.3-1 shows a comparison.

---

*First described by Walter Geffcken in *Interferenzlichtfilter,* Dtsch. Reichspatent 716 153, 8 Dec. 1939.

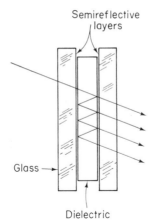

Semireflective
layers

Glass →

Dielectric   **Figure 3.3-9**   Interference filter.

TABLE 3.3-1   TYPICAL VALUES FOR TWO
TYPES OF COLOR FILTERS

	Absorption filter	Interference filter
T_{max}	30%	90%
HW	50 nm	1 Å
Cost	$10	$250

Example

(a) What is the least thickness required of a layer of cryolite ($n = 1.35$) in an interference filter designed to isolate light of 594 nm wavelength?

(b) How will the peak transmittance change if the filter is tilted by 10°?

Solution.   (a) From the condition for maxima in transmitted light, Equation [3.3-1], we find that

$$d = \frac{m\lambda}{2n \cos I} = \frac{(1)(594)}{(2)(1.35)(1)} = \boxed{220 \text{ nm}}$$

(b) The angle of tilt is in effect the angle of incidence, I. But we need the angle *inside* the spacer, that is, the angle of refraction, I'. From Snell's law

$$n \sin I = n' \sin I'$$

$$I' = \arcsin\left(\frac{1.00}{1.35} \sin 10°\right) = 7.39°$$

Substituting this figure for I in Equation [3.3-1] and solving for λ gives

$$\lambda = (2)(1.35)(220 \times 10^{-9})(\cos 7.39°) = 589 \text{ nm}$$

so that

$$\Delta\lambda = 594 - 589 = \boxed{5 \text{ nm}}$$

Note that the transmission of an interference filter, when tilted, always changes toward the *shorter* wavelengths.

Interference filters can also be made for reflection. Such filters are called *dichroic mirrors;* they reflect certain wavelengths and transmit others. Heat reflectors or *hot mirrors* reflect infrared (which on absorption converts into heat) and transmit the visible; *cold mirrors,* used in movie projectors and Klieg lights, transmit infrared and reflect the visible. *Wedge-type interference filters* vary in spacing across the filter. This causes a whole spectrum to be transmitted, from blue at one end of the filter to red at the other. Such filters, combined with a slit for wavelength selection, are a "poor man's monochromator." *Rejection* or *minus filters* eliminate given wavelengths from the spectrum; they are used, for example, in laser safety goggles. Multilayer filters are discussed later in conjunction with antireflection coatings.

ANTIREFLECTION COATINGS

Part of the light passing through a boundary is lost due to reflection. Quantitatively, when the two media forming the boundary have refractive indices n_1 and n_2, the *reflectivity, R,* is

$$R = \left(\frac{n_2 - n_1}{n_2 + n_1}\right)^2 \qquad [3.3\text{-}25]$$

a relationship that we discuss in more detail later (on page 291). For example, at a boundary between air and glass ($n = 1.5$),

$$R = \left(\frac{1.5 - 1.0}{1.5 + 1.0}\right)^2 = 0.04$$

which means that 4% of the light is reflected. While for a single surface this may not be much, for a system that contains several lenses the loss is considerable. There are two ways of reducing such losses.

1. Reflection will not occur if $n_1 = n_2$. But then no refraction will occur either. Also, reflection will not occur if the transition from n_1 to n_2 is gradual, rather than abrupt.
2. Another way of eliminating reflection is by interference, applying a suitable coating to the surface.* Assume that the light is incident on a glass surface coated with a suitable thin film (Figure 3.3-10). Both sides of the film will reflect some of the light. But these reflections will cancel if the two reflected waves are out of phase by 180°, and if they have the same amplitude, $\mathbf{A}_1 = \mathbf{A}_2$.

---

*This discovery was made by Alexander Smakula (1900–   ), German-born physicist, at that time head of the research department at the Carl Zeiss Optical Company in Jena. Smakula later came to the United States to work at Ft. Belvoir, then at MIT. Fa. Carl Zeiss, *Verfahren zur Erhöhung der Lichtdurchlässigkeit optischer Teile durch Erniedrigung des Brechungsexponenten an den Grenzflächen dieser optischen Teile,* Dtsch. Reichspatent 685 767, 1 Nov. 1935.

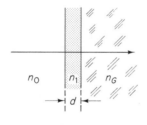

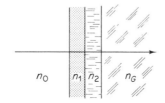

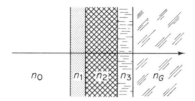

Figure 3.3-10 Antireflection coatings comprising one layer (*top*), two layers (*center*), and three layers (*bottom*).

(a) The amplitude condition, $\mathbf{A}_1 = \mathbf{A}_2$, is met if

$$\left(\frac{n_1 - n_0}{n_1 + n_0}\right)^2 = \left(\frac{n_g - n_1}{n_g + n_1}\right)^2 \qquad [3.3\text{-}26]$$

where n_0 is the outside refractive index, n_1 the index of the coating, and n_g the index of the glass. For $n_0 = 1$,

$$\frac{n_1^2 - 2n_1 + 1}{n_1^2 + 2n_1 + 1} = \frac{n_g^2 - 2n_1 n_g + n_1^2}{n_g^2 + 2n_1 n_g + n_1^2}$$

which reduces to

$$4n_1^3 n_g + 4n_1 n_g = 4n_1^3 + 4n_1 n_g^2$$

Dividing both sides by $4n_1$ and rearranging gives

$$n_1^2 - n_1^2 n_g + n_g^2 - n_g = 0$$

from which

$$\boxed{n_{\text{coating}} \approx \sqrt{n_{\text{glass}}}} \qquad [3.3\text{-}27]$$

(b) The phase condition is met if the two contributions are out of phase by 180°. Both of these contributions are returned at a rare-to-dense boundary (air-to-coating at the first boundary and coating-to-glass at the second), and show a π phase change. But two π changes are equivalent to no phase change. Thus, for minimum reflection to occur, we use Equation [3.3-2] which, for normal incidence and $m = 1$, becomes

$$2n_{coating}d = \tfrac{1}{2}\lambda$$

or

$$\boxed{n_{coating}d = \tfrac{1}{4}\lambda} \qquad\qquad [3.3\text{-}28]$$

With the coating transparent and no light lost to absorption or scattering, the law of conservation of energy tells us that all of the light must go through. Clearly, the coating should also be insoluble and resistant to wear and tear. The best material known today is magnesium fluoride, MgF_2; its refractive index is 1.38.

Multilayer antireflection coatings.   A single-layer coating works well only for a narrow range of wavelengths. The wavelengths chosen are usually near the center of the visible spectrum, causing part of the blue and red to be reflected, which makes the coating appear purple. A much wider coverage is possible with multiple coatings, called *multilayers*. In a *two-layer antireflection (AR) coating,* for example, each layer is made $\tfrac{1}{4}\lambda$ thick. This is called a *quarter-quarter coating* (Figure 3.3-10, center). No reflection will occur if

$$\frac{n_1^2 n_3}{n_2^2} = n_0 \qquad\qquad [3.3\text{-}29]$$

In many *three-layer AR coatings* the center, or "absentee," layer is made $\tfrac{1}{2}\lambda$ thick, the other two layers $\tfrac{1}{4}\lambda$ each. This is called a *quarter-half-quarter coating* (Figure 3.3-10, bottom). Such coatings are widely used; they are effective over most of the visible spectrum. The first, outside layer is often a $\lambda/4$ coating of magnesium fluoride, the next is $\lambda/2$ zirconium dioxide (ZrO_2, index 2.10), and the layer next to the substrate is $\lambda/4$ cerium trifluoride (CeF_3, index 1.65) or aluminum oxide (Al_2O_3, index 1.76). High-performance interference filters, as well as antireflection coatings, can have up to 200 layers of alternating high- and low-index materials. Some others are of the gradient-index type.

Antireflection coatings are of historic interest. During World War II, the Germans tried to develop AR coatings for submarines to avoid detection by enemy radar. Such coatings need to have a magnetic permeability equal in magnitude to their relative dielectric constants. If, in addition, the coating is absorptive for microwaves of a given frequency, that is, if it is ferromagnetic, the target becomes essentially invisible. (Of course, it is easy to change the radar's frequency but difficult to apply a new coating each time the frequency is changed.) So, while correct in theory, this method had little success in practice.

SUGGESTIONS FOR FURTHER READING

O. S. HEAVENS, *Thin Film Physics* (New York: Barnes & Noble, 1970).

Z. KNITTL, *Optics of Thin Films, an Optical Multilayer Theory* (New York: John Wiley & Sons, Inc., 1976).

P. Baumeister and G. Pincus, *"Optical Interference Coatings,"* *Sci. Am.* **223** (Dec. 1970), 58–75.

G. Hernandez, *Fabry–Perot Interferometers* (New York: Cambridge University Press, 1986).

H. K. Pulker, *Coatings on Glass* (Amsterdam: Elsevier, 1984).

PROBLEMS

3.3-1. An oil film, 0.1 μm thick and of index 1.52, rests on a body of water ($n = \frac{4}{3}$).
 (a) How many π phase changes will occur on reflection?
 (b) What is the wavelength reflected by the oil?

3.3-2. A soap bubble, seen at normal incidence and in white light, shows a particularly strong first-order reflection of red (630 nm). If the soapy water has a refractive index of 1.4, how thick is the wall of the bubble?

3.3-3. An oil film (index = 1.47, thickness 0.12 μm) rests on a pool of water. If light strikes the film at an angle of 60°, what is
 (a) The angle of refraction inside the oil?
 (b) The number of π phase changes?
 (c) The wavelength reflected in the first order?

3.3-4. A soap bubble ($n = 1.4$) has a wall 0.36 μm thick. If seen at normal incidence in reflected white light, what color is it?

3.3-5. Two glass plates are in contact at one end and separated by a hair at the other end. If the wedge-like space between the plates is filled with water ($n = \frac{4}{3}$) and if with light of 546 nm wavelength 98 fringes are counted across the length of the plates, how thick is the hair?

3.3-6. Two circular disks of plane glass are laid on top of each other. At one point on their circumference the disks are separated by 0.0027 mm.
 (a) How many dark fringes will be seen in reflected blue light (450 nm)?
 (b) What shape will the fringes have?

3.3-7. Two slabs of glass are in contact at one end and slightly separated at the other. When the gap between the slabs is filled with water ($n = \frac{4}{3}$), 15 fringes per millimeter are seen; when filled with oil, 18 fringes. What is the index of the oil?

3.3-8. A wedge-shaped space between plane plates is filled with water ($n = \frac{4}{3}$) in such a way that a few bubbles of air are trapped between the plates. If 18 fringes are counted within a given distance inside an air bubble, how many fringes, within the same distance, are seen in the water?

3.3-9. What plate separation is needed in a Fabry–Perot interferometer in order to resolve two spectrum lines 0.05 Å apart if the average wavelength is 633 nm and the reflectance 80 percent?

3.3-10. **(a)** What is the order of the center fringe, seen with an etalon 2.5 cm deep and using the green 546-nm mercury line?
 (b) If $r = 0.9$, what is the resolvance?

3.3-11. If in transmitted light a Fabry–Perot interferometer produces very narrow (bright) maxima, what do these maxima look like in *reflected* light?

3.3-12. The two surfaces in a scanning Fabry–Perot interferometer are kept 10 mm apart by a barium titanate spacer which, when a voltage of 380 V is applied, changes in length by one-half of 546 nm.
 (a) When 1.14 kV is applied, and assuming that the response of the spacer is linear, how many fringes pass through the center of the field?
 (b) If $r = 0.8$, how close a doublet (near 546 nm) can be resolved?

3.3-13. When two waves of equal amplitude and frequency are reflected between parallel surfaces and, hence, travel in opposite directions, they form a *standing-wave* pattern. The standing wave illustrated in Figure 3.3-11 is a *time exposure*. Show what the wave looks like:
(a) In a *snapshot* photograph, at an instant of maximum displacement.
(b) $\frac{1}{4}$, $\frac{1}{2}$, and 1 period later.

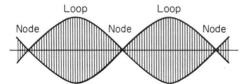

Figure 3.3-11

3.3-14. Standing sound waves, which have a velocity of 331 m s^{-1}, are set up in a cavity 20 cm $\times$ 30 cm $\times$ 50 cm in size. If only *nodes* can be reflected at the walls of the cavity, what are the lowest frequencies possible in each of the three dimensions?

3.3-15. In an experiment on Newton's rings, the diameter of the tenth dark ring formed by yellow sodium light (589 nm) and seen in reflection is 3.6 mm. What is the radius of curvature of the lens surface?

3.3-16. Newton's rings are seen in *transmitted* light of 500 nm wavelength. If the diameter of the twentieth bright ring is 4 mm, what is the radius of curvature of the lens surface?

3.3-17. In a given Newton ring arrangement the diameter of the twentieth ring is 4 mm. What is the diameter of the thirtieth ring?

3.3-18. The diameter of the fourth bright Newton ring is 10 mm. When an unknown liquid is poured into the gap between lens and support, the diameter of this ring shrinks to 8.45 mm. Calculate the liquid's index.

3.3-19. Plot the transmittance as a function of wavelength of:
(a) A *hot mirror* of the dichroic type.
(b) A *cold mirror* of the same type.

3.3-20. A certain interference filter *transmits* the green 546-nm mercury line only. Seen in *reflected* white light, however, the filter looks metallic (silvery).
(a) Why doesn't it look red (complementary to green)?
(b) Plot the transmittance and the reflectance of the filter as a function of wavelength.

3.3-21. What percentage of light is reflected at normal incidence on a surface between air and glass ($n = \frac{5}{3}$)?

3.3-22. The antireflection coating on a lens is made of magnesium fluoride ($n = 1.38$). How thick a coating is needed to produce minimum reflection at 552 nm?

3.3-23. What percentage of the incident light is lost:
(a) On one side of an uncoated glass lens of $n = 1.9044$?
(b) After a coating of optimum refractive index has been applied?

3.3-24. If the first layer of a two-layer antireflection coating is magnesium fluoride ($n = 1.38$) and the substrate is ophthalmic crown ($n = 1.523$), what index should the second layer have?

3.4

Coherence

COHERENCE IS SOMETIMES DEFINED as the condition necessary to produce interference, and interference is defined as an interaction of waves that are "coherent." Nothing much follows from such circular arguments. Moreover, coherence is occasionally thought to be either present or not present, as if light were *either* coherent *or* incoherent. In fact, no light is completely coherent, and none is completely incoherent.

Although the distinction is sometimes blurred, we distinguish two classes of coherence, *spatial coherence* and *temporal coherence*. Spatial coherence refers to the phase relationship between waves traveling side-by-side through space, at the same time but at some distance from one another. Temporal coherence refers to the constancy, and predictability, of phase as a function of time. The two waves now travel through space along the same path but at slightly different times, a characteristic that in essence is the same as *monochromaticity*.

SPATIAL COHERENCE

Consider first the case of *spatial coherence*. Assume that the light comes from an extended source (Figure 3.4-1, left), passes through a double slit (center), and then reaches a screen (right). On the screen a system of interference fringes will be seen, consisting of maxima and minima that are equidistant and parallel to each other and to the slits.

Note that the light comes from an extended source, that is, it comes from an assembly

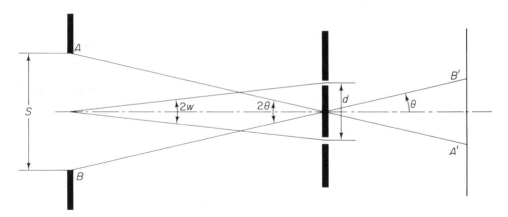

Figure 3.4-1   Young's double-slit experiment and spatial coherence. Vertical dimensions exaggerated.

of groups of atoms that all emit independently from one another. Each group produces maxima whose angular distance θ from the axis is given by the double-slit equation,

$$d \sin \theta = m\lambda \qquad [3.4\text{-}1]$$

If the angles are small (so that $\sin \theta \approx \theta$, in radians), consecutive maxima subtend angular distances

$$\theta = \frac{\lambda}{d}$$

Now consider a slit S that limits the source. As seen from the double slit, the limiting edges of the slit, A and B, subtend an angle 2θ. The double slit, in turn, subtends at the source an angle $2w$. Individually, both A and B produce double-slit interference but the two patterns, as they arrive at the screen, are shifted by 2θ: the center of one pattern lies at A', the other at B'.

The two patterns will cancel whenever a maximum of one falls on a minimum of the other. Since in double-slit interference the minima are halfway between the maxima, that happens when

$$2\theta = \frac{1}{2}\frac{\lambda}{d} \qquad [3.4\text{-}2]$$

In that case, no fringes are seen on the screen.

As slit S is gradually opened further, and points A and B move apart from each other, patterns A' and B' will move apart too and maxima of one pattern fall on the *maxima* of the other and the fringes will reappear. That happens when

$$2\theta = \frac{\lambda}{d}$$

As A and B are moved still farther away from each other, the fringes disappear again, then reappear, and so on, through several cycles.

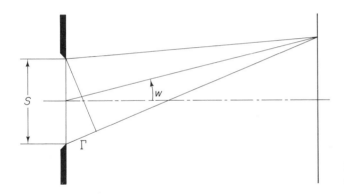

Figure 3.4-2 Deriving the coherence condition.

From the construction in Figure 3.4-2 it follows that the path difference, Γ, between the two contributions is

$$\Gamma = S \sin w \qquad \qquad [3.4\text{-}3]$$

Fringes of high contrast will result if the path difference is small; in fact, the path difference becomes negligible only if it is less than one-half of a wavelength:

$$\Gamma \ll \tfrac{1}{2}\lambda \qquad \qquad [3.4\text{-}4]$$

Combining Equations [3.4-3] and [3.4-4] gives

$$S \sin w \ll \tfrac{1}{2}\lambda \qquad \qquad [3.4\text{-}5]$$

which is the *coherence condition*. It determines the diameter S of a source that, within an angle $2w$, is sufficiently *spatially coherent* to produce fringes of satisfactory contrast.

TEMPORAL COHERENCE

Temporal coherence requires discussion of the fact that light comes in *wavetrains*. Some electromagnetic radiation, such as microwaves and radiowaves, and also sound waves and other mechanical waves, can be generated in the form of waves of almost unlimited length, but light waves cannot. The wavetrains of light are of finite length, each train containing only a limited number of waves (Figure 3.4-3). The length of a wavetrain, Δs, is called the *coherence length*. It is the product of the number of waves, N, contained in the train and of their wavelength, λ:

$$\Delta s = N\lambda \qquad \qquad [3.4\text{-}6]$$

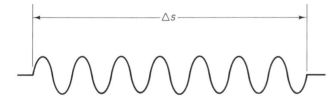

Figure 3.4-3 Schematic representation of a wavetrain.

Since velocity is the distance traveled per unit of time, it takes a wavetrain of length Δs a certain length of time, Δt, to pass a given point, and therefore

$$\Delta t = \frac{\Delta s}{c} \qquad \qquad [3.4\text{-}7]$$

where c is the velocity of light. The length of time Δt is called the *coherence time.*

To measure the temporal coherence, we could again use Young's double slit. We only need to *delay* one of the beams, that is, place a sheet of transparent material over one of the slits. If at a given thickness (of the material) no interference fringes are seen, the path difference introduced by the material exceeds the temporal coherence.

Instead, it is more convenient to use a Michelson interferometer. To obtain high contrast, the two arms of the interferometer are made as equal in length as possible. That is necessary in particular if the wavetrains are fairly short; in that case, the contrast is inversely proportional to the path difference.

Some wavetrains, however, are rather long. Light from the green line of mercury, for example, has a coherence length of about 11 mm, light from the orange krypton line about 80 cm, and light from a laser can have a coherence length of many kilometers. Consequently, with highly coherent laser light we can observe fringes still if one arm of the interferometer is considerably longer than the other (as in Figure 3.4-4); moreover, fringes result even with light that is passing through *atmospheric turbulence.* *

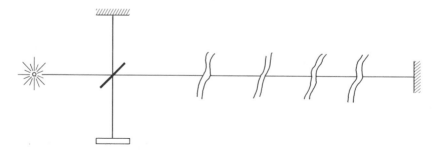

Figure 3.4-4   Unequal-arm Michelson interferometer.

PARTIAL COHERENCE

So far, we have always assumed that two wavetrains, each of finite length Δs, overlap to their full extent. Such complete overlap will produce fringes of the highest contrast. But even if the wavetrains overlap only in part, as in Figure 3.4-5, interference is possible; it is only that the degree of contrast becomes less. The question, then, is not how much the wavetrains must

---

*R. B. Herrick and J. R. Meyer-Arendt, "Interferometry through the Turbulent Atmosphere at an Optical Path Difference of 354 m," *Appl. Opt.* **5** (1966), 981–83. Three years later, an even longer path difference was reached: V. V. Pokasov and S. S. Khmelevtsov, "Interferometry with a Path Difference of up to 500 m through the Turbulent Atmosphere" (in Russian), *Izv. Vyssh. Zaved. Fiz. (Tomsk U.)* **8** (1969), 139–41.

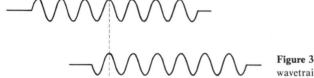

Figure 3.4-5 Partial overlap of two wavetrains.

overlap to produce interference; the question is how much contrast we *need* to see any fringes.

Contrast, as we have shown earlier (on page 187), can be defined as

$$\gamma = \frac{L_{\max} - L_{\min}}{L_{\max} + L_{\min}} \qquad [3.4\text{-}8]$$

but how does contrast relate to coherence? Assume that two points on a screen are illuminated by two bundles of light that produce equal illuminance, E_0. Each of these bundles consists of two parts, A and B. Parts A may be "completely coherent" and cause an illuminance

$$E_A = CE_0 \qquad [3.4\text{-}9]$$

Parts B may be "completely incoherent," both to themselves and with respect to A, and cause an illuminance

$$E_B = (1 - C)E_0 \qquad [3.4\text{-}10]$$

The quantity C is called the *degree of coherence.**

If interference occurs, it is because of parts A. These parts form fringes whose maxima, from Equation [3.2-25], have an illuminance four times as high as the individual contributions. The maximum illuminance, thus, is $4CE_0$. The minimum illuminance, if both bundles are equal, is zero. On this pattern a uniform distribution is superimposed which, because it comes from two sources, has an illuminance twice that of Equation [3.4-10],

$$E_B = 2(1 - C)E_0$$

As a result, the illuminance in the maxima is

$$E_{\max} = 4CE_0 + 2(1 - C)E_0 = 2(1 + C)E_0 \qquad [3.4\text{-}11]$$

and in the minima it is

$$E_{\min} = 2(1 - C)E_0 \qquad [3.4\text{-}12]$$

If Equations [3.4-11] and [3.4-12] are substituted in [3.4-8], we find that

$$\gamma = \frac{2(1 + C)E_0 - 2(1 - C)E_0}{2(1 + C)E_0 + 2(1 - C)E_0} = \frac{4CE_0}{4E_0} = C \qquad [3.4\text{-}13]$$

which shows that the *degree of contrast* of the fringes produced by interference of two waves *is equal to the degree of coherence* between these two waves.

---

*Following P. H. van Cittert, "Degree of Coherence," *Physica* **24** (1958), 505–507.

The highest contrast will result when, following Equation [3.4-8], the illuminance in the minima is zero. Both the contrast, and the degree of coherence, will then be *unity*. Although conceivable in theory, this figure cannot be attained in practice because scattering and diffraction prevent the minima from receiving no light at all. *Complete* coherence, in short, is merely a theoretical limit.

But why can't we have complete incoherence? Because diffraction will cause any image *point* to spread out into a more diffuse image *patch*. These patches are incoherent from patch to patch but, since each patch comes from one group of atoms, are coherent *within* each patch. Since the patches overlap, the image will have a certain degree of coherence and *complete* incoherence is not possible either.

APPLICATIONS

Stellar interferometry. Let us consider some practical applications. Ordinarily, the separation d of the two slits in Young's experiment is a few millimeters at most. We could move the slits farther apart by placing next to them a converging lens. That lens would deflect the light passing through toward the axis, but now the wavefronts in the two contributions subtend a larger angle and the fringes become very closely spaced, requiring a magnifying glass to see them. The lens and the magnifier together then form a *telescope*. Indeed, placing two slits in front of a telescope is a method well known in astronomy for determining the angular separation of binaries (double stars) or the diameter of fixed stars, dimensions too small to measure by direct (noninterferometric) observation.

An extension of this principle is found in *Michelson's stellar interferometer.** Here the two slits are separated even farther, exceeding the diameter of the telescope's aperture. This is done by four mirrors mounted in front of the objective as shown in Figure 3.4-6. The two inner mirrors are fixed, but the two outer mirrors can be moved apart up to several meters; these two mirrors take the place of the double slits.

Again, interference is not seriously affected by atmospheric turbulence. The reason is that the two bundles that produce interference are small in cross section compared to the air cells causing turbulence, and the resulting fringes, though in motion, remain distinct. A conventional image, on the other hand, is produced by light integrated over the telescope's whole aperture, and in turbulent weather may be so blurred as to be worthless.

Lab Experiment. Take two fiber bundles and let the light emerging from them be reflected at a small metal ball. These reflections represent our double star. First we determine, by interferometry, the *distance between the two stars*. At a distance $L = 1.5$ m from the stars, the light passes through a

---

*A. A. Michelson and F. G. Pease, "Measurement of the Diameter of α Orionis with the Interferometer," *Astrophys. J.* **53** (1921), 249–59. The fringes seen disappeared when the outer mirrors were 307 cm apart. Using Rayleigh's criterion (see Chapter 3.5) and assuming a wavelength of 575 nm, that gives an angular diameter

$$\theta = 1.22 \frac{\lambda}{d} = 1.22 \frac{575 \times 10^{-9}}{3.07} = 2.29 \times 10^{-7} \text{ rad} = 0.000013°$$

and a linear diameter of 3.8×10^8 km, more than the diameter of the orbit of Earth (3×10^8 km).

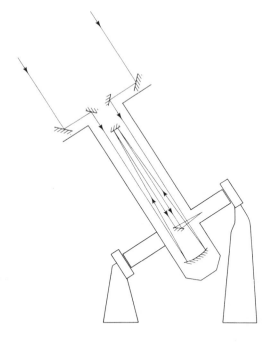

Figure 3.4-6 Michelson's stellar interferometer.

double slit. Individually, either of the reflections produces double-slit interference. We vary the separation of the two slits, d, such that the fringes disappear. Following Equation [3.4-2] that occurs when

$$2\theta = \frac{1}{2}\frac{\lambda}{d}$$

Assume that the wavelength is $\lambda = 600$ nm and the fringes disappear when $d = 2.5$ mm. Then the angular distance between the two reflections is 2θ and the linear distance between them is

$$D = 2\theta L = \frac{1}{2}\frac{\lambda L}{d} = \frac{1}{2}\frac{(600 \times 10^{-9})(1.5)}{0.0025} = \boxed{0.18 \text{ mm}}$$

Next we determine the *width of a single source*. The source has in front of it a slit of variable width. With the slit very narrow, the light passing through, and then passing through a double slit, will produce typical interference. But as the slit is gradually opened wider, the fringes disappear.

Assume that the dimensions are the same as before. Then, to find the width of the slit, S, we take the upper bound of the coherence condition,

$$S \sin w = \tfrac{1}{2}\lambda \qquad\qquad\qquad [3.4\text{-}14]$$

solve for S and, from Figure 3.4-1, substitute

$$\sin w = \frac{1}{2}\frac{d}{L}$$

That gives

$$S = \frac{\lambda L}{d} = \frac{(600 \times 10^{-9})(1.5)}{0.0025} = \boxed{0.36 \text{ mm}}$$

twice the separation of the double star. (The factor 1.22, as it occurs in Michelson's calculations, applies only to a circular source, not to a slit as we use it here.)

Retinal acuity.   Both interference and coherence play an important role in some acuity measurements. That is especially true in cases of *cataract,* where, because of the turbidity of the crystalline lens, conventional acuity measurements are not possible. Indeed, even with a cataractous lens blocking the way, fringes can be seen, as shown schematically in Figure 3.4-7.

But, how can light go through a diffusing medium, and still form distinct fringes? Surely, no distinct *image* could form this way. The answer is that the lines seen by the patient do not pass as ready fringes through the eye; they are *formed on the retina.*

In practice, a beam of light is split into two and projected into the pupil of the eye. The two beams pass through the lens separate, side by side, but despite the turbidity of the lens they retain most of their spatial coherence and still can form fringes that are (nearly) as distinct as without the cataract.

Figure 3.4-7   Fringes of different spatial frequency and orientation as seen by the patient.

Intensity interferometry.   In most any interferometer we add *amplitudes.* In the *intensity interferometer* we add *intensities.* In the first experiment of this kind, the light came from a mercury arc (Figure 3.4-8, top left) and by a beamsplitter was divided to reach two photodetectors, P_1 and P_2, one of them mounted on a sliding support.*

Ordinarily, the amount of current produced by a photodetector is proportional to the amount of light received. But if the steady (dc) component of the current is filtered out, the remaining fluctuations in the two beams are found to be correlated depending on the degree of (spatial) coherence.

The same principle can be applied to astronomy, using two telescopes and two detectors. Sirius, for example, was found to have an angular diameter of 0.0068 arc sec. But even smaller stars could be measured, down to 0.0005 arc sec, about 1/100 of the resolution of Michelson's stellar interferometer.

The largest such radio telescope in operation today is an Y-shaped array of 27 parabolic antennas, each arm 21 km long. But even larger interferometers are being planned: 10 radio antennas extending 8000 km across the United States, providing a resolution comparable to that of a single antenna nearly as big as the diameter of the earth.

---

*R. Hanbury Brown and R. Q. Twiss, "Correlation Between Photons in Two Coherent Beams of Light," *Nature (London)* **177** (1956), 27–29.

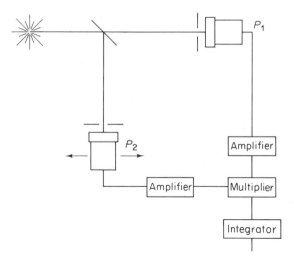

Figure 3.4-8 Determining intensity correlations between two beams of light.

SUGGESTIONS FOR FURTHER READING

M. J. BERAN and G. B. PARRENT, JR., *Theory of Partial Coherence* (Englewood Cliffs, NJ: Prentice-Hall, Inc., 1964).

A. S. MARATHAY, *Elements of Optical Coherence Theory* (New York: John Wiley & Sons, Inc., 1982).

J. W. GOODMAN, *Statistical Optics* (New York: Wiley-Interscience, 1985).

R. HANBURY BROWN, *The Intensity Interferometer, Its Application to Astronomy* (New York: John Wiley & Sons, Inc., 1974).

PROBLEMS

3.4-1.   Red light of 600 nm wavelength is used to produce double-slit interference. If the double slits are separated by 1.5 mm, and if they are 2 m from the source, how narrow a slit should be placed in front of the source to obtain fringes of satisfactory contrast?

3.4-2.   The angular size of the sun, as seen from the earth, is approximately 30 arc min. Considering light of 550 nm coming from opposite points on the circumference of the sun, what should be the separation of two slits whose interference patterns just *cancel?*

3.4-3.   If light of $\lambda = 660$ nm has wavetrains 20λ long, what is its:
(a) Coherence length?
(b) Coherence time?

3.4-4.   Determine the number of waves per wavetrain in light from:
(a) The green mercury line (546 nm).
(b) The orange krypton line (606 nm).
(c) A helium-neon laser (633 nm), assuming that its coherence length is 20 km.

3.4-5.   If the contrast in an interference pattern is 50%, and if the maxima receive 15 units of light, how much do the minima receive?

3.4-6. Two wavetrains overlap to 29% of their length. If the maxima in the resulting interference pattern receive 20 units of light, how much do the minima receive?

3.4-7. If we put a red filter (with a peak transmission at 600 nm) over a slit that is 1 mm wide and mounted in front of a light source, how close to the source could we place a double slit (with a slit separation of 1.2 mm) to observe fringes of satisfactory contrast?

3.4-8. Continue with Problem 3.4-7 and assume that instead of a red filter we now have a *blue* filter (480 nm).

3.4-9. Atomic hydrogen, with its emission at 21 cm wavelength, is the subject of much of *radio astronomy*. What should be the least distance between two dish antennas, for example, if we wish to measure a radio star, its angular diameter believed to be 10 arc min?

3.4-10. When determining by interference the acuity of a patient who is highly *myopic*, the fringes will not be recorded at the same distance from the eye's pupil as if the patient had normal vision. How will that affect the result?

Diffraction

WHEN A POINT LIGHT SOURCE is casting a shadow on a screen, we expect the shadow to be sharp and well defined. But the edges of the shadow are blurred: A certain amount of the light is deflected into the shadow, and another portion is deflected outside the shadow forming brighter and darker fringes (Figure 3.5-1). Such deviation of the light from the predictions of geometrical optics is called *diffraction*.

Huygens' principle. Diffraction fringes that form along the edges of a shadow were first observed by Francesco Grimaldi,* more than a hundred years before the wave nature of light was established by Thomas Young. But, even with the wave concept accepted, how can light deviate from its initial direction and bend around an obstacle? That can be explained by *Huygens' principle.*†

---

*Francesco Maria Grimaldi (1618–1663), Italian mathematician and teacher of rhetoric at the College of the Jesuit Order in Bologna. Grimaldi let sunlight enter a dark room through a hole in a curtain and found that the shadow cast by an object was surrounded by colored fringes. In his book *Physico-mathesis de Lumine, Coloribus et Iride (The Science of Physics of Light, Colors, and Vision),* published by a friend two years after his death, Grimaldi wrote: "Lumen propagatur seu diffunditur non solum directe, refracte ac reflexe, sed etiam quodam quarto modo diffracte" (Light propagates and diffuses not only directly or by refraction or reflection, but also, in a certain fourth manner, by diffraction).

†Christiaan Huygens (1629–1695), Dutch mathematician, physicist, and astronomer. Huygens made noteworthy contributions to many fields, invented the pendulum clock, formulated the laws governing conservation of momentum, centrifugal force, and the moment of inertia. He learned to grind lenses, as many of his contemporaries did, built telescopes of superior quality, found that the "arms" of Saturn were actually a ring, discovered Saturn's sixth moon, Titan, and built a planetarium that still stands today in Leiden. In 1678 he presented the concept of what is now known as *Huygens' principle* to the French Academy of Sciences: C. Huygens, *Traité de la lumière* (Leiden: Van der Aa, 1690).

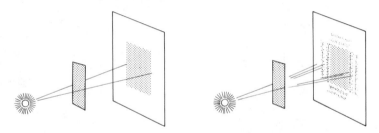

Figure 3.5-1   Shadow in geometrical optics (*left*) and, more correctly, in wave optics (*right*).

Assume that a point light source, on the left in Figure 3.5-2, emits a series of spherical wavefronts. On one of the wavefronts I have marked several points. Each of these points is considered to be the origin of a new, secondary wave, a "wavelet." Then an envelope is drawn over the wavelets; that envelope is the next wavefront. Points on that new wavefront are the source of still other wavelets (not shown), these form another wavefront, and so on, causing the light to proceed.

But now an obstacle is placed in the path of the light. Some distance away from the obstacle, the light proceeds without change. But close to the edge, the wavelet that originates next to it finds no counterpart coming from below the edge; hence, the wavefront normal is turning (clockwise) around the edge.

Depending on the distances involved, we distinguish two types of diffraction, *Fraunhofer diffraction* and *Fresnel diffraction*. Fraunhofer diffraction occurs when the source and the screen are far apart and the light is essentially parallel; it is a matter of *far-*

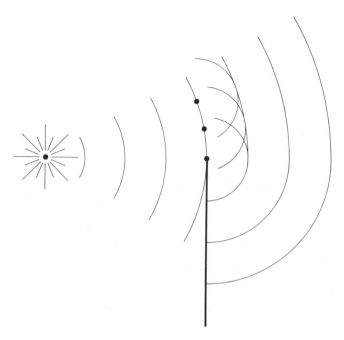

Figure 3.5-2   Huygens' construction of wavefronts bending around an obstacle.

field diffraction. Fresnel diffraction occurs when the source or the screen are close; it is a matter of *near-field diffraction.* Fresnel diffraction is more general; it includes Fraunhofer diffraction as a special case. But Fraunhofer diffraction is so much easier to discuss mathematically that it is customarily presented first.

FRAUNHOFER DIFFRACTION

Single slit.   We begin with *Fraunhofer diffraction on a single slit.* The slit may have a width s (Figure 3.5-3). On a distant screen, at point P_0, the light passing through the slit forms the *zeroth-order maximum.*

Next consider the first diffraction *minimum.* Assume, as shown in Figure 3.5-4, that the light passing through the slit contains three rays, 1–2–3. For a minimum to occur, the light subtends with the axis an angle θ, of a size such that rays 1 and 2 are $\lambda/2$ out of phase, canceling each other. In fact, any ray that passes through the *upper* half of the slit cancels a

Figure 3.5-3   Central maximum in Fraunhofer diffraction.

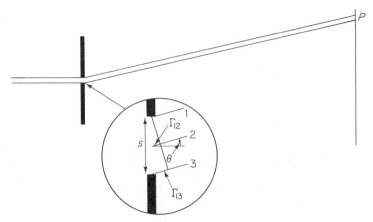

Figure 3.5-4   Fraunhofer diffraction, first minimum. Insert shows path differences between rays.

corresponding ray, a distance $s/2$ apart, that passes through the *lower* half. The first minimum, therefore, occurs where the path difference, Γ, between rays 1 and 2 is

$$\Gamma_{12} = \tfrac{1}{2}\lambda$$

or, between rays 1 and 3,

$$\Gamma_{13} = \lambda \qquad\qquad\qquad [3.5\text{-}1]$$

From the construction it follows that

$$\sin\theta = \frac{\Gamma_{13}}{s} \qquad\qquad\qquad [3.5\text{-}2]$$

and, combining Equations [3.5-1] and [3.5-2], that

$$s\,\sin\theta = \lambda$$

But there will be other minima (whenever Γ_{13} is an integral multiple of λ); thus

$$\boxed{s\,\sin\theta = m\lambda} \qquad m = 1, 2, 3, \ldots \qquad [3.5\text{-}3]$$

which is the equation for *minima in Fraunhofer diffraction on a single slit.**

We note that if the slit is made more narrow, the angle θ becomes larger and the light spreads out wider. We also see that as the wavelength increases, θ increases too; *red light is diffracted more than blue light,* the opposite of what occurs in *re*fraction.

Diffraction maxima.   Now we divide the width of the slit into several narrow strips of equal width Δs. Light passing through each of these strips has an amplitude $\mathbf{A}_i$, and each can be represented by a short phasor (Figure 3.5-5). Since the light is parallel and in phase, these phasors have the same length and direction.

At point P_0, the elemental waves arrive in phase and the resultant amplitude, $\mathbf{A}$, has its *maximum,* the arithmetic sum of the phasors (Figure 3.5-6):

$$\mathbf{A} = \Sigma\,\mathbf{A}_i \qquad\qquad\qquad [3.5\text{-}4]$$

For other points P on the screen, however, the individual phasors, rather than lining up head to toe, will subtend constant angles with one another (Figure 3.5-7).

We call δ the phase angle difference, that is, the angle subtended by the extensions of the first and the last phasor. Thus,

---

*Joseph Fraunhofer (1787–1826), German. After working for a while as a lens grinder and apprentice optician, Fraunhofer became a partner in an optical company that made precision theodolites, professor at the University of Munich, and was knighted by King Maximilian of Bavaria. In his short life (he died of tuberculosis at age 39), Fraunhofer produced large-aperture telescope lenses, exceptionally well corrected for spherical and chromatic aberration, ruled precision gratings and discovered their use for spectroscopy, and found that the spectrum of the sun is crossed by dark lines since named *Fraunhofer lines.* His theory of diffraction appeared first in J. Fraunhofer, "Kurzer Bericht von den Resultaten neuerer Versuche über die Gesetze des Lichtes, und die Theorie derselben," *Ann. Phys.* (3) **14** (1823), 377–78.

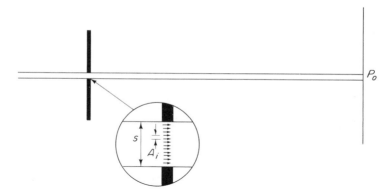

Figure 3.5-5   Fraunhofer diffraction by a single slit. Insert shows arrows inside aperture; these represent phasors.

Figure 3.5-6   Vector representation of amplitudes in the zeroth-order maximum.

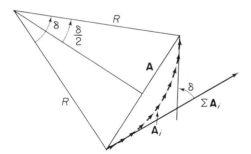

Figure 3.5-7   Vector construction (*vibration curve*) used for finding intensity in single-slit diffraction.

$$\sin \tfrac{1}{2}\delta = \frac{\mathbf{A}/2}{R}$$

$$\mathbf{A} = 2R \sin \tfrac{1}{2}\delta$$

and

$$\delta = \frac{\sum \mathbf{A}_i}{R}$$

Solving the last equation for R and inserting it in the preceding equation gives

$$\mathbf{A} = 2\frac{\sum \mathbf{A}_i}{\delta} \sin \frac{\delta}{2} = \sum \mathbf{A}_i \frac{\sin \delta/2}{\delta/2} \qquad [3.5\text{-}5]$$

Consequently, since intensity is proportional to the square of the amplitude, the intensity at point P_0, at an angular distance θ from the axis, is

$$I_\theta = I_0 \left(\frac{\sin \delta/2}{\delta/2} \right)^2 \qquad\qquad [3.5\text{-}6]$$

where I_0 is the intensity in the zeroth-order maximum. This maximum is the central peak in Figure 3.5-8. At certain angular distances (from the axis), the amplitudes, and the intensities, fall to zero. This will occur whenever

$$\tfrac{1}{2}\delta = \pi, 2\pi, 3\pi, \ldots \qquad\qquad [3.5\text{-}7]$$

The minima, in other words, are equidistant, at least for small angles. The maxima, however, do not fall exactly between the minima; they are slightly displaced toward the center. The first maximum, for instance, is located at $y = 1.4303\,\pi$, the second maximum at $2.4590\,\pi$, the third at $3.4707\,\pi$, and so on, with the higher orders even more nearly halfway between the minima.

The *intensities* in the maxima can be calculated to a good approximation by determining $(\sin^2 \delta/2)/(\delta/2)$ at the halfway positions, that is, where

$$\frac{\delta}{2} = \frac{3\pi}{2}, \frac{5\pi}{2}, \frac{7\pi}{2}, \ldots \qquad\qquad [3.5\text{-}8]$$

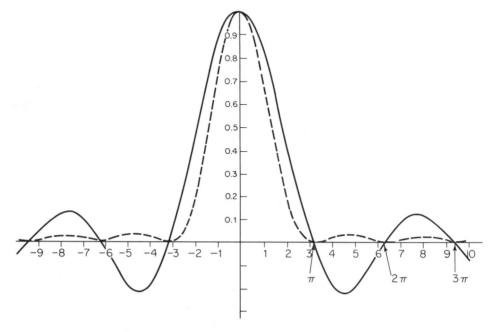

Figure 3.5-8 Amplitude distribution (solid line) and intensity distribution (dashed line) in Fraunhofer diffraction by a single slit.

This gives for the first maximum $4/(9\pi^2)$ or approximately 4.5%, for the second maximum $4/(25\pi^2)$ or 1.6%, for the third maximum $4/(49\pi^2)$ or 0.8%, and so on, of the intensity in the zeroth-order maximum.

Circular aperture.   Fraunhofer diffraction by a circular aperture is of considerable practical interest because most lenses and stops are round. The result is again a series of maxima and minima; but behind a circular aperture these maxima and minima take the form of concentric rings. The bright central maximum is known as *Airy's disk*.* The pattern is very similar to that formed behind a slit, although the dimensions are somewhat different.

The mathematical analysis of diffraction behind a circular aperture is more difficult than diffraction behind a slit. As before, the aperture is divided into a series of narrow strips of equal width. But since these strips are not of equal length, the amplitudes are unequal too. The resultant amplitude is found by integration.

The dimensions, as mentioned, are different. Whereas, for example, the positions of the *minima* behind a slit are given by the simple relationship $s \sin \theta = m\lambda$, we now need to replace m by another factor, J, that derives from *Bessel functions of the first kind*. For the first maximum (behind a circular aperture) we have $J = 1.635$, for the second maximum $J = 2.679$, and for the first three minima we use the values listed in Table 3.5-1.

TABLE 3.5-1   POSITIONS OF MINIMA IN FRAUNHOFER DIFFRACTION

Minimum	Single slit, m	Circular aperture, J
First-order	1	1.220
Second-order	2	2.233
Third-order	3	3.238

Example

Diffraction at a circular *aperture* gives essentially the same results as diffraction at a circular *obstacle*. The reason is that diffraction is essentially an edge effect. Even better to see is diffraction at *multiple* apertures or obstacles. For instance, looking at the sun through the hazy atmosphere, with its vast number of droplets or ice crystals suspended in mid air, we see colored rings around the sun, a larger red ring outside and a smaller blue ring inside. If the observer is standing so that the sun is blocked by the corner of a house, the rings may persist (if caused by the atmosphere), or they may disappear (if caused within the eye).

I have seen rings of the second kind, probably caused by pollen allergy affecting the eye. The blue ring, seen at a distance of 3.2 m from the eye, had a diameter of 40 cm, the red ring 53 cm. The diffraction angle for blue, therefore, was

$$\frac{0.4/2}{3.2} = 0.0625 \text{ rad}$$

*Sir George Biddell Airy (1801–1892), British mathematician and Astronomer Royal. Airy is perhaps best known for the disk just mentioned. He described it in G. B. Airy, "On the Diffraction of an Object-glass with Circular Aperture," *Trans. Cambridge Philos. Soc.* **5** (1835), 283–91.

and for red

$$\frac{0.53/2}{3.2} = 0.0828 \text{ rad}$$

Using $J = 1.635$ for the first maximum showed that the particles causing the diffraction were, for the blue ring,

$$D = 1.635 \times \frac{475 \times 10^{-9}}{0.0625} = 12.4 \ \mu\text{m}$$

in diameter and, for the red ring,

$$D = 1.635 \times \frac{630 \times 10^{-9}}{0.0828} = 12.4 \ \mu\text{m}$$

That means that apparently it was the epithelium cells of the cornea that cause the effect.

Rayleigh's criterion.   Diffraction limits the resolvance (resolving power) of an optical system. For example, consider two nearby stars. Only when their diffraction patterns are separate will the stars appear separate. When the central maxima fuse, the two stars appear as one. When the central maximum of one star coincides with the first minimum of the other, resolution is marginal, a condition called *Rayleigh's criterion* (Figure 3.5-9).

Therefore, from Table 3.5-1, the *minimum angle of resolution,* in radians, provided by a lens of diameter D, at a wavelength λ, is

$$\theta_{\min} \approx 1.22 \frac{\lambda}{D} \qquad [3.5\text{-}9]$$

The lens, even if it could be fully corrected for all aberrations, is still *diffraction-limited*.

Rayleigh's criterion has been criticized for several reasons. Some people claim that it is overgenerous; under favorable conditions details can be resolved that are even smaller in size. Others believe in Rayleigh's criterion as if it were written in tablets of gold by the angels.* Rayleigh himself considered it to be no more than an approximation.

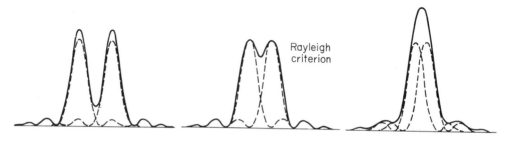

Figure 3.5-9   Rayleigh's criterion. Note the separation of the maxima in the left-hand plot and the close overlap on the right.

---

*As quoted from F. H. Perrin, "Methods of Appraising Photographic Systems," *J. Soc. Motion Pict. Telev. Eng.* **69** (1960), 152.

FRESNEL DIFFRACTION

Fresnel integrals.   *Fresnel diffraction** is not restricted to parallel light. Consider the arrangement shown in Figure 3.5-10. A horizontal slit source, S, emits cylindrical wavefronts. Wavefront W has just reached the center of a slit, oriented parallel to the source and placed at distance a from the source and at distance b from a screen. As before, we divide the wavefront into a series of narrow strips, each strip of width dW and each parallel to the edges of the slit.

For reasons that will soon become apparent, it is practical to choose the width of these strips so that they correspond to path lengths that increase by one-half of a wavelength, $\frac{1}{2}\lambda$, from strip to strip. Thus, while the distance along the axis, from the center of the slit to P_0, is b, the distance from the next higher element is $b + \frac{1}{2}\lambda$, from the next higher, second element $b + 2(\frac{1}{2}\lambda)$, from the third element $b + 3(\frac{1}{2}\lambda)$, and so on. The wavefront elements defined this way are known as *Fresnel half-period zones.* These zones are discussed in more detail when we come to circular apertures and *zone plates* (page 263).

Consider first the disturbance, dy, caused by the element dW which is closest to the edge, acting on point P_0 on the screen. From Equation [3.1-21], this disturbance is

$$dy = A \sin 2(\pi\nu t) \, dW \qquad\qquad [3.5\text{-}10]$$

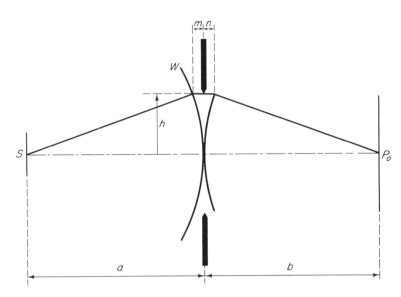

Figure 3.5-10   Fresnel diffraction by a slit. (Vertical dimensions exaggerated.)

---

*Augustin Jean Fresnel (1788–1827), French physicist. Fresnel studied mathematics, then civil engineering, at the French École des Ponts et Chaussées (School of Bridges and Roads); he went into optics later. In his dissertation, "Sur la Diffraction de la lumière, où l'on examine particulièrement le phénomène des franges colorées que présentent les ombres des corps éclairés par un point lumineux," *Ann. Chim. Phys.* (2) **1** (1816), 239–81, Fresnel presented the first rigorous treatment of diffraction. He also worked on more mundane problems, invented flat, weight-saving lenses for use in lighthouses.

Since $\nu = 1/T$, this is equal to

$$dy = A \sin\left(2\pi \frac{t}{T}\right) dW \qquad [3.5\text{-}11]$$

At another point P on the screen, some distance away from P_0, the disturbance is

$$dy = A \sin\left[2\pi\left(\frac{t}{T} - \frac{b}{\lambda}\right)\right] dW \qquad [3.5\text{-}12]$$

The disturbance due to some *other* element dW, farther away from the edge, is

$$dy = A \sin\left[2\pi\left(\frac{t}{T} - \frac{b}{\lambda}\right) + \delta\right] dW \qquad [3.5\text{-}13]$$

where δ is the phase difference between the light from these two elements.

The *total* disturbance at P, due to all elements dW, is found by integration:

$$y = A \int \sin\left[2\pi\left(\frac{t}{T} - \frac{b}{\lambda}\right) + \delta\right] dW \qquad [3.5\text{-}14]$$

If we use the identity $\sin(\alpha + \beta) = \sin\alpha\cos\beta + \cos\alpha\sin\beta$, we can write Equation [3.5-14] as

$$y = A \sin\left[2\pi\left(\frac{t}{T} - \frac{b}{\lambda}\right)\right] \int \cos\delta\, dW + A \cos\left[2\pi\left(\frac{t}{T} - \frac{b}{\lambda}\right)\right] \int \sin\delta\, dW$$

$$[3.5\text{-}15]$$

We then set

$$\mathbf{A} \cos\theta = \mathbf{A} \int \cos\delta\, dW$$

and $\qquad\qquad\qquad\qquad\qquad\qquad\qquad\qquad\qquad\qquad\qquad\qquad [3.5\text{-}16]$

$$\mathbf{A} \sin\theta = \mathbf{A} \int \sin\delta\, dW$$

so that Equation [3.5-15] becomes

$$y = \mathbf{A} \sin\left[2\pi\left(\frac{t}{T} - \frac{b}{\lambda}\right) - \theta\right] \qquad [3.5\text{-}17]$$

This equation represents the disturbance caused by all contributions, at any point, P.

Before we proceed to a determination of the intensity, consider the *path difference*, Γ, and the *phase difference*, δ, between rays (from different wavefront elements). First, we find from Figure 3.5-10, and from the sagitta formula derived on page 63, that

$$2am = h^2 \qquad \text{and} \qquad 2bn = h^2 \qquad [3.5\text{-}18]$$

The path difference, then, is the sum of $m + n$;

$$\Gamma = \frac{h^2}{2a} + \frac{h^2}{2b} = h^2\left(\frac{a + b}{2ab}\right) \qquad [3.5\text{-}19]$$

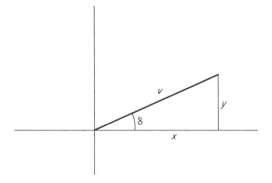

Figure 3.5-11   Part of a vibration curve.

Since path difference and phase difference are connected as $\Gamma/\lambda = \delta/2\pi$, solving for δ and substituting Equation [3.5-19] gives

$$\delta = 2\pi \frac{\Gamma}{\lambda} = \pi h^2 \left(\frac{a+b}{ab\lambda} \right) \qquad [3.5\text{-}20]$$

In order to become independent of the specific values of a, b, and λ (which may vary from experiment to experiment), we introduce a variable, v. This variable is part of the vibration curve, representing the amplitude vector (phasor) of an element of light passing through one of the half-period zones. Numerically, v is defined as

$$\tfrac{1}{2}\pi v^2 = \delta \qquad [3.5\text{-}21]$$

so that by substitution in Equation [3.5-20],

$$\pi h^2 \left(\frac{a+b}{ab\lambda} \right) = \tfrac{1}{2}\pi v^2 \qquad [3.5\text{-}22]$$

or

$$v = h\sqrt{\frac{2(a+b)}{ab\lambda}} \qquad [3.5\text{-}23]$$

The direction of v is given by the angle, δ, it subtends with the $+x$ axis (Figure 3.5-11). But v is only a *short length* on what may be a long and complicated curve and, therefore,

$$\sin \delta = \frac{dy}{dv} \qquad \text{and} \qquad \cos \delta = \frac{dx}{dv} \qquad [3.5\text{-}24]$$

Solving for dx and dy, respectively, and substituting Equation [3.5-21] gives

$$dx = \cos \delta \, dv = \cos \tfrac{1}{2}\pi v^2 \, dv$$

and $\qquad\qquad\qquad\qquad\qquad\qquad\qquad\qquad\qquad\qquad$ [3.5-25]

$$dy = \sin \delta \, dv = \sin \tfrac{1}{2}\pi v^2 \, dv$$

Now, in order to obtain the *intensity*, we square, and add, Equations [3.5-16]. This gives

$$I \propto \mathbf{A}^2 = \left[\int \cos \delta \, dW \right]^2 + \left[\int \sin \delta \, dW \right]^2 \qquad [3.5\text{-}26]$$

We can then substitute Equation [3.5-20] in [3.5-26]:

$$I = \left[\int \cos\left(\pi h^2 \frac{a+b}{ab\lambda} \right) dW \right]^2 + \left[\int \sin\left(\pi h^2 \frac{a+b}{ab\lambda} \right) dW \right]^2 \qquad [3.5\text{-}27]$$

and, using Equation [3.5-22], take the $ab\lambda$ term out in front of the integral signs. Equation [3.5-27] then reduces to

$$I = \tfrac{1}{2}\left(\frac{ab\lambda}{a+b} \right)\left[\left(\int \cos \tfrac{1}{2}\pi v^2 \, dv \right)^2 + \left(\int \sin \tfrac{1}{2}\pi v^2 \, dv \right)^2 \right] \qquad [3.5\text{-}28]$$

The two integrals that occur in this equation are known as *Fresnel's integrals:*

$$\boxed{\begin{aligned} x &= \int \cos \tfrac{1}{2}\pi v^2 \, dv \\[2em] y &= \int \sin \tfrac{1}{2}\pi v^2 \, dv \end{aligned}} \qquad [3.5\text{-}29]$$

If these integrals are solved between certain limits of v, they yield the resultant of the disturbances produced by corresponding elements of the wavefront. In short, we can predict the amount of light reaching the screen.

Cornu spiral. If Fresnel's integrals are solved between a lower limit of $v_1 = 0$ and an upper limit of $v_2 = \infty$,

$$x = \int_0^\infty \cos \tfrac{1}{2}\pi v^2 \, dv$$
$$y = \int_0^\infty \sin \tfrac{1}{2}\pi v^2 \, dv \qquad [3.5\text{-}30]$$

then x and y assume the values listed in Table 3.5-2. If we now plot x versus y, we obtain a curve known as *Cornu's spiral** (Figure 3.5-12). This spiral provides us with a most elegant means for the graphical solution of diffraction problems.

The magnitude of v is the length measured *along the spiral.* Each short fraction of v, dv, represents the amplitude of the light in one wavefront element, dW. In Figure 3.5-11 we saw that the direction of dv is given by the tangent of the angle δ it makes with the $+x$ axis,

$$\tan \delta = \frac{dy}{dx} \qquad [3.5\text{-}31]$$

---

*Marie Alfred Cornu (1841–1902), French physicist, professor of experimental physics at the École Polytechnique in Paris. While working earlier at the École des Mines, Cornu became interested in optics, read Félix Billet's book *Traité d'optique*, repeated all the experiments described therein. He determined the speed of light by Fizeau's method and made precise measurements of the wavelengths of certain lines in the spectrum of hydrogen. He is probably best known for the spiral he described in A. Cornu, "Méthode nouvelle pour la discussion des problèmes de diffraction dans le cas d'une onde cylindrique," *J. Phys.* **3** (1874), 5–15, 44–52.

TABLE 3.5-2 FRESNEL INTEGRALS

v	x	y	v	x	y
0.00	0.0000	0.0000	4.50	0.5261	0.4342
0.10	0.1000	0.0005	4.60	0.5673	0.5162
0.20	0.1999	0.0042	4.70	0.4914	0.5672
0.30	0.2994	0.0141	4.80	0.4338	0.4968
0.40	0.3975	0.0334	4.90	0.5002	0.4350
0.50	0.4923	0.0647	5.00	0.5637	0.4992
0.60	0.5811	0.1105	5.05	0.5450	0.5442
0.70	0.6597	0.1721	5.10	0.4998	0.5624
0.80	0.7230	0.2493	5.15	0.4553	0.5427
0.90	0.7648	0.3398	5.20	0.4389	0.4969
1.00	0.7799	0.4383	5.25	0.4610	0.4536
1.10	0.7638	0.5365	5.30	0.5078	0.4405
1.20	0.7154	0.6234	5.35	0.5490	0.4662
1.30	0.6386	0.6863	5.40	0.5573	0.5140
1.40	0.5431	0.7135	5.45	0.5269	0.5519
1.50	0.4453	0.6975	5.50	0.4784	0.5537
1.60	0.3655	0.6389	5.55	0.4456	0.5181
1.70	0.3238	0.5492	5.60	0.4517	0.4700
1.80	0.3336	0.4508	5.65	0.4926	0.4441
1.90	0.3944	0.3734	5.70	0.5385	0.4595
2.00	0.4882	0.3434	5.75	0.5551	0.5049
2.10	0.5815	0.3743	5.80	0.5298	0.5461
2.20	0.6363	0.4557	5.85	0.4819	0.5513
2.30	0.6266	0.5531	5.90	0.4486	0.5163
2.40	0.5550	0.6197	5.95	0.4566	0.4688
2.50	0.4574	0.6192	6.00	0.4995	0.4470
2.60	0.3890	0.5500	6.05	0.5424	0.4689
2.70	0.3925	0.4529	6.10	0.5495	0.5165
2.80	0.4675	0.3915	6.15	0.5146	0.5496
2.90	0.5624	0.4101	6.20	0.4676	0.5398
3.00	0.6058	0.4963	6.25	0.4493	0.4954
3.10	0.5616	0.5818	6.30	0.4760	0.4555
3.20	0.4664	0.5933	6.35	0.5240	0.4560
3.30	0.4058	0.5192	6.40	0.5496	0.4965
3.40	0.4385	0.4296	6.45	0.5292	0.5398
3.50	0.5326	0.4152	6.50	0.4816	0.5454
3.60	0.5880	0.4923	6.55	0.4520	0.5078
3.70	0.5420	0.5750	6.60	0.4690	0.4631
3.80	0.4481	0.5656	6.65	0.5161	0.4549
3.90	0.4223	0.4752	6.70	0.5467	0.4915
4.00	0.4984	0.4204	6.75	0.5302	0.5362
4.10	0.5738	0.4758	6.80	0.4831	0.5436
4.20	0.5418	0.5633	6.85	0.4539	0.5060
4.30	0.4494	0.5540	6.90	0.4732	0.4624
4.40	0.4383	0.4622	6.95	0.5207	0.4591

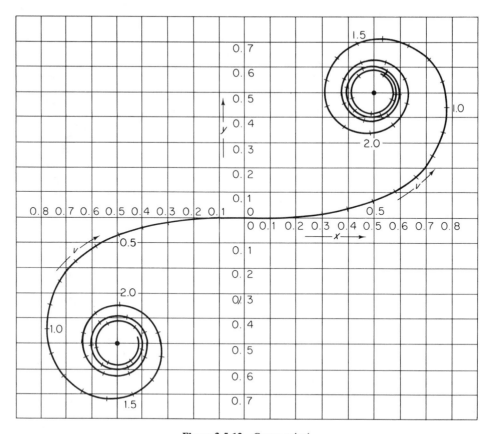

Figure 3.5-12 Cornu spiral.

But this angle, $\delta = \frac{1}{2}\pi v^2$, Equation [3.5-21], is proportional to the *square* of v. Therefore, as we move away from the origin and v *increases,* the *steepness of the curve increases even faster*—which accounts for the spiral shape.

Of particular interest are the two endpoints, the *eyes* of the spiral. At these points $v \to \pm\infty$. Using the identity

$$\int_0^\infty \sin ax^2 \, dx = \int_0^\infty \cos ax^2 \, dx = \frac{1}{2}\sqrt{\frac{\frac{1}{2}\pi}{a}} \qquad [3.5\text{-}32]$$

we substitute $\frac{1}{2}\pi$ for a, and v for x, and obtain

$$\int_0^\infty \sin \tfrac{1}{2}\pi v^2 \, dv = \int_0^\infty \cos \tfrac{1}{2}\pi v^2 \, dv = \frac{1}{2}\sqrt{\frac{\frac{1}{2}\pi}{\frac{1}{2}\pi}} = \frac{1}{2} \qquad [3.5\text{-}33]$$

which means that the upper eye of the spiral has the coordinates

$$x = y = (+0.5, \, +0.5) \qquad [3.5\text{-}34]$$

and the lower eye,

$$x = y = (-0.5, \, -0.5) \qquad [3.5\text{-}35]$$

As with any vibration curve, the amplitude due to a wavefront element, dW, corresponds to an element of length, dv, *on* the curve. But the total amplitude corresponds to the shortest distance *between* points; it corresponds to the *chord*. The square of this chord represents the intensity.

Applications.   First we apply Cornu's spiral to diffraction at a *knife edge*. Consider point P_0, located at the geometric limit of the shadow (Figure 3.5-13). The amplitude at P_0 is proportional to the length of the chord which extends from the lower eye of the spiral to the origin, as shown.

As point P moves *into* the shadow, the chord on the spiral retracts (Figure 3.5-14). The lower end of the chord remains anchored to the lower eye, but the upper end retraces the path of the spiral downward, becoming gradually shorter.

Conversely, as P moves *outside* the shadow, the upper end of the chord moves into the upper branch of the spiral (Figure 3.5-15), while its lower end remains anchored to the lower eye. But the *length of the chord undergoes periodic variations,* passing through a series of maxima and minima, rather than changing monotonically as before.

At certain points the amplitude is even higher than the amplitude with no obstacle in the path. In short: *Inside* the shadow the intensity drops off gradually, and *outside* the shadow there are fringes (Figure 3.5-16). At the edge of the shadow, only one-half of the wave contributes; thus the amplitude has fallen to one-half, and the intensity to one-fourth, of the unobstructed wave.

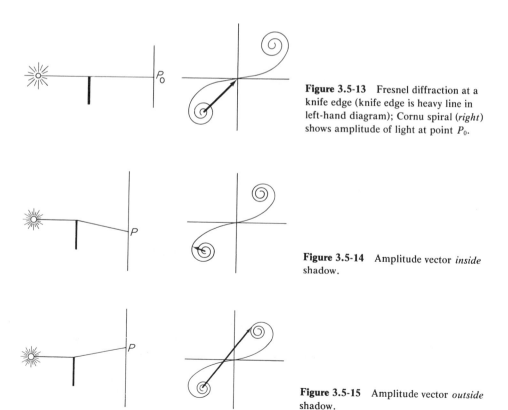

Figure 3.5-13   Fresnel diffraction at a knife edge (knife edge is heavy line in left-hand diagram); Cornu spiral (*right*) shows amplitude of light at point P_0.

Figure 3.5-14   Amplitude vector *inside* shadow.

Figure 3.5-15   Amplitude vector *outside* shadow.

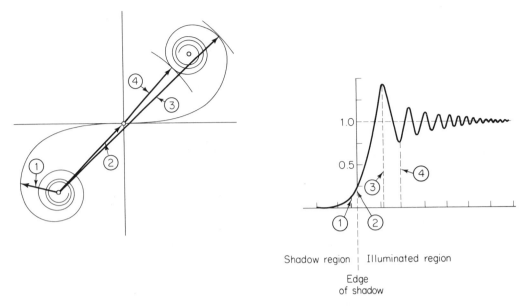

Figure 3.5-16   Fresnel diffraction at a knife edge. Amplitude in Cornu spiral (*left*) and intensity distribution on screen (*right*). Note corresponding numbers on left and right.

Example

Whereas Cornu's spiral illustrates the principle very well, higher accuracy is obtained when the tabulated values of Fresnel's integrals are used directly. For example, calculate the relative intensity at a point:

(a) Where $v = -1.00$, which is *inside* the shadow.
(b) Where $v = +1.00$, which is *outside* the shadow.

Solution.   From Table 3.5-2,

$$v = 1.00, \qquad x = 0.7799, \qquad y = 0.4383$$

(a) *Inside* the shadow, the top of the phasor lies in the same (lower left-hand) quadrant as the lower eye of the spiral to which it is anchored. Thus the phasor extends from the coordinates $(-0.5; -0.5)$ to $(-0.7799; -0.4383)$ and the absolute values of the x and y components, respectively, must be *subtracted*:

$$\Delta x = 0.5 - 0.7799 = |0.2799|$$

and

$$\Delta y = 0.5 - 0.4383 = 0.0617$$

We need not solve for the actual length of the phasor (which has to be squared anyway to obtain the intensity). However, we recall that Cornu's construction gives a value of two for the unobstructed wave, and hence the relative intensity at the specified point is

$$I = \tfrac{1}{2}(0.2799^2 + 0.0617^2)I_0 = \boxed{0.041 I_0}$$

(b) *Outside* the shadow, the phasor extends from the lower eye to the first (upper right-hand) quadrant; thus the x and y components must be *added:*

$$\Delta x = 0.5 + 0.7799 = 1.2799$$

and

$$\Delta y = 0.5 + 0.4383 = 0.9383$$

The relative intensity then is

$$I = \tfrac{1}{2}(1.2799^2 + 0.9383^2)I_0 = \boxed{1.26 I_0}$$

Zone plates.   We now come to Fresnel diffraction by a *circular aperture*. For simplicity, assume that the light comes from infinity. The wavefronts that pass through the aperture in Figure 3.5-17, therefore, are *plane*. We divide the aperture into half-period zones, which from zone to zone are $\tfrac{1}{2}\lambda$ farther away from point P_0. If b is the distance from the center of the aperture to P_0, then $b + \tfrac{1}{2}\lambda$ corresponds to a (first) ring around the center of the aperture, $b + 2(\tfrac{1}{2}\lambda)$ to a second, wider ring, $b + 3(\tfrac{1}{2}\lambda)$ to the third ring, and $b + m(\tfrac{1}{2}\lambda)$ to the mth ring. These distances refer to *boundaries* between zones, rather than to the zones as such.

From Pythagoras' theorem we find that the radius of the first boundary is

$$R_1 = \sqrt{(b + \tfrac{1}{2}\lambda)^2 - b^2} \qquad [3.5\text{-}36]$$

and the radius of the mth boundary

$$R_m = \sqrt{(b + \tfrac{1}{2}m\lambda)^2 - b^2} \qquad [3.5\text{-}37]$$

Furthermore, the *area* of the first *zone* is

$$A_1 = \pi R_1^2 = \pi[(b + \tfrac{1}{2}\lambda)^2 - b^2] = \pi b\lambda + \tfrac{1}{4}\pi\lambda^2$$

or, since λ is small compared with b,

$$A_1 \approx \pi b\lambda \qquad [3.5\text{-}38]$$

The area of the second zone is

$$A_2 = \pi[(b + \lambda)^2 - b^2] - \pi b\lambda \approx 2\pi b\lambda - \pi b\lambda = \pi b\lambda \qquad [3.5\text{-}39]$$

and so on. The individual zones all have approximately the same area. Therefore, the contributions from these zones are very nearly equal; but since the zones differ by one-half of a wavelength, the resultant at P_0 is zero. However, any odd number of zones will give a bright center (while any even number will give a dark center).

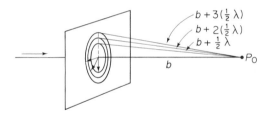

Figure 3.5-17   Fresnel's half-period zones on a plane wavefront.

If we now block out all odd-numbered zones, then only zones 2, 4, 6, . . . will contribute to the light reaching P_0, and P_0 will be bright; it will becomes the *focus* of a *zone plate**
(Figure 3.5-18).

In order to derive the focal length of a zone plate, we square Equations [3.5-37] and [3.5-36] and divide one by the other:

$$\frac{R_m^2}{R_1^2} = \frac{(b + \frac{1}{2}m\lambda)^2 - b^2}{(b + \frac{1}{2}\lambda)^2 - b^2}$$

Canceling b and λ gives

$$R_m^2 + \tfrac{1}{2}R_m^2\lambda = m(R_1^2 b + \tfrac{1}{2}R_1^2\lambda)$$

and, since $\lambda \ll R$,

$$R_m = R_1\sqrt{m} \qquad\qquad [3.5\text{-}40]$$

which means that the boundaries between zones have radii that are proportional to the square root of the natural numbers.

If we again square Equation [3.5-37] and rearrange the terms, we obtain

$$b^2 = (b + \tfrac{1}{2}m\lambda)^2 - R_m^2 = b^2 + bm\lambda + (\tfrac{1}{2}m\lambda)^2 - R_m^2$$

and, since $\lambda \ll b$,

$$bm\lambda \approx R_m^2$$

The *focal length,* therefore, is

$$f = b \approx \frac{R_m^2}{m\lambda} \qquad\qquad [3.5\text{-}41]$$

Figure 3.5-18   Center portion of a zone plate.

---

*A zone plate, or *zone lens* because of its focusing properties, is the only image-forming element for which there is no prototype in nature. The first to make a zone plate, in 1871, was Lord Rayleigh. Four years later, J. L. Soret described its properties, "Ueber die durch Kreisgitter erzeugten Diffractionsphänomene," *Ann. Phys.* (6) **6** (1875), 99–113. Today, zone plates are more than a curiosity. They are used for image formation in regions of the spectrum (X rays, microwaves) where conventional lenses do not exist.

Note the wavelength λ in this equation. It shows that a zone plate has severe chromatic aberration. (A conventional lens has chromatic aberration too, because its refractive index is a function of wavelength. This variation, though, is much smaller because n varies very slowly.)

A zone plate acts both as a positive and a negative lens; its focal length is actually $\pm f$. This is because light diffracted at the boundaries is directed not only *toward* the axis but also *away* from it. Also unlike an ordinary lens, a zone plate has several foci. This is because of the higher orders that occur in addition to the first order. The most intense, first-order focus is farthest away. The higher-order foci are located at distances $\frac{1}{3}, \frac{1}{5}, \frac{1}{7}, \ldots$ of that of the first order.

If a zone plate has completely opaque and completely clear zones, it is called a *Fresnel zone plate*. If the zone plate has a sinusoidal transmissivity distribution, it is a *Gabor zone plate*—which, oddly enough, has no higher-order foci.

Gauss' thin-lens equation $1/s + 1/f = 1/s'$, applies as well to a zone plate. Rayleigh's criterion, $\theta_{\min} \approx 1.22\lambda/D$, also applies. If we replace θ by d/f and f by Equation [3.5-41] then

$$d_{\min} \approx 1.22 \frac{R_m^2}{mD}$$

But D is twice the radius, and thus

$$d_{\min} \approx 1.22 \frac{R_m}{2m} \qquad [3.5\text{-}42]$$

Interestingly, λ has dropped out: The wavelength, although it causes much chromatic aberration, has no effect on the resolution. Finally, we see from Equation [3.5-42] that to make a better zone plate (which will resolve smaller details), it (a) must be as small as possible and (b) must have as many zones as possible, two requirements that call for high-resolution photographic material on which to reproduce such zone plates.

Drawing a zone plate by computer.   We start out with Equation [3.5-40], $R_m = R_1\sqrt{m}$, written in BASIC:

$$\text{RM} = \text{R1} * \text{SQR (M)}$$

Using these radii, RM, we draw a set of concentric circles:

$$\text{CIRCLE (160,100), RM, 2,,, .86}$$

The last number, .86, is the *aspect,* which, in general, describes the shape of an ellipse; here, this figure has been determined empirically to make the circles as round as possible.

The space between alternate concentric circles is then filled using the **PAINT** command. The coordinates specified with PAINT refer to a pixel (*any* pixel) inside this space. The next number is the color of the paint; that color must match the color of the circle. In Program 3.5-1, ZONEPA, we make the innermost radius so large, and use so many rings, that they fill the screen and are still sufficiently separate from one another; the printout is then reduced photographically.

PROGRAM 3.5-1. CONSTRUCTION OF A ZONE PLATE

```
10      REM       Drawing a Zone Plate
20      REM       Program ZONEPA
30      REM       Radius of the Innermost Boundary = R1
40      REM       Radius of the Mth Boundary = RM
50      REM       Total Number of Boundaries = TOTAL
60      CLS
70      R1 = 28
80      TOTAL = 13
90      FOR M = 1 TO TOTAL
100     RM = R1 * SQR (M)
110     SCREEN 1
120     CIRCLE (160,100), RM, 2,,, .86
130     NEXT M
140     PAINT (161,100), 2
150     FOR L = 2 TO (TOTAL-1) STEP 2
160     RING2 = R1 * SQR (L) + 160
170     RING3 = R1 * SQR (L+1) + 160
180     MEDIAN = (RING2 + RING3) / 2
190     PAINT (MEDIAN, 100), 2
200     NEXT L
220     END
```

Concluding phenomena.   There are a few more diffraction phenomena worth mentioning.

If instead of an (open) aperture we have a (solid) circular obstacle, Fresnel's theory and experimental observation show that in the center of the shadow there is a bright point of light, called *Poisson's spot** (Figure 3.5-19). The reason is that in diffraction by a circular disk or circular aperture, there are only two rays (from diametrically opposite points) diffracted to any given off-axis point. But there are an infinite number of rays (from the whole circumference) that converge on the axis.

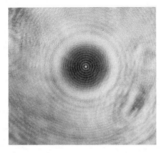

Figure 3.5-19   Poisson's spot. Note the bright spot in the center of the shadow.

---

*Named after Siméon Denis Poisson (1781–1840), French physicist. As a member of the committee which was to judge Fresnel's dissertation, Poisson concluded that, if Fresnel were right, a central bright spot would appear in the shadow of a circular obstacle, a result which he offered as a *reductio ad absurdum,* that is, evidence that Fresnel's theory is absurd. He did not know that the bright spot in question had been discovered by Maraldi and Deslisle some 100 years earlier. After the objection raised by Poisson, Arago immediately tried the experiment using a disk 2 mm in diameter and rediscovered Maraldi's spot. Poisson's name became immortal, either despite his objection, or because of it, and the phenomenon is since called *Poisson's spot.*

If the disk is illuminated by two point sources, two Poisson spots appear. If there is an array of many point sources, there are as many Poisson spots. And if the many point sources are replaced by a complex object, such as a photographic slide illuminated from behind, a real image of the slide results, having used as the "objective lens" nothing but a solid metal sphere.

Two diffracting objects are said to be complementary to each other if one is the photographic negative of the other. The (Fraunhofer) diffraction patterns generated by two such objects, except for the zeroth order, are the same, an effect known as *Babinet's principle.*

Neither Huygens' principle nor Fresnel's theory considers the fact that the secondary wavelets cannot have equal amplitudes in all directions. If they did, they would move forward as well as backward. This dilemma was solved by Kirchhoff, who introduced the *obliquity factor.*[*]

The obliquity factor makes the amplitude vary as a function of the angle θ by which the light departs from the forward direction. The amplitude then takes the form

$$\mathbf{A} = \tfrac{1}{2}\mathbf{A}_0(1 + \cos \theta) \tag{3.5-43}$$

In the forward direction, where $\theta = 0$, the obliquity factor reaches its highest value, unity. At $\theta = 90°$, the amplitude falls off to $\tfrac{1}{2}$, and the intensity to $\tfrac{1}{4}$. In the backward direction, $180°$, the obliquity factor and the amplitude and intensity all become zero. If the amplitude and intensity are plotted in three-dimensional space, a figure results that has the shape of a cardioid (Figure 3.5-20).

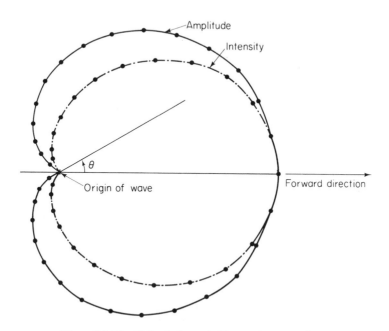

Figure 3.5-20 Obliquity factor in Huygens' construction.

*For Kirchhoff footnote, see page 381. The fundamental paper on Kirchhoff's diffraction theory is G. Kirchhoff, "Zur Theorie der Lichtstrahlen," *Ann. Phys.* (Neue Folge) **18** (1883), 663–95.

SUGGESTIONS FOR FURTHER READING

R. B. HOOVER and F. S. HARRIS, JR., "Die Beugungserscheinungen: A Tribute to F. M. Schwerd's Monumental Work on Fraunhofer Diffraction," *Appl. Opt.* **8** (1969), 2161–64.

M. V. KLEIN and TH. E. FURTAK, *Optics,* 2nd edition, pp. 337–468 (New York: John Wiley & Sons, Inc., 1986).

M. BORN and E. WOLF, *Principles of Optics,* 6th edition, pp. 370–458 (Oxford: Pergamon Press Ltd., 1980).

J. B. KELLER, "Geometrical Theory of Diffraction," *J. Opt. Soc. Am.* **52** (1962), 116–30.

PROBLEMS

3.5-1. How many wavelengths wide must a single slit be if the first Fraunhofer diffraction minimum occurs at an angular distance of 30° from the optic axis?

3.5-2. A slit 0.14 mm wide is illuminated by monochromatic light. If on a screen 2 m away from the slit, the two second-order minima are 3 cm apart, what is the wavelength of the light?

3.5-3. When you are driving at night, the headlight of a motorcycle following you, seen through the foggy rear window of your car, is surrounded by colored rings. Which color is inside, which outside?

3.5-4. When looking at the sun with the sky slightly overcast, a colored ring is seen around the sun that is red outside, and therefore due to diffraction. How can you decide whether the diffraction takes place in the atmosphere or in your eyes?

3.5-5. If white light is passing through a small aperture that has the shape of an equilateral triangle, a regular pentagon, or a regular hexagon, respectively, how many light *fans* will occur because of diffraction?

3.5-6. An opaque mask contains a great many small open rectangles, all of the same shape and orientation but distributed at random. With the mask placed in a parallel beam of monochromatic light, compare the resulting diffraction pattern with that produced by a single rectangle.

3.5-7. Lycopodium seeds, which are of spherical shape and nearly uniform size, are dusted on a glass plate. If with parallel light of 640 nm wavelength the angular radius of the first diffraction maximum is 2°, how large are the seeds?

3.5-8. Seen through a cloudy medium, a light source 5 m away appears surrounded by a ring 50 cm in diameter. If the wavelength of the light is 500 nm, how large are the particles that cause the ring?

3.5-9. If the headlights of an approaching car are 122 cm apart, what is the maximum distance at which the eye can resolve them? Assume pupils 4 mm in diameter and light of 500 nm wavelength.

3.5-10. How large a target can a telescope resolve that has an aperture 3 cm in diameter, is aimed at an object 5 km away, and uses light of 600 nm wavelength?

3.5-11. A diffraction-limited telescope with a 7.6-cm aperture is aimed at a target 12.5 km away. Assuming light of 500 nm wavelength and neglecting air turbulence, what size details can be resolved by the telescope?

3.5-12. **(a)** What is the resolution of a lens that is 25 mm in diameter, has 130.4 cm focal length, and is used with light of 550 nm wavelength?

(b) If the lens is focused for infinity, what is the linear resolution in the image plane?

3.5-13. From the table of Fresnel integrals, determine the relative intensity at a point:

(a) $v = -0.7$ *inside* the shadow.

(b) $v = +1.4$ *outside* the shadow behind a knife edge.

3.5-14. A wire, 0.3 mm in diameter, is placed 50 cm away from a source and an equal distance in front of a screen. If $\lambda = 500$ nm, determine the value of v on Cornu's spiral and the relative intensity at the edge of the geometrical shadow.

3.5-15. What should be the radius of the twentieth boundary of a zone plate that, with light of 500 nm wavelength, has a focal length of 160 cm?

3.5-16. If a zone plate with 100 boundaries, when used with light of 500 nm wavelength, has 2 m focal length, what is the least distance that it can resolve?

3.5-17. (a) What should be the diameter of a zone plate that has 55 zones and can resolve details 10 μm in size?

(b) What wavelength is necessary?

3.5-18. Determine the longitudinal chromatic aberration for two wavelengths, 450 nm and 600 nm, of the zone lens described in Problem 3.5-17.

3.5-19. In the lecture hall we show *Poisson's spot* on a screen 2.25 m away from an obstacle 2 mm in diameter. Why can't we see Poisson's spot also during a solar eclipse? (Angular size of moon $\approx$ $\frac{1}{2}°$.)

3.5-20. A wavefront, which has an amplitude of $\mathbf{A} = 1$ arbitrary unit, is advancing due east. Considering the *obliquity factor*, what is the ratio of the intensities of secondary Huygens' waves traveling west, northwest, north, northeast, and east?

3.6

Diffraction Gratings

A DIFFRACTION GRATING IS AN EXTENSION of Young's double slit and as such is based on both diffraction and interference. In fact, the same equations apply to both double-slit interference and to interference produced by a diffraction grating. It is only that gratings, and their applications, have developed very much into a technology all their own, that I present diffraction gratings in a separate chapter.

THE GRATING EQUATION

In its most elementary form, a diffraction grating consists of multiple parallel, equidistant slits in an opaque screen. As in double-slit interference (see Figure 3.2-2), d is the distance between the centers of any two adjacent slits, θ the angle through which the light is diffracted, m the order, and λ the wavelength. Then

$$\boxed{d \sin \theta = m\lambda} \qquad m = 0, 1, 2, \ldots \qquad \text{[3.6-1]}$$

which is the *grating equation,* the same equation that we had derived for double-slit maxima. If the light is incident on the grating at an angle I, then

$$d(\sin \theta - \sin I) = m\lambda \qquad \text{[3.6-2]}$$

The principal use of diffraction gratings is in spectroscopy. A *grating spectrograph,* as we see from Figure 3.6-1, is similar to a prism spectrograph. The light passes first through a

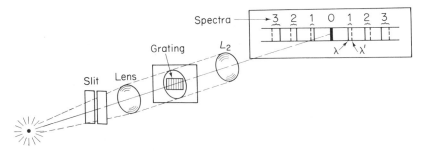

Figure 3.6-1 Schematic diagram of grating spectrograph. Light source is assumed to emit two wavelengths, $\lambda < \lambda'$.

combination of a slit and a collimating lens. It then reaches the grating, set with its rulings parallel to the slit. Another lens, L_2, focuses the light on the screen. Light of different wavelengths, as a consequence of Equation [3.6-1], is diffracted at different angles, θ, and for each order, m, is drawn out into a spectrum, the blues nearer to the optic axis and the reds farther away from it. The zeroth order retains the composite color of the source and therefore is easy to identify.

The number of orders possible is limited by the separation of the slits. A coarse grating, with relatively few slits per unit width, will give many orders. A finely ruled grating may give only one or two orders, but their spectra are spread out much wider than those formed by a coarse grating.

Example

When looking at a hydrogen discharge tube, which emits light of 656 nm wavelength, we see two red lines, to either side of the zeroth order. If at a distance of 90 cm from the grating, the two lines are separated by 62.5 cm, how many lines per millimeter does the grating have?

Solution. First we determine the angle θ that the lines subtend with the optic axis. From the construction in Figure 3.6-2 we find that

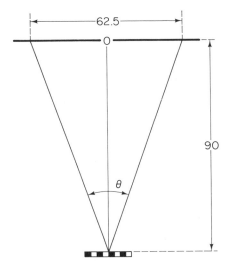

Figure 3.6-2

$$\theta = \arctan \frac{62.5/2}{90} = 19.15°$$

Then we use the grating equation, solve it for d, set $m = 1$, and insert the values given:

$$d = \frac{\lambda}{\sin \theta} = \frac{656 \times 10^{-6} \text{ mm}}{0.328} = 0.002$$

The grating, therefore, has

$$\frac{1}{0.002} = \boxed{500 \text{ lines per millimeter}}$$

Grating theory (continued).   So far we have considered only the separation, d, of the slits. But the slits also have a *finite width, s* (Figure 3.6-3).

This significantly changes the intensity distribution behind the grating. Consider the light that passes through a *single slit*. The path difference between peripheral rays, touching the edges of the slit, from Equation [3.5-2], is

$$\Gamma = s \sin \theta$$

Following Equation [3.3-9], this is equivalent to a phase difference

$$\delta = \frac{2\pi}{\lambda} \Gamma = \frac{2\pi}{\lambda} s \sin \theta \qquad [3.6\text{-}3]$$

The intensity distribution behind a single slit, from Equation [3.5-6], is

$$I_\theta = I_0 \left(\frac{\sin \delta/2}{\delta/2} \right)^2 \qquad [3.6\text{-}4]$$

We now substitute Equation [3.6-3] in [3.6-4],

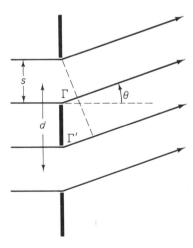

Figure 3.6-3 Diagram of diffraction grating showing slit width, s, and slit separation, d.

$$I_\theta = I_0 \frac{\sin^2\left(\dfrac{\pi}{\lambda} s \sin\theta\right)}{\left(\dfrac{\pi}{\lambda} s \sin\theta\right)^2} \qquad [3.6\text{-}5]$$

which is the *diffraction contribution*. It is due to the *finite width* of the slits.

Now we consider *several, N, consecutive slits*. The path difference between rays that pass through (equivalent points in) adjacent slits, from Equation [3.2-2], is

$$\Gamma' = d \sin\theta$$

and the phase difference

$$\phi = \frac{2\pi}{\lambda} d \sin\theta \qquad [3.6\text{-}6]$$

The intensity distribution behind N slits could be found by integration; however, adding the complex amplitudes is more convenient. The amplitude of the light passing through each slit, from Equation [3.1-11], is

$$\mathbf{A} = Ae^{i\phi}$$

But since, from slit to slit, there is a constant phase difference δ', consecutive amplitudes vary as

$$\mathbf{A}_1 = Ae^{i\delta'}, \qquad \mathbf{A}_2 = Ae^{i2\delta'}, \qquad \mathbf{A}_3 = Ae^{i3\delta'}, \ldots \qquad [3.6\text{-}7]$$

The sum of these amplitudes is that of a *geometric series*. In general,

$$a + ar + ar^2 + ar^3 + \cdots + ar^{N-1} = a\frac{1 - r^N}{1 - r} \qquad [3.6\text{-}8]$$

and therefore,

$$Ae^{i\delta'} = A[1 + e^{i\delta'} + e^{i2\delta'} + \cdots + e^{i(N-1)\delta'}] = A\frac{1 - e^{iN\delta'}}{1 - e^{i\delta'}} \qquad [3.6\text{-}9]$$

To find the intensity, we multiply the amplitude with its complex conjugate. This gives

$$\mathbf{A}^2 = A^2\left[\frac{(1 - e^{iN\delta'})(1 - e^{-iN\delta'})}{(1 - e^{i\delta'})(1 - e^{-i\delta'})}\right] \qquad [3.6\text{-}10]$$

But

$$(1 - e^{iN\delta'})(1 - e^{-iN\delta'}) = 1 - \cos N\delta' \qquad [3.6\text{-}11]$$

and likewise,

$$(1 - e^{i\delta'})(1 - e^{-i\delta'}) = 1 - \cos \delta' \qquad [3.6\text{-}11A]$$

Using the trigonometric relationship

$$1 - \cos\alpha = 2\sin^2(\tfrac{1}{2}\alpha) \qquad [3.6\text{-}12]$$

we obtain

$$\mathsf{A}^2 = A^2 \frac{\sin^2(N\frac{1}{2}\delta')}{\sin^2(\frac{1}{2}\delta')} \qquad\qquad [3.6\text{-}13]$$

and substituting Equation [3.6-6] in [3.6-13], we have

$$I_\theta = I_0 \frac{\sin^2\left(N\dfrac{\pi}{\lambda} d \sin\theta\right)}{\sin^2\left(\dfrac{\pi}{\lambda} d \sin\theta\right)} \qquad\qquad [3.6\text{-}14]$$

which is the *interference contribution*. It is due to the *multiplicity* of slits.

The total, actual pattern behind the grating is found by multiplying Equations [3.6-5] and [3.6-14]:

$$I_\theta = I_0 \frac{\sin^2\left(\dfrac{\pi}{\lambda} s \sin\theta\right)\sin^2\left(N\dfrac{\pi}{\lambda} d \sin\theta\right)}{\left(\dfrac{\pi}{\lambda} s \sin\theta\right)^2 \sin^2\left(\dfrac{\pi}{\lambda} d \sin\theta\right)} \qquad\qquad [3.6\text{-}15]$$

For simplicity, call D the pertinent term $(\pi s \sin\theta)/\lambda$ in the diffraction contribution and I the term $(\pi d \sin\theta)/\lambda$ in the interference contribution. The total pattern produced by the grating is then

$$I_\theta = I_0 \frac{\sin^2 D}{D^2} \frac{\sin^2 NI}{\sin^2 I} \qquad\qquad [3.6\text{-}16]$$

which means that one contribution is being *modulated* by the other contribution.

The numerator, $\sin^2 NI$, in the interference contribution becomes zero whenever

$$NI = k\pi \qquad k = 1, 2, 3, \ldots \qquad\qquad [3.6\text{-}17]$$

The denominator, $\sin^2 I$, similarly, becomes zero whenever

$$I = 0, \pi, 2\pi, \ldots \qquad\qquad [3.6\text{-}18]$$

Since the ratio 0/0 is indeterminate, Equation [3.6-17] is the condition for minima for all values of k *except* where $k = 0, N, 2N, 3N, \ldots, mN$. Substituting $k = mN$ in Equation [3.6-17], and replacing I by $(\pi d \sin\theta)/\lambda$, gives

$$N \frac{\pi d \sin\theta}{\lambda} = mN\pi \qquad\qquad [3.6\text{-}19]$$

This reduces to

$$d \sin\theta = m\lambda \qquad\qquad [3.6\text{-}1]$$

which again is the grating equation. It refers to the *principal maxima*. Between any two adjacent (principal) maxima, there will be $N-1$ loci of zero intensity (see Figure 3.6-4). Between these minima there are *secondary maxima*, but their intensities are much less than those in the principal maxima.

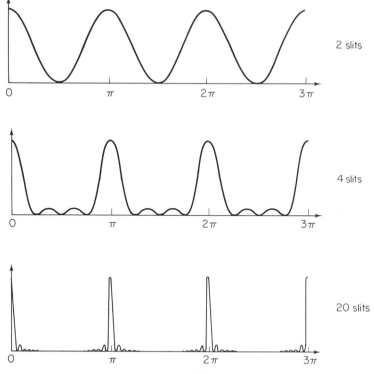

Figure 3.6-4 Intensity distribution behind a grating with 2 slits (*top*), 4 slits (*center*), and 20 slits (*bottom*).

Finally, note the $\sin^2 N$ term in Equation [3.6-16]. It shows that the intensity in the principal maxima is proportional to the *square* of the number of slits. If there are *more* slits, the principal maxima become higher and more narrow and the secondary maxima between them are suppressed; these orders are "missing."

Example

A grating has slits that are 0.15 mm wide and separated, center to center, by 0.6 mm. Which of the higher-order maxima are missing?

Solution.   We divide the grating equation (which relates to interference *maxima*),

$$d \sin \theta = m\lambda$$

by the single-slit equation for *minima*,

$$s \sin \theta = m'\lambda$$

This gives

$$\frac{d}{s} = \frac{m}{m'} = \frac{0.6 \text{ mm}}{0.15 \text{ mm}} = 4$$

Setting m' successively equal to 1, 2, 3, . . . , the missing orders are

$$m = (m')(4) = \boxed{4, 8, 12, \ldots}$$

Resolvance.   The *resolvance* ("resolving power") of a grating describes its ability to separate closely adjacent spectrum lines. First we take the grating equation,

$$d \sin \theta = m\lambda$$

and differentiate θ with respect to λ. Since $\Delta\lambda \ll \lambda$, this gives

$$\frac{d\theta}{d\lambda} = \frac{m}{d \cos \theta} \qquad [3.6\text{-}20]$$

The term $d\theta/d\lambda$ is the *angular dispersion* of the grating; it increases as the order, m, increases. Angular dispersion by itself, however, does not make two closely spaced lines more distinct; that is a matter of *resolvance*.

Next consider that the light contains two wavelengths, λ and $(\lambda + \Delta\lambda)$. Whereas for the first wavelength the grating equation holds,

$$d \sin \theta = m\lambda$$

for the second wavelength it becomes

$$d \sin \theta = m(\lambda + \Delta\lambda) \qquad [3.6\text{-}21]$$

To satisfy Rayleigh's criterion, the second wavelength must be no closer to the first wavelength than the minimum adjacent to λ, in the same order:

$$d \sin \theta = \left(m + \frac{1}{N}\right)\lambda \qquad [3.6\text{-}22]$$

Setting Equations [3.6-22] and [3.6-21] equal by eliminating $d \sin \theta$ gives

$$\left(m + \frac{1}{N}\right)\lambda = m(\lambda + \Delta\lambda)$$

and therefore,

$$\boxed{\frac{\lambda}{\Delta\lambda} = Nm} \qquad [3.6\text{-}23]$$

which is the resolvance of the grating.

We note that the resolvance, $\lambda/\Delta\lambda$, is proportional to the number of lines, N, in the grating, and to the order, m, in which it is used. To obtain higher resolution, we may either use a grating with more lines or go to a higher order. The width of the grating is immaterial. In practice, grating lines are ruled very close to each other in order to keep the aperture of the system to a reasonable size.

Example

The sodium D doublet has wavelengths of 589 nm and 589.6 nm.
(a) If only a grating with 400 rulings is available, what is the lowest order possible in which the D lines are resolved?
(b) How wide does the grating have to be?

Solution.   (a) From Equation [3.6-23],

$$m = \frac{\lambda}{\Delta\lambda N} = \frac{589 \times 10^{-9}}{(0.6 \times 10^{-9})(4 \times 10^2)} = 2.45$$

But orders must be *whole numbers* and therefore the lowest possible order in which the D lines are resolved is

$$m = \boxed{3}$$

(b) This is $\boxed{\text{immaterial.}}$

Types of gratings.   We distinguish *transmission gratings* (which let the light pass through, as in Figures 3.6-1 and 3.6-2) from *reflection gratings* (which reflect it; Figures 3.6-5 and 3.6-6). Either type can be made with more than 1000 lines per millimeter. In contrast, *Ronchi grids** have only a few lines per millimeter. Such grids are used to obtain moiré patterns, to be discussed soon, and are based on geometric optics; the wavelength plays no role.

A grating with completely opaque bars and clear intervals is called an *amplitude grating*. If the bars do not block the light but merely retard its phase, we have a *phase grating*. If the light distribution across the bars and intervals is sinusoidal, rather than "square-wave," all the light is diffracted into the two first-order spectra.

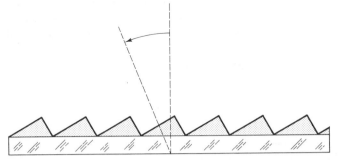

Figure 3.6-5 Profile of a blazed reflection grating. Angle shown is *blaze angle.*

---

*Named after Vasco Ronchi (1897–   ), Italian physicist, long-time director of the Istituto Nazionale di Ottica in Florence. After graduating from the University of Pisa, Ronchi devoted his life to optics, writing 20 books and more than 700 papers on various topics, most notably optical testing and interferometry, the history of science, and physiological/psychological aspects of the design of optical systems. When asked why he worked so intensively, Ronchi replied: "I don't do it for glory or for remuneration or for personal prestige, I do it because I enjoy it. I only regret that I can't work more, to do things which only I can do."

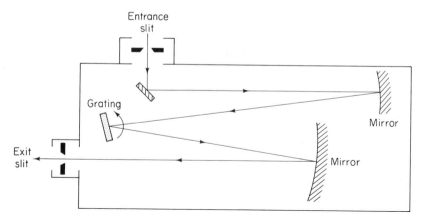

Figure 3.6-6 Czerny–Turner mount: diffraction grating and two concave mirrors.

Actually, there are no "slits" or "lines" in a grating. Modern gratings are *blazed*. First a plate of low-expansion glass is coated with a suitable metal such as aluminum or gold. Then the coating is ruled with a diamond tool, moved across in repetitive, parallel, equidistant strokes and with just enough pressure to make the metal flow aside. After the diamond has passed, the metal hardens into the desired profile (Figure 3.6-5). Whereas conventional gratings direct most of the light into the zeroth order, the advantage of blazed gratings is that they direct much more light into some of the higher orders, making more efficient use of the light available.

Gratings can be made with high precision, in sizes up to 50×75 cm^2, using interferometrically and electronically controlled ruling engines. From original *masters,* plastic *replicas* can be produced. But even the best gratings sometimes have periodic errors that give rise to spurious lines in the spectrum, called *ghosts*.

Like prisms, diffraction gratings can be mounted between a collimating lens and a focusing lens. If the grating is of the reflecting type, one lens is sufficient, as in the *Littrow mount* (not shown). No lens at all is required with a *concave reflection grating*. But curved gratings are difficult to rule. So, after all, it is better to use a plane grating, often together with two concave mirrors as in the *Czerny–Turner mount* shown in Figure 3.6-6.

THREE-DIMENSIONAL GRATINGS

The gratings discussed so far are one-dimensional. A two-dimensional or *cross-grating* results if two gratings are laid on top of each other, their rulings oriented at right angles. A zone plate could also be considered a two-dimensional grating, of rotational symmetry (and space-varying frequency). I will refer again to two-dimensional gratings in Chapter 4.5. For now, we proceed to *three-dimensional gratings*.

X-ray diffraction. For a while after Röntgen's discovery of X rays it was not known whether X rays were particles (like cathode rays) or waves of exceedingly short length. It occurred to Max von Laue that a crystal may act on X rays in a way similar to a grating on

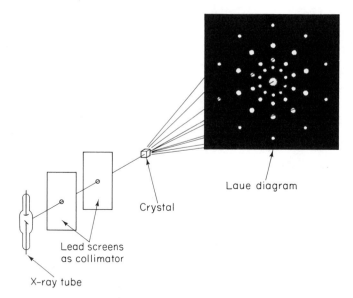

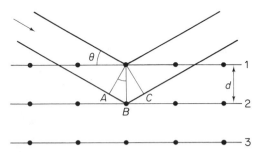

Laue diagram

Crystal

Lead screens
as collimator

X-ray tube

Figure 3.6-7 Laue diagram (top right)
produced by X rays diffracted in a crystal.

light. Together with W. Friedrich and P. Knipping, von Laue tried crystals of zinc sulfide
and copper sulfate. Both gave fine X-ray diffraction patterns, called *Laue diagrams* (Figure
3.6-7), which established the wave nature of X rays.*

A crystal is a three-dimensional lattice of atoms, or groups of atoms, that is built out of
repeating fundamental units of structure, called *unit cells*. X-ray diffraction on such units
may be visualized as reflections (Figure 3.6-8). Let a beam of monochromatic X rays incident on a crystal make an angle θ with the surface.† Some of the incident rays are reflected at
the top layer, 1, but other rays penetrate to the deeper layers, 2, 3, and so on. The planes
have an interplanar spacing d. If the wave reflected at 1 is reinforced by the wave reflected at
2, the path difference Γ between the two rays must be $\Gamma = AB + BC = \lambda$ or it must be a
multiple of λ, $\Gamma = m\lambda$.

Figure 3.6-8 Deriving Bragg's law.

---

*Max Theodor Felix von Laue (1879–1960), German physicist, professor of theoretical physics and director of
the Kaiser Wilhelm Institute, the present Max Planck Institute, in Berlin. W. Friedrich, P. Knipping, and M. von
Laue, "Interferenzerscheinungen bei Röntgenstrahlen," *Ann. Phys.* (4) **41** (1913), 971–88. In 1914, von Laue received the Nobel Prize in physics.

†Note that angles in X-ray crystallography are measured from the surface and not from the surface *normal* as
elsewhere in optics.

Since

$$\sin \theta = \frac{AB}{d}$$

$$AB = BC = d \sin \theta$$

Therefore, in order to produce a maximum, the path difference,

$$\Gamma = 2d \sin \theta \qquad\qquad [3.6\text{-}24]$$

must be equal to an integral multiple of a wavelength,

$$\boxed{2d \sin \theta = m\lambda} \qquad\qquad [3.6\text{-}25]$$

which is *Bragg's law.** The same relationship holds if the X rays are transmitted through the crystal, rather than reflected by it.

MOIRÉ FRINGES

When two window screens, or other sets of repetitive lines, are laid on top of one another, intersection of their lines produces another sequence of lines, called *moiré fringes.*†

There are several ways of producing such fringes. One way is to use two Ronchi rulings held in contact. Assume that the separation of the lines in the two original grids is d, and that one of the grids has been turned through an angle θ with respect to the other. Then from the construction in Figure 3.6-9 it follows that

$$\sin \frac{\theta}{2} = \frac{d/2}{M}$$

---

*Named after (William) Lawrence Bragg (1890-1971), British crystallographer, the son of William (Henry) Bragg (1862-1942). When in 1904 the older Bragg, at that time professor of mathematics and physics at Adelaide, Australia, and president of Section A of the Australasian Association for the Advancement of Science, had to deliver the presidential address in Dunedin, New Zealand, he choose radioactivity as his subject. He was intrigued by Marie Curie's results but, in his meticulous way, felt that some more experiments were needed; these established his scientific reputation. Four years later he returned to England. When he and his son saw von Laue's photographs, the younger Bragg, still a student at Cambridge, thought that perhaps the Laue spots could be explained by diffraction, established the law named after him, and thus laid the foundations of X-ray crystallography. He succeeded Rutherford as head of the Cavendish Laboratory and, like his father before him, became president of the Royal Institution. Father and son shared the Nobel Prize in physics in 1915. W. L. Bragg and W. H. Bragg, "The Reflection of X-Rays by Crystals," *Proc. Roy. Soc. (London)* **88** (1913), 428.

†The term "moiré" comes from the French; it refers to the wavy finish of textiles made from the wool of angora goats, in English it is "mohair." The moiré effect in optics was first noted by Lord Rayleigh, "On the Manufacture and Theory of Diffraction-gratings," *Philos. Mag.* (4) **47** (1874), 81–93. See also Y. Nishijima and G. Oster, "Moiré Patterns: Their Application to Refractive Index and Refractive Index Gradient Measurements," *J. Opt. Soc. Am.* **54** (1964), 1–5; and G. Oster and Y. Nishijima, "Moiré Patterns," *Sci. Am.* **208** (May 1963), 54–63.

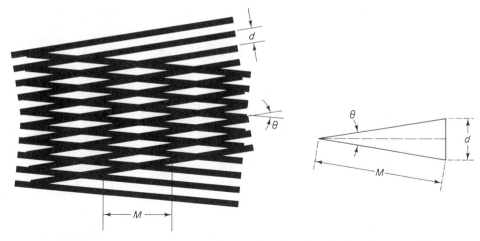

Figure 3.6-9 Moiré pattern formed by two grids of spacing d, subtending an angle θ. Distance M is spacing of moiré fringes.

where M is the spacing of the moiré fringes. Solving for M gives

$$M = \frac{d}{2\sin(\theta/2)}$$ [3.6-26]

which shows that the more parallel the two grids, the wider the spacing of these fringes.

No such rotation is needed in *moiré topography*. Moiré topograms can be obtained either by placing a single grid next to a (preferably white) surface and observing or photographing the shadow cast on the surface through the same grid. The shadow, and the grid through which the shadow is seen, superimpose, producing the moiré effect. That is called the *shadow method*.

Or the image of one grid can be projected onto the surface and the image be observed through another grid. That is called the *projection method;* it is illustrated in Figure 3.6-10.

Example

Fringes obtained by either the shadow method or the projection method are easy enough to see, but how can we use them to draw quantitative conclusions? Assume that the test object is a sphere 16 mm in diameter, its curvature comparable to that of the cornea of the eye, with a small defect in its surface. On a photographic print we see that the defect has caused some of the fringes to become distorted by $\frac{1}{5}$ of their spacing. How deep is the defect?

Solution. First we measure the diameter of the sphere as it appears on the print. If that diameter is 36 mm, the print is magnified $\frac{36}{16}$ times. Then we measure the spacing of the fringes. If the spacing is 2.25 mm, the fringes (on the actual sphere) are separated by

$$2.25 \times \tfrac{16}{36} = 1 \text{ mm}$$

Therefore, the depth of the defect is

$$1 \times \tfrac{1}{5} = \boxed{0.2 \text{ mm}}$$

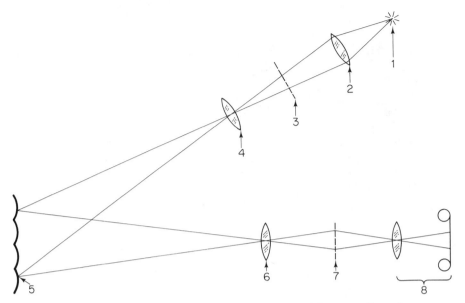

Figure 3.6-10 Moiré topography, projection method. 1, Light source; 2, condenser lens; 3, first grid; 4, projection lens; 5, surface; 6, relay lens; 7, second grid; 8, camera.

well within the accuracy of the method (estimated to be 0.01 mm). To determine the depth of a defect, in short, we need a mere four measurements, three of them taken on a photograph, without knowing the dimensions of the actual experiment, the magnification, or the rotation of one grid relative to the other.

SUGGESTIONS FOR FURTHER READING

S. P. Davis, *Diffraction Grating Spectrographs* (New York: Holt, Rinehart and Winston, 1970).

M. C. Hutley, *Diffraction Gratings* (New York: Academic Press, Inc., 1982).

V. Ronchi, "Forty Years of History of a Grating Interferometer," *Appl. Opt.* **3** (1964), 437–51.

P. S. Theocaris, *Moiré Fringes in Strain Analysis* (Oxford: Pergamon Press Ltd., 1969).

G. Oster, "Optical Art," *Appl. Opt.* **4** (1965), 1359–69.

PROBLEMS

3.6-1. A grating, used in the second order, diffracts light of 400 nm wavelength through an angle of 30°. How many lines per millimeter does the grating have?

3.6-2. A grating has 8000 slits ruled across a width of 4 cm. What is the color of the light whose two fifth-order maxima are 90° apart?

3.6-3. As seen through a diffraction grating, red light of 632 nm wavelength is displaced 20 cm from the center of a meter stick mounted 60 cm in front of the grating. How many lines per millimeter does the grating have?

3.6-4. Blue light of wavelength 470 nm is diffracted by a grating ruled with 500 lines per millimeter. Considering that no light can be diffracted more than 90° away from the axis, what is the highest order possible?

3.6-5. If you compare photographs of a certain line spectrum, one taken with a prism spectrograph and the other with a grating spectrograph, how can you tell which is which?

3.6-6. A transmission grating is mounted on the inner surface of a glass tank and illuminated with parallel light from a high-intensity light source. The spectra are observed on a sheet of cardboard mounted on the opposite side. Do the spectra change, and if so, how do they change, when the tank is filled with water?

3.6-7. Collimated light containing the wavelengths 600 nm and 610 nm is diffracted by a plane grating ruled with 60 lines to the millimeter. If a lens of 2 m focal length is used to focus the light on a screen, what is the linear distance between these two lines in the first order?

3.6-8. A spectrum of white light is obtained with a grating ruled with 2500 lines/cm. What is the angular separation between the green (520 nm) and red (630 nm) in the second order?

3.6-9. The spectrum of mercury contains, among others, a blue line of 435.8 nm and a green line of 546.1 nm. If these two lines are to have an angular separation of at least 7°, and if the grating available has 500 lines/mm, in which order must the grating be used?

3.6-10. Red light of 630 nm wavelength, diffracted by a grating in a given order, overlaps blue light of 475 nm wavelength, diffracted by the same grating in the next-higher order. What are the two orders?

3.6-11. Two slits 0.05 mm wide are moved apart until the third maximum produced by interference from the two slits on a screen 2 m away coincides with the first minimum in the diffraction pattern produced by either one of the slits. These lines of coincidence are 1.8 cm on each side from the zeroth-order maximum. What is the wavelength of the light, and what is the separation of the slits?

3.6-12. A grating is mounted at distance b from a meter stick on which a certain spectrum line is observed at distance a from the center (Figure 3.6-11). Derive a general equation for the wavelength of the light and check with the example on page 271.

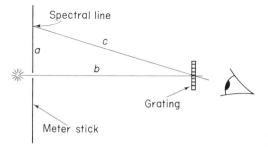

Figure 3.6-11

3.6-13. What is the least separation between wavelengths that can be resolved near 640 nm in the second order, using a diffraction grating that is 5 cm wide and ruled with 32 lines per millimeter?

3.6-14. How many lines must a grating have, used in the second order and near 550 nm, to resolve two lines 0.1 Å apart?

3.6-15. X rays of wavelength $\lambda = 1.3$ Å are diffracted in a Bragg spectrograph at an angle, in the first order, of 22°. What is the interplanar spacing of the crystal?

3.6-16. How far apart are the diffracting planes in a NaCl crystal for which X rays of wavelength 1.54 Å in the first order make a glancing angle of 15.9°?

3.6-17. Two identical grids are laid on top of each other. What angle must one grid subtend with the other so that moiré lines result that have twice the spacing in the original grids?

3.6-18. Continuing with Problem 3.6-17, what angle will give 13 times the spacing?

Light
as an Electromagnetic
Phenomenon

Virtually all interference can be explained by the wave nature of light. Early in the history of physics, light waves were thought to be longitudinal, the same as sound waves that already were known to be longitudinal. But polarization experiments showed that they are transverse. Moreover, light waves consist of varying electric and magnetic "fields." These fields are scattered by molecules as well as by larger particles, they can be polarized, and can be manipulated by various types of transformations important in optical data processing and holography.

MAXWELL'S EQUATIONS

Light waves are often compared to mechanical waves. But light is fundamentally different. For example, light requires no medium for propagation, a fact that for a while posed much difficulty when trying to explain the nature of light. The problem was solved by Clerk Maxwell, who showed that light waves are not mechanical but an electromagnetic phenomenon that can be described by two vectors, the amplitude of the electric field strength, **E**, and the amplitude of the magnetic field strength, **H**. These vectors oscillate at right angles to each other and to the direction of propagation (Figure 4.1-1). They cannot be separated. They can be expressed in the form of four fundamental equations known as *Maxwell's equations.* *

---

*James Clerk Maxwell (1831–1879), Scottish mathematician and physicist. At age 15, Clerk Maxwell (his last name!) presented a paper "On the Description of Oval Curves and Those Having a Plurality of Foci" before the Royal

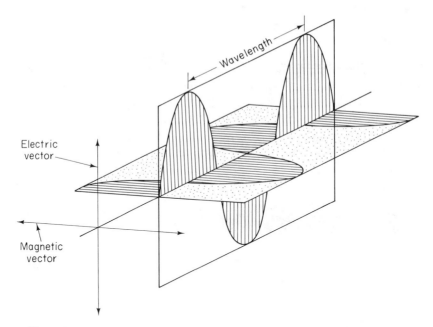

Figure 4.1-1   Electromagnetic wave represented by electric and magnetic field vectors.

Maxwell's first equation can be derived from *Coulomb's law.* Coulomb's law refers to the force, **F**, between two electric charges *at rest,*

$$\mathbf{F} = \frac{1}{4\pi\epsilon_0} \frac{q_1 q_2}{R^2} \hat{\mathbf{R}}$$ [4.1-1]

Here q_1 and q_2 are the charges, separated by a distance R, $\hat{\mathbf{R}}$ is a unit vector indicating the direction from one charge to the other, and ϵ_0 is the electric permittivity of free space, approximately $8.85 \times 10^{-12} \, C^2 \, N^{-1} \, m^{-2}$.

Society, at 16 became interested in optics when he had the chance of visiting Nicol, who gave him a pair of polarizing prisms. Using these, Clerk Maxwell went to work on photoelasticity, made major contributions also to refraction in nonhomogeneous media, color vision, gas dynamics, heat transfer, the theory of servomechanisms (governors), and correctly explained the rings of Saturn. In 1864 he read a paper before the Royal Society, published a year later: J. Clerk Maxwell, "A Dynamical Theory of the Electromagnetic Field," *Philos. Trans. Roy. Soc. London* **155** (1865), 459–512. Near the end of part I, on page 466, Clerk Maxwell says that "light itself . . . is an electromagnetic disturbance in the form of waves propagated through the electromagnetic field according to electromagnetic laws."

Some 22 years later, in 1887, Heinrich Rudolf Hertz (1857–1894), German physicist, professor of physics at the Technische Hochschule in Karlsruhe, found the waves that Clerk Maxwell had predicted. During one of his lecture demonstrations on electromagnetic induction, Hertz was surprised to see that the spark produced by the discharge of a Leyden jar could be detected some distance away. The detector he used was a loop of wire with a small gap in it: discharge of the Leyden jar caused a tiny spark to jump across that gap. Since the distance was too large for direct induction, the only alternative seemed to be that the secondary spark was produced by some kind of radiation propagating through space. With a setup as simple as this, Hertz found that the waves were transverse, that their velocity was the same as that of light, and that they could be reflected, refracted, polarized, and brought to interference, just like light waves. In 1901, Guglielmo Marconi (1874–1937), Italian scientist and businessman, transmitted them across the Atlantic Ocean. H. Hertz, *Wiedemann's Ann. Phys.* **31** (1887), 421.

From Coulomb's law we derive *Gauss' law.* It describes the relationship between the electric flux through a closed surface and the charge enclosed by the surface. Expressed in differential form, Gauss' law is in fact *Maxwell's first equation,*

$$\nabla \cdot \mathbf{E} = \frac{\rho}{\epsilon_0} \qquad [4.1\text{-}2]$$

where $\nabla \cdot$ is the divergence operator, $\mathbf{E}$ the electric field strength, and ρ the charge density.

Next we consider *moving charges,* that is, a *current.* A current, i, in an element of wire dW, will set up a *magnetic induction field,* $\mathbf{B}$. As before, R is the distance from dW to the test point (Figure 4.1-2).

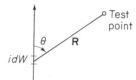

Figure 4.1-2   Measuring the magnetic field produced by a current.

The moving charge per wire element, $i\,dW$, makes a contribution dB,

$$dB = \mu_0 \frac{1}{4\pi} \frac{1}{R^2} i\,dW \sin\theta$$

or, in vector notation,

$$d\mathbf{B} = \mu_0 \frac{1}{4\pi} \frac{1}{R^2} i\,d\mathbf{W} \times \hat{\mathbf{R}} \qquad [4.1\text{-}3]$$

The total induction, $\mathbf{B}$, is found by integration:

$$\mathbf{B} = \mu_0 \frac{1}{4\pi} i \int \frac{1}{R^2} d\mathbf{W} \times \hat{\mathbf{R}} \qquad [4.1\text{-}4]$$

This is *Biot–Savart's law* where μ_0 is the magnetic permeability of free space, $4\pi \times 10^{-7}$ Wb A^{-1} m^{-1}. The direction of $\mathbf{B}$ is given by the vector cross product $d\mathbf{W} \times \mathbf{R}$; it points in a direction *normal* to the plane defined by $\mathbf{W}$ and $\mathbf{R}$.

In analogy to ρ, we replace the current i by the current density, $\mathbf{j} = i/A$. If $\mathbf{j}$ is constant, $\nabla \cdot \mathbf{j} = 0$, then all that is left is $\mathbf{j} \cdot \nabla \times (\hat{\mathbf{R}}/R)$, which becomes zero, so that

$$\nabla \cdot \mathbf{B} = 0 \qquad [4.1\text{-}5]$$

which is Maxwell's second equation.

While Maxwell's first equation deals with a *steady* electric field, and the second equation with a steady magnetic field, we now come to *fields that vary as a function of time.* Assume that a bar magnet is thrust through a loop of wire (see footnote, page 318). This will cause an electromotive force, EMF, to occur in the wire. Its magnitude is proportional to the time rate of change of the magnetic flux, ϕ,

$$\text{EMF} = -\frac{d\phi}{dt} \qquad [4.1\text{-}6]$$

which is *Faraday's law of induced electricity.* From Faraday's law we can derive Maxwell's third equation,

$$\nabla \times \mathbf{E} = -\frac{\partial \mathbf{B}}{\partial t} \qquad [4.1\text{-}7]$$

which means, the same as Faraday's law, that whenever a magnetic field changes with time, it generates an electric field.

Maxwell's fourth equation is the final link. It is the reverse of Faraday's law, stating that a changing electric field generates a magnetic field.

Maxwell's fourth equation can be derived from Biot–Savart's law, extending it to fluctuating currents. Consider a capacitor with charge flowing in that then accumulates on one of the plates. The plate, as shown in Figure 4.1-3, is surrounded by a closed surface. Now, while the capacitor is being charged, current *i* is going in and apparently none is coming out but, to assure continuity, Clerk Maxwell reasoned that the same amount of *displacement current* must come out. That leads to Maxwell's fourth equation,

$$\nabla \times \mathbf{B} = \mu_0 \left(\mathbf{j} + \epsilon_0 \frac{\partial \mathbf{E}}{\partial t} \right) \qquad [4.1\text{-}8]$$

Clearly, displacement current exists only whenever the electric field *changes* ($\partial \mathbf{E}/\partial t$); for steady currents, the last term drops out.

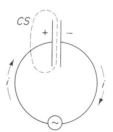

Figure 4.1-3 Deriving concept of displacement current.

In conclusion, Maxwell's first equation, as well as Coulomb's law, show that a charge distribution generates a steady electric field. Maxwell's second equation, and Biot–Savart's law, refer to a steady magnetic field. His third equation, and Faraday's law, tell us that an electric field can also be generated by a time-varying magnetic field. And Maxwell's fourth equation extends to a time-varying electric field.

From the vector cross products in Maxwell's third and fourth equations we also see that the two fields are normal to each other and, as we will find out shortly, normal to the direction of propagation. Most significantly, the electric field arises directly from the (time-varying) magnetic field, and the magnetic field directly from the (time-varying) electric field. No loop of wire, or *any other medium,* is necessary: *Electromagnetic waves propagate through space entirely on their own.*

It should be easy to assume that electromagnetic waves carry energy: consider only that light, even after it has traveled from a distant star to the eye of the observer, still can initiate the process of vision. The *Poynting vector of energy flow,* **S**, is given by the cross product of the electric field **E** and the magnetic field **H**,

$$\mathbf{S} = \mathbf{E} \times \mathbf{H} \qquad [4.1\text{-}9]$$

Presumably, the total energy contained in an electromagnetic wave is equally shared by the two fields; however, the **E** field seems to be more effective than the **H** field. Most phenomena caused by light, from the photoelectric effect to the dissociation of silver halides in photographic film to the response of the retina, are due to the **E** field; the only exception is probably the *pressure* exerted by electromagnetic radiation.

The *velocity of propagation* of electromagnetic energy in free space is

$$c = \frac{1}{\sqrt{\epsilon_0 \mu_0}}$$

[4.1-10]

an important relationship that defines the *velocity of light in terms of two electrical constants*.

FRESNEL'S EQUATIONS

The term *Fresnel's equations* refers to a set of four equations which describe the reflection, transmission, and state of polarization of radiation at a boundary between two media. Assume that the electric field, **E**, oscillates in one of two directions, either *parallel* to the plane of incidence (as in Figure 4.1-4), which is called p polarization (for parallel) and the field given the subscript $\parallel$, or *normal* to the plane of incidence, which is called s polarization (for the German "senkrecht," meaning perpendicular) and the field given the subscript $\perp$. In addition, the incident light is given the subscript i, the reflected light r, and the transmitted (refracted) light t.

Hence, there are four coefficients, two *amplitude reflection coefficients*, $r_\parallel$ and $r_\perp$, and two *amplitude transmission coefficients*, $t_\parallel$ and $t_\perp$. The squares of these coefficients are

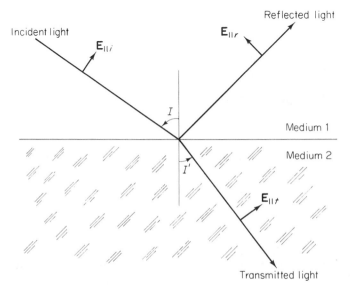

Figure 4.1-4 Incident, reflected, and transmitted light oscillating *parallel* to the plane of the paper.

the *reflectivities*, $R_{\parallel}$ and $R_{\perp}$, and the *transmissivities*, $T_{\parallel}$ and $T_{\perp}$, respectively.*

Next we assume that on crossing the boundary there is no loss of energy and, there-fore, the Poynting vector of the incident light is equal to the sum of the Poynting vectors of the reflected and the transmitted light:

$$\mathbf{S}_i = \mathbf{S}_r + \mathbf{S}_t \qquad [4.1\text{-}11]$$

Vectors $\mathbf{S}_i$ and $\mathbf{S}_r$ are in medium 1 (index n_1) and vector $\mathbf{S}_t$ is in medium 2 (n_2); thus, since energy is proportional to the square of the amplitude,

$$n_1 \mathbf{E}_i^2 = n_1 \mathbf{E}_r^2 + n_2 \mathbf{E}_t^2$$

$$n_1 (\mathbf{E}_i^2 - \mathbf{E}_r^2) = n_2 \mathbf{E}_t^2$$

and, in terms of cross sections of the individual bundles,

$$n_1 (\mathbf{E}_i^2 - \mathbf{E}_r^2) \cos I = n_2 \mathbf{E}_t^2 \cos I' \qquad [4.1\text{-}12]$$

Now consider separately the $\parallel$ and $\perp$ components. First we replace the term in paren-theses on the left-hand side of Equation [4.1-12] by $(\mathbf{E}_i + \mathbf{E}_r)(\mathbf{E}_i - \mathbf{E}_r)$. For the $\parallel$ compo-nents to be equal and continuous across the boundary, we divide Equation [4.1-12] by

$$\mathbf{E}_{\parallel i} - \mathbf{E}_{\parallel r} = \mathbf{E}_{\parallel t}$$

This gives

$$n_1 (\mathbf{E}_{\parallel i} + \mathbf{E}_{\parallel r}) \cos I = n_2 \mathbf{E}_{\parallel t} \cos I' \qquad [4.1\text{-}13]$$

We then eliminate $\mathbf{E}_{\parallel t}$, and use Snell's law to eliminate n_2/n_1:

$$\left(\frac{\mathbf{E}_r}{\mathbf{E}_i} \right)_{\parallel} = \frac{\sin I \cos I - \sin I' \cos I'}{\sin I \cos I + \sin I' \cos I'}$$

Thus, the amplitude reflection coefficient for parallel light is

$$r_{\parallel} = \left(\frac{\mathbf{E}_r}{\mathbf{E}_i} \right)_{\parallel} = \frac{\tan(I - I')}{\tan(I + I')} \qquad [4.1\text{-}14]$$

This is *Fresnel's first equation.*

If instead we eliminate $\mathbf{E}_{\parallel r}$ in Equation [4.1-13], we obtain

$$\left(\frac{\mathbf{E}_t}{\mathbf{E}_i} \right)_{\parallel} = \frac{2 \cos I \sin I'}{\sin I \cos I + \sin I' \cos I'}$$

and

$$t_{\parallel} = \left(\frac{\mathbf{E}_t}{\mathbf{E}_i} \right)_{\parallel} = \frac{2 \cos I \sin I'}{\sin(I + I') \cos(I - I')} \qquad [4.1\text{-}15]$$

which is *Fresnel's second equation.*

---

*Terms ending in -*ion,* such as *reflection,* describe a process. Words ending in -*ivity,* such as *reflectivity,* refer to the general property of a material. Terms ending in -*ance,* such as *reflectance,* refer to the properties of a given object.

If we divide Equation [4.1-12] by

$$\mathbf{E}_{\perp i} - \mathbf{E}_{\perp r} = \mathbf{E}_{\perp t}$$

we obtain

$$n_1(\mathbf{E}_{\perp i} - \mathbf{E}_{\perp r})\cos I = n_2\mathbf{E}_{\perp t}\cos I'$$

Eliminating $\mathbf{E}_{\perp r}$ yields

$$\left(\frac{\mathbf{E}_r}{\mathbf{E}_i}\right)_{\perp} = \frac{\sin I \cos I' - \sin I' \cos I}{\sin I \cos I' - \sin I' \cos I}$$

and

$$r_{\perp} = \left(\frac{\mathbf{E}_r}{\mathbf{E}_i}\right)_{\perp} = -\frac{\sin(I - I')}{\sin(I + I')} \qquad \text{[4.1-16]}$$

This is *Fresnel's third equation.*

If we eliminate $\mathbf{E}_{\perp r}$, we obtain

$$\left(\frac{\mathbf{E}_t}{\mathbf{E}_i}\right)_{\perp} = \frac{2 \cos I \sin I'}{\cos I \sin I' + \sin I \cos I'}$$

and

$$t_{\perp} = \left(\frac{\mathbf{E}_t}{\mathbf{E}_i}\right)_{\perp} = \frac{2 \cos I \sin I'}{\sin(I + I')} \qquad \text{[4.1-17]}$$

which is *Fresnel's fourth equation.*

Now consider the *reflectivities.* At *normal incidence,* there is no difference between $\parallel$ and $\perp$ light. But we cannot simply set $I = 0$, because substituting this in Fresnel's first or third equation would give an indeterminate result. Instead, we combine Equations [4.1-14] and [4.1-16],

$$\left(\frac{\mathbf{E}_r}{\mathbf{E}_i}\right)_{\parallel} = \frac{\tan(I - I')}{\tan(I + I')} = -\left(\frac{\mathbf{E}_r}{\mathbf{E}_i}\right)_{\perp} = \frac{\sin(I - I')}{\sin(I + I')}$$

$$= \frac{\sin I \cos I' - \cos I \sin I'}{\sin I \cos I' + \cos I \sin I'}$$

Then, dividing both numerator and denominator by $\sin I'$ and, applying Snell's law, replacing $\sin I/\sin I'$ by n_2/n_1, we obtain

$$\frac{\mathbf{E}_r}{\mathbf{E}_i} = \frac{n_2 \cos I' - n_1 \cos I}{n_2 \cos I' + n_1 \cos I} \qquad \text{[4.1-18]}$$

At normal incidence, $I = I' = 0$, the cosine terms drop out and the ratio $\mathbf{E}_r/\mathbf{E}_i$ squared is the percentage of radiation reflected at the boundary between two media, the *reflectivity, R,*

$$\boxed{R = \left(\frac{n_2 - n_1}{n_2 + n_1}\right)^2} \qquad \text{[4.1-19]}$$

But how does the *reflectivity change as a function of the angle of incidence?* We distinguish between light incident from the side of the rarer medium (as in Figure 4.1-5, left) and light incident from the side of the denser medium (right). In *external* reflection (where $n_1 < n_2$), we find, if we square Fresnel's third equation, that the reflectivity of the ⊥ light gradually increases until at grazing incidence it reaches 100 percent.

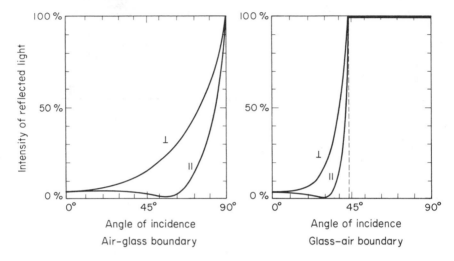

Figure 4.1-5   Reflectivity as a function of the angle of incidence.

The reflectivity of the ‖ light is determined by Fresnel's first equation. As the incident light reaches an angle such that $I + I' = 90°$, the denominator in Equation [4.1-14] becomes infinite; thus, at that angle no ‖ light is being reflected (see the dip to zero in the left-hand plot) and consequently, all of the reflected light is ⊥ light. That occurs at *Brewster's angle*.

In *internal* reflection the ⊥ reflectivity again gradually increases. However, it reaches 100 percent sooner, at the minimum angle of total internal reflection. The ‖ light, likewise, reaches the Brewster angle sooner (Figure 4.1-5, right).

Finally, we note that Fresnel's third equation has a minus sign. This means that in external reflection there is a phase change of 180° or π radians (as we have already seen in Figure 3.2-9, page 215). In internal reflection there is none.

SUGGESTIONS FOR FURTHER READING

F. L. PEDROTTI and L. S. PEDROTTI, *Introduction to Optics* (Englewood Cliffs, NJ: Prentice-Hall, Inc., 1987).

R. K. WANGSNESS, *Electromagnetic Fields* (New York: John Wiley & Sons, Inc., 1986).

PROBLEMS

4.1-1. (a) What is the force between two small spheres separated by 2 cm, one sphere being charged to $+8$ nC, the other to -4 nC?

(b) What is the force if the spheres are first brought into contact and then are separated and moved 2 cm apart?

4.1-2. A long straight wire carries a steady current of 6 A. Using Biot–Savart's law, determine the magnetic induction caused by the current within 1 cm length of wire, at a point 20 cm from the wire.

4.1-3. Since the magnetic permeability of free space is *defined* as $4\pi \times 10^{-7}$ Wb A^{-1} m^{-1}, take the velocity of light, using Equation [1.1-3], and determine the electric permittivity.

4.1-4. A point light source is placed:

(a) In the focus of a collimating lens.

(b) In the focus of a paraboloid mirror.

(c) In one focus of a closed ellipsoid cavity, concentrating the light into the other focus.

Does the system, because of radiation pressure, experience any recoil?

4.1-5. Linearly polarized light is incident at $30°$ on a boundary between glass ($n = 1.5$) and air. Determine the ratio of the amplitudes, and energies, of the reflected light to those of the incident light, assuming that the **E** fields are oriented normal to the plane of incidence.

4.1-6. Light linearly polarized in the p orientation is internally reflected at an angle of $35°$ at the boundary between heavy flint ($n = 1.72$) and air ($n = 1.00$). Find the amplitude reflection coefficient and indicate whether or not there will be a phase change.

4.1-7. Calculate Brewster's angle of maximum polarization for *internal* reflection at a surface between glass ($n = 1.52$) and oil ($n = 1.81$).

4.1-8. What is the least angle at which a π phase change occurs in linearly polarized light, the **E** vector oscillating parallel to the plane of incidence and the light internally reflected at a boundary between crown ($n = 1.52$) and carbon disulfide ($n = 1.63$)?

4.1-9. When light strikes a water surface ($n = \frac{4}{3}$) covered with a film of oil ($n = 1.65$), how much light is lost due to reflection on both sides of the oil?

4.1-10. If two mirrors of 99% reflectance each are facing each other, and if a beam of light is reflected 50 times back and forth between the mirrors ($= 100$ reflections), what percentage of the light is lost?

4.1-11. (a) If the refractive index n_2 in Figure 4.1-6 is higher than both n_1 and n_3, or, in shorthand notation, if $1 < 2 > 3$, and if the light is incident from the left, at which of the two boundaries, 1–2 and 2–3, will a 180° phase change occur on reflection?

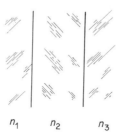

n_1 n_2 n_3 **Figure 4.1-6**

(b) At which boundaries will a phase change occur if 1 < 2 < 3?

(c) At which boundaries will a phase change occur if 1 > 2 < 3?

(d) At which boundaries will a phase change occur if 1 > 2 > 3?

4.1-12. Prepare a table listing (1) external reflection and (2) internal reflection, each with respect to p- and s-oriented linearly polarized light, and indicate whether or not there is a π phase change for angles:

(a) Less than Brewster's angle.

(b) Larger than Brewster's angle.

4.2

Light Scattering

"IT IS NOW, I BELIEVE, GENERALLY ADMITTED that the light which we receive from the clear sky is due in one way or another to small suspended particles which divert the light from its regular course." With these words John William Strutt, later Lord Rayleigh, began his first paper on *light scattering*. Light scattering is a phenomenon that occurs widely in nature and is of great utility. It accounts for the blue of the sky, as well as the white of the clouds, and is used as an analytical tool in chemistry and biology. We distinguish two major classes of light scattering, *Rayleigh scattering* and *Mie scattering*.

RAYLEIGH SCATTERING

Phenomenology. Letting a beam of white light pass through air or a variety of gases, Tyndall observed that a faint blue light emerges sideways out of the beam. He thought that the light, of "a colour rivalling that of the purest Italian sky," was due to the presence of small particles; he considered pure gases "optically empty."*

---

*John Tyndall (1820–1893), Irish physicist. Working as a surveyor, railroad engineer, and science teacher, Tyndall obtained a Ph.D. in mathematics, studied chemistry under Bunsen, met Faraday, became professor of physics at the Royal Institution, and, when Faraday retired, succeeded him as superintendent. Tyndall wrote some 30 books, is noted for his work on light scattering, designed foghorns and both emission and absorption spectrophotometers, discovered the effect of penicillin on bacteria, and studied the mechanics of glacier motion, the latter no doubt because he was an accomplished mountain climber who, in 1861, made the first ascent of the Weisshorn, 4512 m, near Zermatt, in the Swiss Alps. J. Tyndall, "On the Blue Colour of the Sky, the Polarization of Skylight, and on the Polarization of Light by Cloudy matter generally," *Philos. Mag.* (4), **37** (1869), 384–94.

At first Rayleigh agreed. Later he recognized that it is the molecules (of the air), rather than other particles, that account for the blue of the sky. Since then, *Rayleigh scattering* has become synonymous with *molecular scattering.* The approximate upper limit of the size of the particles that cause such scattering is *one-tenth of the wavelength of light.**

Scattering as dipole radiation. Consider light that is incident on a single molecule. Without ionization, a molecule contains an equal number of positive and negative charges. Under the influence of the light's electric field, the positive charges are displaced in one direction and the negative charges are displaced in the opposite direction, forming an *electric dipole* (Figure 4.2-1).

Figure 4.2-1 Electric dipole consists of two charges of opposite sign, separated by distance d.

It is convenient to describe the properties of a dipole in terms of its *dipole moment, p*:

$$\mathbf{p} = q\mathbf{d} \qquad [4.2\text{-}1]$$

where q is the charge and d the distance between the charges. With an external electric field **E**, the induced dipole moment is

$$\mathbf{p} = \alpha\mathbf{E} \qquad [4.2\text{-}2]$$

where α is the *polarizability* of the molecule. If the field varies sinusoidally as a function of time, the dipole moment will vary as well:

$$\mathbf{p}_t = \mathbf{p}_0 \sin(\omega t) \qquad [4.2\text{-}3]$$

The external field, in short, will make the dipole *oscillate,* in synchrony with the impinging field.

The positive charges in a molecule are relatively heavy (they represent the atom's nuclei) and, under the influence of the field, move very little. The negative charges (the electrons) are much lighter and move much farther. Thus the induced oscillations affect mainly the electrons. But electrons that undergo oscillatory accelerations emit electromagnetic

*John William Strutt, third Baron Rayleigh (1842-1919), British physicist, professor and later chancellor of Cambridge University, and successor of Clerk Maxwell as director of the Cavendish Laboratory. His first papers on light scattering were J. W. Strutt, "On the Light from the Sky, its Polarization and Colour," *Philos. Mag.* (4) **41** (1871), 107-120, 274-79; and "On the Scattering of Light by small Particles," *ibid.* (4) **41** (1871), 447-54. Later, in Lord Rayleigh, "On the Transmission of Light through an Atmosphere containing Small Particles in Suspension, and on the Origin of the Blue of the Sky," *Philos. Mag.* (5) **47** (1899), 375-84, he corrected himself saying that "even in the absence of foreign particles we would still have a blue sky." Rayleigh also made important contributions to photography, resolving power, zone plates, blackbody radiation, color vision, and acoustics. On his estate at Terling, he had several darkrooms, one of them painted black with a mixture of soot and beer. It was here that he discovered argon, which earned him the 1904 Nobel Prize in physics. In 1899, Rayleigh was admiring the sight of Mt. Everest from a place near Darjeeling, 160 km away. At that distance, the outline of the mountain could barely be seen in the haze. From the degree of visibility and the refractive index of air, Rayleigh calculated the number of molecules per milliliter of air, and found 3×10^{19}, a figure close to today's value.

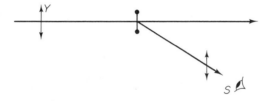

Figure 4.2-2 Linearly polarized light incident from the left, forming dipole. Observer at *S*.

waves, just as a dipole antenna emits radio waves; hence, the incident light is almost instantaneously *reradiated*. Light, in other words, does not simply "pass through" transparent matter; instead, it forms dipoles, these dipoles radiate, and this reradiation appears as light.

Assume the incident field is oscillating along the *y* axis, as shown in Figure 4.2-2. Consequently, the induced dipoles also oscillate along the *y* axis. An observer at *S* will see these oscillations as light; in fact, the light is seen from any direction, from straight ahead, from the side, and from the rear, as long as it is in a direction *perpendicular to the axis of the dipole*.

But now let the observer look from a direction *in the plane of the drawing* (the paper plane). Call θ the angle subtended by the forward direction and the direction of scattering. In the forward direction where $\theta = 0$, and in the backward direction where $\theta = 180°$, the scattered light is well visible. But when looking at the dipole from directly above or directly below (where $\theta = 90°$), there will be no light. That is because electromagnetic waves are transverse, rather than longitudinal, the same as with a dipole antenna which radiates in all directions, *except* in the direction of its own length.

In the horizontal plane, in short, the amplitude of the scattered light is the same all around but in the vertical plane it varies as a function of $\cos \theta$; the intensity, following $I \propto A^2$, varies as $\cos^2 \theta$. Combining the effects in three-dimensional space, a plot of the intensity looks like a doughnut (Figure 4.2-3).

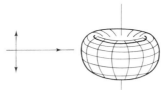

Figure 4.2-3 Three-dimensional plot of linearly polarized light scattered by a dipole.

Rayleigh scattering as a function of wavelength.   Now let a bundle of light pass through a volume of scatterers of thickness Δx. We call I_0 the intensity of the light incident on the volume and I' the intensity passing through. A certain fraction of the light, i, is scattered out of the path and lost:

$$I_0 - I' = i \qquad [4.2\text{-}4]$$

The magnitude of this loss is proportional to I_0, to the thickness Δx, and to a constant of proportionality called *turbidity*, τ.

$$I' - I_0 = -i = I_0 \tau \, \Delta x \qquad [4.2\text{-}5]$$

Turbidity, in other words, is the *fractional loss* of light, the ratio i/I_0, per unit thickness; it is a term comparable to *absorptivity* (see Chapter 5.4) although I hasten to add that in pure scattering there is no absorption, or irretrievable loss of light: all the light removed from the incident beam is reemitted again almost instantaneously in the form of scattered light.

The *total* turbidity, that is, the total amount of light scattered forward or backward, is found by integrating over the surface of a hemisphere of radius R:

$$\tau = 2\pi \int_0^\pi \frac{i}{I_0} R^2 \sin\theta \, d\theta = \frac{8\pi}{3} \frac{i}{I_0} R^2 \qquad [4.2\text{-}6]$$

The amplitude of the induced oscillation (of a dipole) increases as the driving frequency (of the incident light) approaches the natural frequency of oscillation of the molecule. Since the natural frequency of a typical molecule is comparable to the frequency of ultraviolet radiation, more light is scattered at shorter wavelengths.

More specifically, assume that the scattering medium may be a gas that contains N molecules per unit volume. The amplitude of the scattered light is inversely proportional to the *square* of the wavelength and, since $I \propto A^2$, the intensity is inversely proportional to the *fourth power* of the wavelength,

$$i = I_0 \frac{8\pi^4 N\alpha^2}{\lambda^4 R^2} (1 + \cos^2\theta) \qquad [4.2\text{-}7]$$

The essential point of Rayleigh scattering, in short, is that

$$\boxed{i \propto \frac{1}{\lambda^4}} \qquad [4.2\text{-}8]$$

a relationship that is illustrated in Figure 4.2-4.

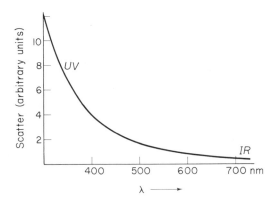

Figure 4.2-4 Rayleigh scattering: plot of intensity as a function of wavelength.

Example

How much more Rayleigh scattering is produced by light of 528 nm wavelength than by light of 628 nm wavelength?

Solution.   From Equation [4.2-8],

$$i \propto \frac{1}{\lambda^4}$$

we find that

$$\frac{i_1}{i_2} = \left(\frac{\lambda_2}{\lambda_1}\right)^4 = \left(\frac{628}{528}\right)^4 = \boxed{2.0}$$

Applications.   Light scattering can be used to determine the weight of almost any particles, from simple molecules to colloids, polymers, and proteins and other substances of biochemical interest. Conversely, if the composition is known, the number of molecules per unit volume can be found.

Observations are usually made of the intensity of the scattered light as a function of the angle θ of scattering relative to the forward direction. The quantity actually measured is the *Rayleigh ratio*, R_θ, a parameter that is defined as

$$R_\theta = \frac{i_\theta}{I_0} R^2 \qquad\qquad [4.2\text{-}9]$$

and that, as I have indicated by the subscripts, is a function of θ. But, from Equation [4.2-5], $i/I_0 \propto \tau$, and hence R_θ is also proportional to τ_θ.

Most instruments for measuring light scattering follow the outline shown in Figure 4.2-5. Light comes from a source (left) and is collimated by a lens, L. After passing through an optional color or polarizing filter, F, the light is incident on a cell, C, containing the scattering medium. Preferably the cell can be quickly exchanged by an opal-glass reference standard, to permit calibration. After passing through the cell, the primary light is absorbed in a light trap, T.

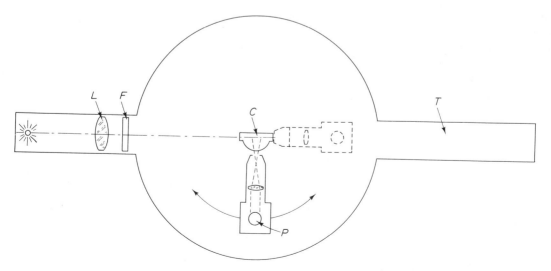

Figure 4.2-5   Schematic diagram of light-scattering apparatus.

Scattered light from the cell or from the reference standard goes through another slit-lens assembly and perhaps another polarizing filter to a photodetector, P, mounted so that it can be swung around with the cell as the pivot. Its output is plotted versus the angle of rotation, the dashed outline in the figure indicating the forward position where $\theta = 0$.

MIE SCATTERING

Rayleigh scattering versus Mie scattering.   When the particles are larger than about one-tenth of a wavelength, the light scattered from one point on the particle may well be out of phase with light scattered from another point. The two contributions then interfere and the scattered-light distribution is no longer symmetrical as in the Rayleigh case. With increasing particle size, in fact, the scattered light becomes more concentrated in the forward direction (Figure 4.2-6). The difference between forward and back scatter is called *dissymmetry,* which is often defined as the ratio $i_{45°}/i_{135°}$. Dissymmetry can be used to determine the average size of particles, provided that their shape is known. For larger particles, however, the angular dependence of the scattering becomes quite complicated, showing a number of maxima and minima, and cannot be adequately described by a single parameter such as the dissymmetry.

Mie's theory[*] takes into account the size of the particles but also their refractive index, and the refractive index of the surrounding medium, as well as the shape, dielectric constant, and absorptivity of the particles. It is based on a formal solution of Clerk Maxwell's equations, leading to a series, in theory an infinite number, of partial waves. The amplitudes of these waves decrease rapidly as the particles become smaller. In Rayleigh scattering, we need to consider only the first few terms, and may neglect all others. Mie scattering, by contrast, requires many more terms; it is more general and includes Rayleigh scattering as a special case.

Mie scattering parameters.   Assume that light of cross section A is incident on a volume V of thickness Δx, $V = A\,\Delta x$. If there are N particles per unit volume, the

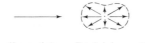

Small particles: Rayleigh scattering

Large particles: Mie scattering

Figure 4.2-6 Angular distribution in Rayleigh and Mie scattering.

---

[*]Gustav Mie (1868–1957), German physicist, professor of physics at the University of Freiburg. G. Mie, "Beiträge zur Optik trüber Medien, speziell kolloidaler Metallösungen," *Ann. Phys.* (4) **25** (1908), 377–445.

actual volume contains $NA \Delta x$ particles. If r is the radius of a particle, each particle has a cross section πr^2 and together they have a cross section

$$\Sigma = NA \Delta x \, \pi r^2 \qquad [4.2\text{-}10]$$

provided that no one particle lies in the shadow of another particle; otherwise, the cross section is less. In most cases, the distances between particles are much larger than their radii, so the actual cross section is about the same as the maximum possible.

But then it turns out that the cross section of the light affected by a particle is not necessarily equal to the cross section of the particle. The reason is the refractive index difference between particles and the medium between them. The ratio of these two cross sections is the *extinction factor, K*. This factor is generally larger than unity, which means that the extinction by a particle is *higher* than what could be expected from its physical size. The ratio i/I_0 in Mie scattering, hence, is

$$\frac{-i}{I_0} = \frac{NA \Delta x \, K\pi r^2}{A} \qquad [4.2\text{-}11]$$

Combining this equation with our earlier Equation [4.2-5], canceling A and Δx, and replacing τ by the *Mie extinction coefficient, μ,* gives

$$\mu = NK\pi r^2 \qquad [4.2\text{-}12]$$

which shows that μ is equal to the product of the number of particles times $K\pi r^2$, a term called the *extinction cross section.* If, as in most cases, the particles are of continuous size distribution, with radii ranging from r_1 to r_2, then

$$\mu = \pi \int_{r_1}^{r_2} N(r)K(r, n)r^2 \, dr \qquad [4.2\text{-}13]$$

The essential point is that there is *no λ^{-4} relationship.* Indeed, with increasing particle size the exponent 4 in the denominator in Equations [4.2-7] and [4.2-8] becomes less and gradually approaches zero. For sufficiently large particles there is no wavelength dependency at all. This is one of the characteristics that distinguish Mie scattering from Rayleigh scattering; it is, of course, the reason why most clouds are white. These clouds, and fog and mist and aerosol sprays, are composed of droplets at least 10 μm in diameter, 200 times the $\lambda/10$ limit referred to before.

LIGHT SCATTERING AND METEOROLOGY

Light scattering accounts for some of the most beautiful color effects that we see in nature. Some examples:

Blue sky.   The blue of the sky is a matter of *Rayleigh scattering:* light of shorter wavelength is scattered more than light of longer wavelength. Indeed, blue is *scattered out of* the light we receive from the sun and therefore the sun appears *yellow.* On the other hand, when we look at the sky in any direction but at the sun, some of the light is *scattered in* (into the line of sight) and the sky appears *blue*. As seen from a high mountain, where there is not much air, or from the moon, where there is none, the sky is black.

The atmosphere may also contain aerosols, colloidal particles both natural and man-made, that are considerably larger than molecules. Scattering by aerosols does not follow the λ^{-4} law; it is a matter of *Mie scattering*. As seen through polluted air, the sky is not as deeply blue anymore; it looks more pale.

Airport lights.   Navigational lights seen at an airport are generally green, white, or red. But the lights lining the taxiways are *blue*. Seen from high up in the air, blue is scattered out and all but invisible; thus, the pilot is not misled to land there.

Red sunset.   With the sun high above the horizon and the path through the atmosphere relatively short, only blue is scattered out and the sun appears yellow. With the sun close to the horizon, the light traverses a much longer path, all but red is scattered out and the sun appears *red*.

Green flash.   The green flash is due to a combination of refraction, dispersion, selective absorption, and scattering. Because of refraction in the gradient-index atmosphere, light from the sun reaches the observer on Earth in a curved path. The light is dispersed: blue is refracted more, reaching the observer in a slightly steeper trajectory than red. The blue image of the sun's disk, therefore, appears *above* the red image, separate by about $\frac{1}{60}$ of the sun's diameter. However, the blue and the green segment are so narrow that, with the sun still above the horizon, they are barely visible at all. But as the sun disappears below the horizon, the red disappears too, the orange and yellow are attenuated by absorption in water vapor, oxygen, and ozone, and the blue is lost to Rayleigh scattering. That leaves the green which sometimes, very briefly, is seen as the *green flash*.

SUGGESTIONS FOR FURTHER READING

H. C. van de Hulst, *Light Scattering by Small Particles* (New York: Dover Publications, Inc., 1981).

M. Kerker, *The Scattering of Light and Other Electromagnetic Radiation* (New York: Academic Press, Inc., 1969).

C. F. Bohren and D. R. Huffman, *Absorption and Scattering of Light by Small Particles* (New York: John Wiley & Sons, Inc., 1983).

D. K. Lynch, *Atmospheric Phenomena*. Readings from *Scientific American* (San Francisco: W. H. Freeman and Company, Publishers, 1980).

R. Greenler, *Rainbows, Halos, and Glories* (New York: Cambridge University Press, 1980).

PROBLEMS

4.2-1.   Plot the angular distribution of Rayleigh scattering for incident light linearly polarized *normal* to the plane of the drawing.

4.2-2.   Unpolarized light is incident on a medium causing molecular scattering. If the turbidity at right angles to the incident light is 5 arbitrary units, what is the turbidity at 45°?

4.2-3. How much more Rayleigh scatter is produced by the 587-nm krypton line than by the 760-nm line?

4.2-4. If light of 635 nm wavelength causes a certain amount of Rayleigh scatter, light of what wavelength will give 10 times as much scatter?

4.2-5. At a wavelength of 546 nm the scattering intensity of pure carbon tetrachloride is found to be 5.9 arbitrary units. What is the intensity at 436 nm?

4.2-6. Light of 760 nm and 810 nm wavelength is passed through a turbid medium. If 20 percent of the shorter wavelength is lost due to Rayleigh scattering, what percentage is lost of the longer wavelength?

4.2-7. If Rayleigh scattering causes 1.85% of 1.06 μm radiation to be lost when passing through 10 km of atmosphere, what percentage is lost of radiation of 694.3 nm?

4.2-8. A certain amount of light of 503.3 nm wavelength is scattered out of a beam passing through a volume of clean air at atmospheric pressure. What percentage of that amount is scattered at 632.8 nm wavelength at one-half the pressure?

4.2-9. Some cars are equipped with yellow foglights, assuming that yellow light penetrates a dense fog (composed of relatively large water droplets) better than white light. Discuss the merits of such a scheme in terms of Rayleigh and Mie scattering.

4.2-10. Light scattering is one of several methods that can be used for building a *smoke detector* for the home. If you were given a small light source of low power consumption, such as a light-emitting diode, and a photocell as the receiver, how would you arrange these components?

4.3

Polarization of Light

THE TERM POLARIZATION HAS actually two meanings. First, it refers to a property of light, which may be linear, elliptical, circular, or partial polarization. But it also refers to the production of polarized light, by means of scattering, reflection, selective absorption, or double refraction. These various properties and techniques of production I will discuss in this chapter. We begin with a summary of different *types of polarized light*.

TYPES OF POLARIZED LIGHT

1. Most light as it comes from the sun or from another incandescent, or fluorescent, source is *unpolarized*. It is a mixture of light polarized in different ways and different directions and polarized to different degrees. These differences result because light is emitted by atoms, or groups of atoms, and these groups of atoms oscillate independently from one another. While within each wavetrain the electric field oscillates in the same mode and direction, from train to train the modes and directions of oscillation vary; the result, therefore, is a sequence of oscillations all *oriented at random*.

2. If the electric field oscillates in a given, constant orientation, the light is said to be *linearly polarized* (Figure 4.3-1). The term "plane polarized" is deprecated, for reasons that will become apparent shortly.

Figure 4.3-1 Linearly polarized light. Projection of wave on a plane intercepting the axis of propagation gives a *line*, hence the term *linear polarization*.

3. If linearly polarized light contains an additional component of natural (unpolarized) light, it is called *partially* (linearly) *polarized* (Figure 4.3-2). Such light could come from a perfectly polarizing filter with some holes in it or from a low-quality filter.

Figure 4.3-2   Partially linearly polarized light seen head-on.

4. In *circular polarization,* the electric vector no longer oscillates in a plane, as in linear polarization. Instead, the vector retains its magnitude and proceeds in the form of a helix *around,* rather than *through,* the axis of propagation (Figure 4.3-3). Within one wavelength, the vector completes one revolution. If the tip of the vector, when seen looking toward the light source, rotates clockwise, the light is said to be right-circularly polarized; if counterclockwise, left.

Figure 4.3-3   Right-circularly polarized light.

5. *Elliptical polarization* stands between circular and linear polarization. It is the most general type of polarization; linear and circular are the two extremes of elliptical polarization (Figure 4.3-4). The tip of the vector now proceeds in the form of a *flattened* helix; the vector rotates and, at the same time, it changes in magnitude.

The term "plane polarization," as opposed to "linear," is based on a three-dimensional convention. Circularly polarized light need then be called "helically polarized." To avoid awkward terms such as "circular helical" and "elliptical helical," I prefer a two-dimensional convention with the terms *linear, elliptical,* and *circular.*

Figure 4.3-4   Circular, elliptical, and linear polarization.

6. *Optical rotation* (optical activity) is entirely different. It is a property of matter, rather than a property of light. We will discuss the molecular basis of optical activity later (on page 317).

PRODUCTION OF POLARIZED LIGHT

Light can be polarized in various ways. We distinguish the following.

Polarization by scattering. Consider a bundle of unpolarized light that passes through an assembly of small particles in suspension (Figure 4.3-5). As we have seen before on page 297, an observer looking at the particles from a direction at right angles to the direction of propagation finds the light to be *linearly polarized*. The direction of oscillation of the scattered light is *normal* (perpendicular) to the plane defined by the direction of propagation and the direction of observation.* (Clearly, the light scattered toward *S* cannot be polarized *parallel* to the plane because that would require the incident light to oscillate back and forth along the axis, "longitudinally," and there is no such light.)

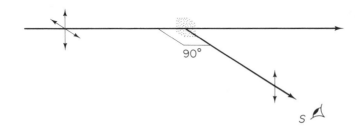

Figure 4.3-5 Light scattered at right angles to the direction of propagation is linearly polarized.

Polarization by reflection. Assume that light is reflected at the surface of matter. Actually, the light penetrates a short distance into the matter where it induces molecular oscillations. As in polarization by scattering, the reflected light can oscillate only *normal* to the plane of incidence and hence, the reflected light is *linearly polarized* (Figure 4.3-6).

Maximum polarization occurs when the reflected light and the refracted light (not shown) subtend an angle of 90°. If that is the case,

$$I + 90° + I' = 180°$$

$$I' = 90° - I$$

Since

$$\sin(90° - I) = \cos I$$

Snell's law, $n \sin I = n' \sin I'$, becomes

$$\frac{\sin I}{\cos I} = \frac{n'}{n}$$

---

*The polarization of light by scattering was first observed by Lord Rayleigh who used colloidal sulfur as the scattering medium, adding acetic acid to a weak solution of photographic fixer (sodium thiosulfate). A more convenient way to produce a colloidal suspension is by adding to water, and stirring, a few grains of nonfat dry milk.

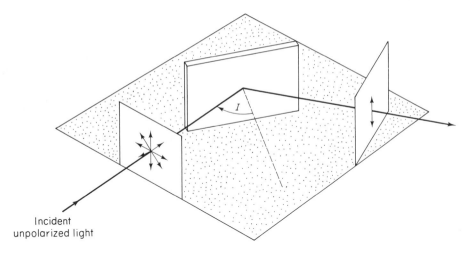

Figure 4.3-6   Polarization by reflection. The plane of incidence, defined by the incident light and the surface normal, is horizontal.

and, since

$$\frac{\sin I}{\cos I} = \tan I$$

$$\tan I = \frac{n'}{n}$$

[4.3-1]

which is the tangent of *Brewster's angle.** Light reflected at any other angle but Brewster's is partially linearly polarized.

Polarization by transmission.   Each time the light is reflected at Brewster's angle, the reflected component is completely linearly polarized. Consequently, in the light transmitted through the surface, that component is missing. If the process is repeated often enough, using a stack of plates as shown in Figure 4.3-7, the transmitted light becomes more and more purely linearly polarized. This method of polarization is of interest in the infrared where thin films of selenium take the place of the glass.

---

*Named after Sir David Brewster (1781–1868), Scottish physicist, professor of physics at St. Andrews College. Initially a minister in the Church of Scotland, Brewster became interested in optics, found the angle named after him, contributed also to dichroism, absorption spectra, and stereophotography, invented the kaleidoscope, and wrote a book about it. A prolific writer, Brewster edited several journals, published a 526-page *Treatise on Optics,* the *Life of Sir Isaac Newton,* several more books, and some 315 journal articles. Brewster's law, in his own words, states that "when a ray of light is polarised by reflexion, the reflected ray forms a right angle with the refracted ray." D. Brewster, "On the laws which regulate the polarisation of light by reflexion from transparent bodies," *Philos. Trans. Roy. Soc. London* **105** (1815), 125–59.

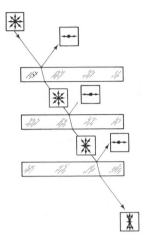

Figure 4.3-7   Polarization by transmission.

Polarization by selective absorption.   Polarization by selective absorption is an important case which can be best understood by the example of the *wire-grid polarizer*.

A wire-grid polarizer is a grid of parallel wires. In 1888, Heinrich Hertz used such grids as polarizers to test the properties of radiowaves he had discovered the year before. The electric field impinging on the grid may be resolved into two orthogonal components, one oscillating parallel to the wires and the other normal to them. The parallel component drives the conduction electrons in the wires along their lengths, the current encounters resistance, heats up the wire, and thereby loses energy. The normal component, however, has no electrons to drive very far and hence passes through without much loss. In short, it is the *normal* component that goes through (contrary to what one might think).

What the wire-grid polarizer does for longer waves can be accomplished for visible light by a *dichroic crystal.* A good example is *tourmaline,* an aluminoborosilicate containing Al_2O_3, B_2O_3, and SiO_2. On passing through the crystal, the light is split into two components, the horizontal component in Figure 4.3-8 being attenuated, the vertical component

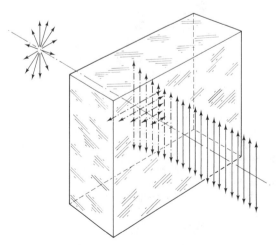

Figure 4.3-8   Polarization by selective absorption.

going through. The light emerging from the crystal, therefore, is linearly polarized. (The transmitted light, because of the natural color of tourmaline, is green; in orthogonally polarized light the crystal appears black, hence the term "dichroic" meaning "two colors.")

The same dichroic effect is seen with crystals of quinine sulfate periodide, called *herapathite*.* These crystals are very small, and cannot be used for polarization. However, after a method was found to align the crystals, an efficient and economical polarizing material was produced, now known as *Polaroid*.†

The most common type of a polarizer used today is *H sheet*, produced by stretching a sheet of synthetic polyvinyl alcohol (PVA, —CH$_2$—CHOH—), thereby aligning its very long molecules (polymers). The sheet is then impregnated with iodine which attaches itself to the polymers. The conduction electrons associated with the iodine move freely along the molecules, the same as the electrons in a wire-grid polarizer, and thus absorb the electric field oscillating parallel to them. Perpendicular to the molecules there is little conduction and the field oscillating that way will go through. *K sheet* is a polyvinylene polarizer optically similar to H sheet but more resistant to high temperatures and humidity.

Polarization by double refraction.   Take a pen and make a dot on a sheet of paper. Place a crystal of calcite or "Iceland spar," calcium carbonate, CaCO$_3$, over the dot and look through. You will see two dots instead of one. Rotate the crystal and observe the dots. One of them will remain in place while the other moves around it. This was how Bartholin‡ discovered *double refraction,* or *birefringence*. Crystals such as calcite, quartz, mica, and ice which show this effect are called *anisotropic*. In contrast, most noncrystalline solids such as unstressed glass are *isotropic*.

Calcite is a good example of a system of anisotropic crystals called *hexagonal*. In order to visualize the shape of a hexagonal crystal, think first of an orthogonal (right-angled) block

---

*Named after William Bird Herapath (1820–1868), physician and surgeon at Queen Elizabeth's Hospital in Bristol, England. Herapath worked on a variety of subjects, from chlorophyll and the identification of bloodstains by microspectroscopy to home sanitation. His best known discovery came when his pupil, a Mr. Phelps, dropped iodine into the urine of a dog that had been fed quinine and saw that tiny emerald-green crystals formed there. Herapath, examining them under a microscope, noticed that the crystals were in some places light where they overlapped and in some places they were dark, an observation which he published under the slightly laborious title "On the Optical Properties of a newly-discovered Salt of Quinine, which crystalline substance possesses the power of polarizing a ray of Light, like Tourmaline, and at certain angles of Rotation of depolarizing it, like Selenite," *Philos. Mag.* (4) **3** (1852), 161–73.

†Invented by Edwin Herbert Land (1909–   ), American physicist, inventor, and businessman. While an undergraduate at Harvard College, Land set out to develop a better way of producing polarized light, hoping to replace the single crystals that were so difficult to grow. Clearly, he reasoned, a great many small crystals would do the same if only they could be aligned properly. He took a suspension of herapathite and placed it in a strong magnetic field. It worked: he had the first man-made polarizer for light. Later he used an electric field to align the crystals. He also suspended them in a thermoplastic matrix, extruding the plastic, while soft, through a narrow slit. The result was "J sheet" polarizer; it contains a great many herapathite crystals all aligned parallel to one another. The first patent on a synthetic sheet polarizer was issued to E. H. Land and J. S. Friedman, *Polarizing refracting bodies,* U.S. Patent 1,918,848, July 18, 1933. For more details, see E. H. Land, "Some Aspects of the Development of Sheet Polarizers," *J. Opt. Soc. Am.* **41** (1951), 957–63.

‡Erasmus Bartholin (1625–1698), Danish physician and professor of mathematics at the University of Copenhagen. He described his discovery in a 60-page publication, *Experimenta crystalli Islandici disdiaclastici quibus mira et insolita refractio delegitur* (Hafniae, 1670). Bartholin, however, was not aware of what we now call polarization; he thought his observation to be an unusual case of refraction, *double refraction*.

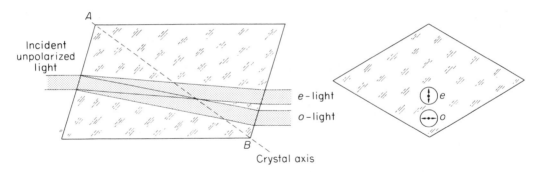

Figure 4.3-9   Double refraction and polarization of light in calcite.

of material. Two diagonally opposite corners may be called A and B, as in Figure 4.3-9. Slightly compress the block along the line A–B: The faces that before were rectangular now become rhombic and corners A and B become blunt; that is, the angles adjacent to them grow larger, the angles away from them acute.

The line A–B is called the *crystal axis*, not to be confused with the optic axis of the system. In the direction specified by this axis the crystal is isotropic and its refractive index, the *ordinary* refractive index, ω, is constant (1.6584 for calcite and sodium light). In other directions, the index changes with the angle subtended with the crystal axis. In a direction 90° from the axis, the index is least (1.4864). This is the *extraordinary* refractive index, ϵ.

If unpolarized light is incident on the crystal, in a direction different from that of its axis, the light is split into two components, the *ordinary* or *o ray,* and the extraordinary or *e ray.* The *o* ray advances at the same velocity in all directions. If we could set up a point source of light within the crystal, Huygens' wavefronts would advance outward in the form of *spheres* (Figure 4.3-10).

The extraordinary ray, however, has different velocities in different directions and the wavefronts advance in the form of *ellipsoids of revolution.* In some crystals, such as quartz, the ellipsoidal wavefronts lie within the spherical wavefronts because the velocity of the *e* ray is less, and the ϵ index is higher, than that of the *o* ray. Such crystals are called *positive* uniaxial. In others, such as calcite, it is the other way around; these are called *negative* uniaxial. The two bundles emerging from the crystal are both linearly polarized, at right

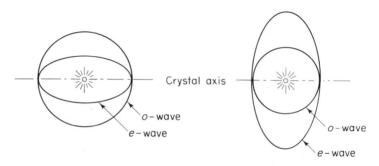

Figure 4.3-10   Huygens' wavefronts in a positive uniaxial crystal (*left*), negative uniaxial crystal (*right*).

angles to each other. Hence, such crystals make good polarizers if we could only eliminate one of the bundles. This is done in the *Nicol prism*.

Nicol prism.   The Nicol prism* is made out of a crystal of Iceland spar, cut into two and cemented together with a thin layer of Canada balsam in between (Figure 4.3-11). The two endfaces are polished down from 71° to 68° as shown. The *o* light is refracted more (downward) and is incident on the Canada balsam at an angle slightly larger than the *e* light. But since the refractive index of Canada balsam (1.526) lies between ω and ϵ for calcite (1.6584 and 1.4864, respectively), the *e* light will pass through but the *o* light will not. The *o* light undergoes total internal reflection and is absorbed in black paint on the sides of the prism, and only the *e* light will emerge from it.

Nicol prisms are good polarizers, but they are expensive and have a limited field of view (28°). This is because, if the light is not nearly parallel, either the *o* light will be incident on the balsam at an angle less than the critical angle and go through, or the *e* light will be reflected out of the prism too and be lost.

There are other prisms besides the Nicol type which are useful for special applications. The *Glan–Thompson prism* has a wider angular aperture (40°), but it is wasteful of calcite and hence even more expensive. The *Glan–Foucault prism* has no cement (but a narrow field) and thus is less likely to be damaged at high power densities. In the *Rochon prism* and the *Wollaston prism* the *o* and the *e* rays are transmitted side by side.

Types of birefringence.   The numerical difference between the ϵ and the ω index of refraction provides us with one definition of birefringence. But since $\epsilon - \omega$, in effect, is a measure of *path difference*, Γ, this path difference *per unit path length*, *L*, gives us another definition; thus

$$\text{birefringence} \equiv \epsilon - \omega = \frac{\Gamma}{L} \qquad\qquad [4.3\text{-}2]$$

If a transparent isotropic substance such as plastic or glass is subject to mechanical stress, it becomes temporarily birefringent. Such birefringence is the basis of *photoelastic stress analysis,* a method widely used in the engineering world. For testing purposes, either a

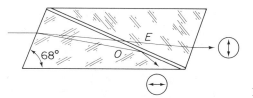

Figure 4.3-11   Nicol prism.

---

*Named after William Nicol (1768–1851), Scottish geologist and physicist. A lecturer at the University of Edinburgh, Nicol published his first paper at age 58. His interests were primarily in the fields of crystallography, mineralogy, and paleontology. In 1828 he invented his prism, and described it in the article "On a Method of So Far Increasing the Divergency of the Two Rays in Calcareous Spar That Only One Image May Be Seen at a Time," *Edinburgh New Philos. J.* **6** (1829), 83–84.

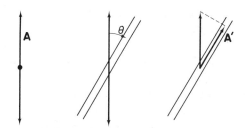

Figure 4.3-12   Malus' law.

transparent replica of the structure is made (and examined in transmitted light) or a plastic coating is applied to the real structure (and examined in reflected light). In either case, the birefringence reveals the strains induced in the structure.

Malus' law.   Consider a beam of linearly polarized light, coming toward you. The electric vector of the light may oscillate in the vertical direction, as shown in Figure 4.3-12, left. The light may then be incident on a polarizer whose *plane of transmission* subtends an angle θ with the direction of oscillation (center). How will the amplitude of the light change?

Draw a normal from the tip of vector **A** to the plane of transmission (right). If $\theta = 0$, all the light will go through and $\mathbf{A} = \mathbf{A}'$. If $\theta = 90°$, none will go through and $\mathbf{A}' = 0$. In order to find the amount of light transmitted at intermediate angles, note that $\mathbf{A}'/\mathbf{A} = \cos \theta$. Therefore, since the intensity of light is proportional to the *square* of the amplitude, the intensity transmitted, I', relates to the intensity incident, I_0, as

$$\boxed{I' = I_0 \cos^2 \theta}$$ [4.3-3]

where θ is the angle through which one of the polarizers has been turned with respect to the other. This relationship is known as *Malus' law.**

THE ANALYSIS OF LIGHT OF UNKNOWN POLARIZATION

If two high-quality polarizers are set with their planes of transmission perpendicular to each other (if the polarizers are "crossed"), *no* light will go through: the minimum transmission is zero. But if some unpolarized light were present in the light reaching the second polarizer

---

*Étienne Louis Malus (1775–1812), French army officer and engineer. One evening in 1808 while standing near a window in his home in the Rue d'Enfer in Paris, Malus was looking through a crystal of Iceland spar at the setting sun reflected in the windows of the Palais Luxembourg across the street. As he turned the crystal about the line of sight, the two images of the sun seen through the crystal became alternately darker and brighter, changing every 90° of rotation. After this accidental observation Malus followed it up quickly by more solid experimental work, described his observation in "Sur une propriété de la lumière réfléchie," *Mem. Phys. Chim. Soc. Arcueil, Paris* **2** (1809), 143–58. He concluded that the light, by reflection on the glass, became *polarized* but, by chance, he defined the "plane of polarization" of the reflected light as being the same as the plane of incidence. Today we know that the **E** vector oscillates *normal* to the plane of incidence; thus Malus' "plane of polarization" is perpendicular to the plane of oscillation.

(which would make the light partially polarized), at no setting would the light be extinguished completely. The same would happen if circularly polarized light were present.

How, then, can we distinguish between the two? This is possible by using a relatively simple device known as a *quarter-wave plate*. Such a plate is made out of birefringent material which, as we have seen, makes light of orthogonal polarizations traverse the material at different velocities. One of the wavefronts will be ahead of the other by a distance that depends on the thickness of the plate (Figure 4.3-13). If one component is ahead by 90°, we have a *quarter*-wave plate. If the thickness is such that the path difference is 180°, we have a *half*-wave plate.

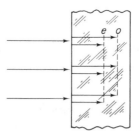

Figure 4.3-13   Retardation of light in a (positive) birefringent crystal such as quartz.

Example

Quarter-wave plates are often made of thin sheets of mica. Although mica is *biaxial* and has *three* refractive indices, we are justified in using 1.5936 for one index and 1.5977 for the other. How thick must the mica be to provide $\frac{1}{4}\lambda$ retardation, using sodium light, 589 nm?

Solution.   From Equation [4.3-2] we find the *path* difference,

$$\Gamma = L(\epsilon - \omega)$$

and from Equation [3.3-9], the *phase* difference,

$$\delta = \left(\frac{2\pi}{\lambda}\right)\Gamma = \left(\frac{2\pi}{\lambda}\right)L(\epsilon - \omega)$$

Then, for the phase difference to be $\delta = \frac{1}{4}\lambda = \frac{1}{2}\pi$,

$$\tfrac{1}{2}\pi = \left(\frac{2\pi}{\lambda}\right)L(\epsilon - \omega)$$

$$\tfrac{1}{2}\lambda = 2L(\epsilon - \omega)$$

$$L = \frac{\lambda}{4(\epsilon - \omega)}$$

and thus

$$L = \frac{589 \times 10^{-9}}{(4)(1.5977 - 1.5936)} = \boxed{0.036 \text{ mm}}$$

Let me illustrate how a quarter-wave plate works, using graphical construction. A linearly polarized, sinusoidal wave, $\mathbf{A}_1$ in Figure 4.3-14, left, rises up from zero at a time $t = 0$. A second wave of the same amplitude and frequency, $\mathbf{A}_2$, is delayed by one-fourth of a

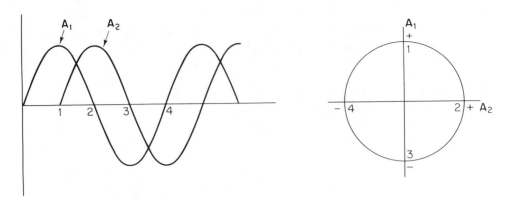

Figure 4.3-14 Adding two linear polarizations of $\lambda/4$ phase difference gives circular polarization.

wavelength, 90°. We consider four points evenly spaced in time, 1–2–3–4, and plot the amplitudes of these two waves, at these times, on coordinates labeled A_1 and A_2 (right). At point 1, the A_1 wave has reached a positive maximum and A_2 is zero; at point 2, $A_1 = 0$ and A_2 has reached a positive maximum; at point 3, A_1 is at a negative maximum and $A_2 = 0$; and at point 4, $A_1 = 0$ and A_2 is at a negative maximum. The plot is that of a circle; thus, the resulting light is *circularly polarized*.

The same conclusion can be reached analytically, and in a more general way. The first wave in Figure 4.3-14 is oscillating in the *y-z* plane and begins at $t = 0$; thus, we have

$$y_1 = A_y \sin(\omega t) \qquad [4.3\text{-}4]$$

Another sinusoidal wave, not shown, may be oscillating in the *x-z* plane, orthogonal to the first wave, and hence is given by

$$x = A_x \sin(\omega t) \qquad [4.3\text{-}5]$$

The second wave shown in Figure 4.3-14 is delayed by a phase difference δ and, therefore,

$$y_2 = A_y \sin(\omega t + \delta) \qquad [4.3\text{-}6]$$

Since

$$\sin(\alpha + \beta) = \sin\alpha\cos\beta + \cos\alpha\sin\beta$$

we can write Equation [4.3-6] in the form of

$$y = A_y \sin\omega t \cos\delta + A_y \cos\omega t \sin\delta \qquad [4.3\text{-}7]$$

We eliminate $\sin\omega t$, substituting Equation [4.3-5], as well as $\cos\omega t$, using

$$\sin^2\alpha + \cos^2\alpha = 1$$

and therefore

$$\cos\omega t = \sqrt{1 - \sin^2\omega t} = \sqrt{1 - \frac{x^2}{A_x^2}}$$

Equation [4.3-7] then becomes

$$y = \mathbf{A}_y \frac{x}{\mathbf{A}_x} \cos \delta + \mathbf{A}_y \sqrt{1 - \frac{x^2}{\mathbf{A}_x^2}} \sin \delta \qquad [4.3\text{-}8]$$

Transposing and squaring both sides gives

$$\left(y - \mathbf{A}_y \frac{x}{\mathbf{A}_x} \cos \delta\right)^2 = \mathbf{A}_y^2 \left(1 - \frac{x^2}{\mathbf{A}_x^2}\right) \sin^2 \delta \qquad [4.3\text{-}9]$$

which reduces to

$$\sin^2 \delta = \frac{x^2}{\mathbf{A}_x^2} + \frac{y^2}{\mathbf{A}_y^2} - 2\frac{xy}{\mathbf{A}_x\mathbf{A}_y} \cos \delta \qquad [4.3\text{-}10]$$

If there is *no* phase difference, $\delta = 0$, between the two contributions,

$$\frac{x^2}{\mathbf{A}_x^2} + \frac{y^2}{\mathbf{A}_y^2} - 2\frac{xy}{\mathbf{A}_x\mathbf{A}_y} = 0$$

and

$$y = \frac{\mathbf{A}_y}{\mathbf{A}_x} x \qquad [4.3\text{-}11]$$

which is the equation of a *straight line:* The resulting light is linearly polarized.

If the phase difference $\delta = 90°$,

$$\frac{x^2}{\mathbf{A}_x^2} + \frac{y^2}{\mathbf{A}_y^2} = 1 \qquad [4.3\text{-}12]$$

which is the equation of an *ellipse.* If, in addition, the two amplitudes are equal, $\mathbf{A}_x = \mathbf{A}_y$, the resulting light is *circularly* polarized. But that is true also in the opposite direction: when two circularly polarized waves of the same amplitude but opposite sense of rotation, one right-handed and the other left-handed, interfere with one another, the result is linearly polarized light.

Such interference follows *Fresnel–Arago's laws:**

1. Two bundles of light, polarized in the same direction, do interfere.
2. Two bundles, linearly polarized in orthogonal directions, do not interfere.
3. Two bundles, derived from orthogonal components of unpolarized light and subsequently brought into the same plane of oscillation, do not interfere.

---

*Dominique François Jean Arago (1786–1853), French physicist. Working at the Paris Observatory, Arago's most noteworthy contributions were to the wave theory of light and to the magnetic effect of electric current. By running a current through a copper wire, Arago in 1820 disproved the then held theory that iron was necessary to produce electromagnetism. When Napoleon made himself emperor in 1852, Arago refused an oath of allegiance and resigned; the new emperor did not accept his resignation and let Arago stay on anyway. A. Fresnel and F. Arago, "Sur l'Action que les rayons de lumière polarisés exercent les uns sur les autres," *Ann. Chim. Phys.* (2) **10** (1819), 288–305. (For Fresnel footnote, see page 255.)

4. Two bundles, derived from coherent portions of light and subsequently brought into orthogonal planes, do interfere. In this case, however, varying degrees of ellipticity result, as we have seen.

Returning to our initial question, by using a polarizer and a quarter-wave plate we can analyze almost any type of light of unknown polarization. The proper steps are listed in Table 4.3-1.

TABLE 4.3-1   ANALYSIS OF LIGHT OF UNKNOWN POLARIZATION

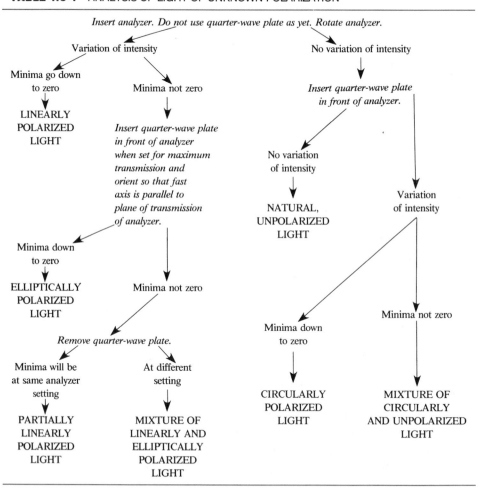

OPTICAL ACTIVITY

When linearly polarized light passes through a crystal of quartz, the light remains linearly polarized but its direction of oscillation changes: It *rotates* (Figure 4.3-15). Such rotation is called *optical activity*. To an observer looking toward the light source, the rotation may be

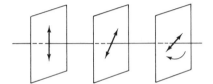

Figure 4.3-15   Optical rotation.

either clockwise or right-handed (as shown) and the substance be *dextrorotatory,* or it may be counterclockwise or left-handed and the substance be *levorotatory.*

The reason for the rotation is the asymmetry of the molecular structure. Even if two molecules contain the same atoms and bonds in the same sequential order, they still may differ in their orientation relative to the coordinate system. Such molecules are called *isomers.* If the isomers are exact mirror images of one another, they are called *enantiomers.**

Optical activity is a matter of *circular birefringence:* an optically active medium has an index of refraction for right-circularly polarized light that is different from the index for left-circularly polarized light. Since linearly polarized light is the resultant of two circular polarizations of the same amplitude but opposite sense of rotation, linearly polarized light passing through the medium will show a rotation of its direction of polarization. The degree of rotation is proportional to the difference between the two indices and to the length of path traversed.

The degree of rotation, for a given wavelength, caused by a 1-mm-thick solid is called *specific rotation.* For liquids, specific rotation is defined as the rotation caused by a solution of 1 g of substrate in 1 ml of solvent through a path 10 cm long. This dependence on concentration is of practical interest, for example when testing syrups or urine for the amount of sugar present. At different wavelengths, the rotation is generally different, an effect called *rotatory dispersion.*

ELECTROOPTICS AND MAGNETOOPTICS

Certain materials, when exposed to an electric field, change their polarization characteristics and become birefringent. Such materials are called *electrooptic materials.* Others change when exposed to a magnetic field; these are called *magnetooptic materials.* Both types are of considerable theoretical and practical interest. They make possible ultrafast shutters and light modulators that can be operated at rates up to many GHz (10^9 Hz), as they are used in data processing, optical computers, and laser communication systems. We distinguish the following.

---

*The link between optical rotation and asymmetry was found by Louis Pasteur (1822–1895), French chemist. When Pasteur saw that some crystals of sodium ammonium tartrate were mirror images of others, painstakingly, using a pair of tweezers, he separated them into two groups. Checking a solution of each in a polarimeter, he found that one was dextrorotatory and the other levorotatory. But optical activity occurs also in solutions, where no crystals exist; hence, he concluded, the asymmetry must be a property of the molecules, rather than the crystals. Pasteur also disproved the doctrine of spontaneous generation (of life) and found that infectious diseases such as anthrax and rabies were transmitted by microorganisms, opening the way to vaccination.

Faraday effect.   Historically the first experimental proof of an interaction between light and an electromagnetic field, *Faraday rotation* is induced by a magnetic field.* Similarly to optical activity, the field changes the index of refraction for right-circular polarization differently than for left-circular polarization.

But the two types of rotation are different. In optical activity, the sense of the rotation depends on the direction of the light, in Faraday rotation it depends on the direction of the field. Light passing through an optically active substance, and reflected back through it, shows no net rotation because the rotations cancel. In Faraday rotation, the total rotation is cumulative; it is twice the one-way rotation. The angle of rotation ϕ is proportional to the magnetic field strength $\mathbf{H}$ and to the distance L the light travels in the substance,

$$\phi = \mathbf{H}LV \qquad\qquad [4.3\text{-}13]$$

where V is called *Verdet's constant.*

Kerr effect.   In the *Kerr electrooptic effect* it is the electric field that causes the substance to become birefringent.† The (induced) difference between the two refractive indices, parallel and normal to the field, is proportional to the *square* of the field, $\mathbf{E}$,

$$\Delta n = K\mathbf{E}^2 \qquad\qquad [4.3\text{-}14]$$

where K is the *Kerr constant.*

A *Kerr cell* contains two parallel plates immersed in a suitable liquid such as nitrobenzene (Figure 4.3-16). Enclosed between crossed polarizers, ordinarily no light is passing through the cell, but with an electric field of several kV cm^{-1} applied, at 45° to the direction of oscillation of the entering light, the light becomes elliptically polarized and the shutter *opens.*

Pockels effect.   The *Pockels effect,* the same as the Kerr effect, is caused by the electric field.‡ In contrast to the Kerr effect, however, which is quadratic, the Pockels effect is a linear function of the field strength. And, what is more important, the voltage needed with a typical Pockels cell is at least an order of magnitude less than with a Kerr cell.

---

*Named after Michael Faraday (1791–1867), British physicist. The son of a blacksmith, Faraday began his career as a bookbinder's apprentice. Reading some of the books that passed through his hands, he became interested in chemistry and worked for a while as Sir Humphry Davy's lab assistant. As a young man, Faraday proclaimed that women were nothing in his life, and even published a poem ridiculing love. At age 30, he met, fell in love with, and married Sarah Barnhard, the 21-year-old daughter of a silversmith. Soon after Oerstedt's 1819 discovery that an electric current produced a magnetic field, Faraday, in 1821, placed a current-carrying wire near a magnetic pole and succeeded in making it rotate, opening the way to electric *motors.* Seeing this, he wondered whether the reverse might be true—whether a magnet could be made to produce a current. After casual experimentation over several years (during which time he liquefied gases, discovered benzene, and found a better way to make steel), he returned to his idea. For days he tried without success until, one day in 1831, in desperation he plunged a magnet down a coil: Suddenly there was an electric pulse. Why hadn't he seen that before? Clearly, it is the *change* of the magnetic field that did it. For this discovery, which opened the way to the electric *generator,* the whole scientific world honored him. In 1845, Faraday discovered the magnetic rotation of polarized light.

†John Kerr (1824–1907), Scottish physicist. In 1875 Kerr used a block of glass (still on display at the Kelvin Museum of the University of Glasgow), with two wires from an induction coil attached to it and placed between crossed Nicol prisms. With the current on, the glass becomes positive uniaxial birefringent and light passes through.

‡Named after German physicist Friedrich Carl Alvin Pockels (1865–1913).

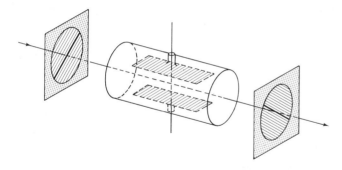

Figure 4.3-16 Kerr cell.

The electric field may either be applied parallel to the optical axis *(longitudinal Pockels effect)* or normal to it *(transverse Pockels effect)*. Transverse-effect cells need lower voltages, longitudinal-effect cells allow particularly high switching rates. Materials often used are crystals of potassium dihydrogen phosphate, KH_2PO_4 (KDP); potassium dideuterium phosphate, KD_2PO_4 (KD*P); and lithium niobate, $LiNbO_3$. As before, the electric field makes the crystal birefringent and linearly polarized light becomes elliptical. With the proper path length, the direction of oscillation changes by 90° and the cell opens, the same as with a Kerr cell but at a much lower voltage. For many applications, this has made the Pockels cell the preferred type of a light modulator.

SUGGESTIONS FOR FURTHER READING

W. A. SHURCLIFF, *Polarized Light, Production and Use* (Cambridge, MA: Harvard University Press, 1962).

J. M. BENNETT and H. E. BENNETT, "Polarization," in W. G. Driscoll and W. Vaughan, editors, *Handbook of Optics*, Sec. 10, pp. 1–164 (New York: McGraw-Hill Book Company, 1978).

W. NESSE, *Introduction to Optical Mineralogy* (New York: Oxford University Press, Inc., 1986).

L. D. BARRON, *Molecular Light Scattering and Optical Activity* (New York: Cambridge University Press, 1983).

R. M. A. AZZAM and N. M. BASHARA, *Ellipsometry and Polarized Light* (New York: North-Holland, 1986).

L. J. PINSON, *Electro-Optics* (New York: John Wiley & Sons, Inc., 1985).

PROBLEMS

4.3-1. Unpolarized light is incident on a small volume of a scattering medium. What is the state of polarization of the light scattered at angles of 0°, 45°, and 135°, measured from the forward direction?

4.3-2. Circularly polarized light is passing through a colloidal suspension. Describe the state of polarization of the light scattered in directions of 0°, 45°, 90°, and 180° from the transmitted light?

4.3-3. What is the angle of incidence for complete polarization to occur on reflection at the boundary between water ($n = \frac{4}{3}$) and glass ($n = 1.589$) assuming the light comes from the side of:
(a) The water?
(b) The glass?

4.3-4. A light source immersed in a liquid of index 1.3972 is reflected at the plane face of a diamond ($n = 2.42$). If the angle of incidence is chosen for maximum polarization, what is the angle of *refraction?*

4.3-5. Sunlight is reflected from the Great Salt Lake ($n = 1.38$). If the reflected light is 100% polarized and the humid air above the lake has an index of 1.0026, *how high above the horizon* is the sun?

4.3-6. Light reflected by the polished surface of a block of glass has maximum polarization when the angle of *refraction* is 33.69°. What is the index of the glass?

4.3-7. Glare produced by wet pavement can be suppressed by Polaroid sunglasses. In which direction must the Polaroid molecules be oriented?

4.3-8. Polaroid filters can be made with the molecules forming concentric circles. If such an *axis finder* is placed in a beam of linearly polarized light, obtained by reflection at Brewster's angle, a dark band is seen extending across the filter in a direction normal to the plane of incidence at the Brewster surface. Therefore, are the molecules in the filter oriented *radial* or *tangential* with respect to the optic axis of the light passing through the filter?

4.3-9. When light passes through two polarizers whose axes of transmission are parallel, a photodetector reads 30 units. If one of the polarizers is then turned through 30°, what will the detector read?

4.3-10. Two Nicol prisms are set for maximum transmission. If one of the prisms is rotated through 60°, how much light is going through?

4.3-11. A bundle of unpolarized light passes through two polarizers whose planes of transmission are parallel to each other. Through what angle(s) must one of the polarizers be turned to reduce the amount of light to one-half of its initial value?

4.3-12. Two polarizers are crossed at 90° so that no light is going through. If a third polarizer, oriented at 45°, is placed between the two polarizers, what percentage of light will emerge from the system? (Try it; there *is* light going through.)

4.3-13. Following the example outlined in the text (page 314), plot the resultant polarization of two orthogonal, linearly polarized waves of equal amplitude which differ in phase by 30°.

4.3-14. A beam of unpolarized light passes through a linear polarizer and a quarter-wave plate, is reflected by a plane mirror, and passes again through the quarter-wave plate and the polarizer. What happens to the light?

4.3-15. Ice is a positive uniaxial crystal with indices of refraction of 1.309 and 1.310. How thick must the ice be to act as a quarter-wave plate for light of 600 nm wavelength?

4.3-16. A thin plate of calcite is cut with its axis parallel to the plane of the plate. What is the minimum thickness required to produce a $\lambda/4$ path difference for sodium light (589 nm)?

4.3-17. The specific rotation of quartz in light of 589 nm wavelength is 21.7°/mm. What thickness of quartz, cut perpendicularly to its axis and inserted between *parallel* polarizers, will cause no light to be transmitted?

4.3-18. A glucose solution of unknown concentration is contained in a 12-cm-long tube and seen to rotate linearly polarized light by 2.5°. Since the specific rotation of glucose is 52°, what is the concentration?

4.3-19. If the Verdet constant for monobromonaphthalene, in yellow D light, is 0.104 arc min/A and if a magnetic field of 2.6×10^5 A m^{-1} is applied and the light traverses a path 10 cm long, through what angle will the plane of oscillation turn?

4.3-20. If the Kerr constant at a wavelength $\lambda_1 = 450$ nm is of a certain magnitude, how large is it at a wavelength $\lambda_2 = 600$ nm?

Fourier Transform
Spectroscopy

TRANSFORMATION MEANS TO CHANGE from a given figure, or set of expressions, to an equivalent figure, or set of expressions. Transformations include many fundamental processes that occur in optics. They are the basis of all image formation. They also are of much interest in the formation of *spectra*. Fourier transform spectroscopy, in fact, is the one major advance in the technique of spectroscopy since Bunsen invented the spectroscope more than a hundred years ago. Fourier spectroscopy may not be the most convenient way to obtain a spectrum but it has an undisputed advantage over any prism or grating instrument: It is less wasteful of light.

FRINGE CONTRAST VARIATIONS

Return for a moment to double-slit interference. If the light were perfectly monochromatic, the resulting fringes would extend from the zeroth order out to infinity on both sides. But there is no such light, and hence the contrast farther away from the zeroth order becomes less (Figure 4.4-1a). To scan across these fringes, we move a slit with a photodetector behind it (b), which results in the trace shown in (c).

   If the light had been *more* monochromatic, the fringes would have extended farther, and an envelope drawn across the peaks of the trace would have been wider. If the light had been *less* monochromatic, the envelope would have been more narrow. Obviously, there exists some relationship between the degree of monochromaticity of the light and the shape of the envelope.

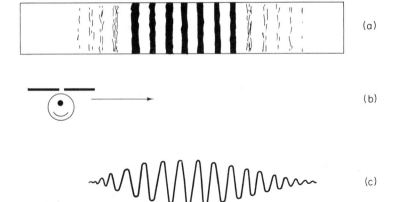

Figure 4.4-1   Interference fringes becoming less distinct farther away from the center (a); scanning across fringes (b) gives tracing (c).

But now assume that we use sodium D light and observe the fringe pattern produced by it in a Michelson interferometer. When one of the mirrors is moved through a considerable distance, we notice that at some times the fringes are more visible than at other times. In fact, the fringes go through periodic variations of contrast, reaching maximum contrast every 980th fringe. Fizeau, as early as 1862, followed such variations through 52 cycles; he then concluded that yellow sodium light has two components of about equal intensity, that it is a *doublet* (i.e., two lines) with a separation of $1/980$ of the average wavelength.* How did he arrive at this conclusion?

Think of two waves of slightly different wavelengths, λ_1 and λ_2. One wave is represented by the upper bar pattern in Figure 4.4-2, the other wave by the lower bar pattern. The difference between them is

$$\lambda_2 - \lambda_2 = \Delta\lambda \qquad [4.4\text{-}1]$$

On the left in the diagram the two waves coincide; on the right they coincide again. But obviously there is one more wave in the upper row. Then, if N is the number of waves between any two coincidences,

$$(N + 1)\lambda_1 = N\lambda_2$$

$$N\lambda_1 + \lambda_1 = N\lambda_2$$

$$\lambda_1 + \frac{\lambda_1}{N} = \lambda_2$$

$$\lambda_2 - \lambda_1 = \frac{\lambda_1}{N}$$

---

*H. Fizeau, "Recherches sur les modifications que subit la vitesse de la lumière dans le verre et plusieurs autres corps solides sous l'influence de la chaleur." *Ann. Chim. Phys.* (3) **66** (1862), 429–82.

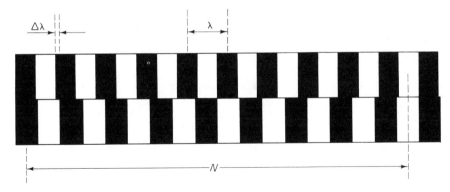

Figure 4.4-2 Vernier scale shows coincidences of spacing N, comparable to *beats* in acoustics and to the interference of doublet lines.

and, since $\lambda_1 \approx \lambda_2$,

$$N = \frac{\lambda}{\Delta\lambda}$$ [4.4-2]

the same equation that defines the *resolvance* of a diffraction grating, in the first order (page 276).

If the light were perfectly monochromatic, such light would have wavetrains of infinite length, with an unlimited number, N, of waves in each train, and the *bandwidth*, $\Delta\lambda$, would be zero. Needless to say—there is no such light.

Consider again the *coherence length* of the light,

$$\Delta s = N\lambda$$

and substitute for N Equation [4.4-2]. That results in

$$\Delta s = \frac{\lambda^2}{\Delta\lambda}$$ [4.4-3]

an expression which connects coherence length, wavelength, and bandwidth.

In Fizeau's observation, the wavelength difference between the two lines, from Equation [4.4-2], was found to be

$$\Delta\lambda = \frac{\lambda}{N} = \frac{589 \text{ nm}}{980} = 0.6 \text{ nm}$$ [4.4-4]

This agrees with the facts known today: The sodium D_2 line has a wavelength of 588.9953 nm and the D_1 line 589.5923 nm.

Fizeau's observation is of historic interest. It marks the first time that interferometry was used to reveal the doublet structure of a spectral line which had been believed to be monochromatic. From these beginnings, this technique has developed into a field of its own. Stated simply, we project light into an interferometer, obtain an interferogram, and convert the interferogram into a spectrogram, using Fourier transformation. This is the principle of *Fourier transform spectroscopy*.

Example

A Michelson interferometer is used for measuring the separation of a doublet known to have an average wavelength of 677 nm. Periodic variations of the fringe contrast are seen whenever one of the mirrors is moved through 0.44 mm. What are the two wavelengths?

Solution.   From the Michelson interferometer equation, page 214,

$$m = \frac{2d}{\lambda} = \frac{(2)(4.4 \times 10^{-4})}{677 \times 10^{-9}} = 1300$$

But $m = N$ and hence, from Equation [4.4-4],

$$\Delta\lambda = \frac{\lambda}{N} = \frac{677}{1300} = 0.52 \text{ nm}$$

If both lines have the same intensity, then $\lambda = 677 \mp 0.26$ nm, and thus

$$\boxed{\lambda_1 = 676.74 \text{ nm}, \qquad \lambda_2 = 677.26 \text{ nm}}$$

MICHELSON INTERFEROMETER SPECTROSCOPY

We now turn to the practical side of Fourier transform spectroscopy. Take a Michelson interferometer and let light from a source S be incident on the beamsplitter (Figure 4.4-3). The light proceeds to the two mirrors. One of the mirrors is stationary while the other is made to oscillate sinusoidally. If the incident light were monochromatic, the center of the field would alternately become bright and dark, and the photodetector would produce a sinusoidal alternating-current (ac).

The frequency of the signal depends both on the frequency of oscillation of the mirror and on the wavelength. Not only is it essential how fast the mirror moves but also how many wavelengths fit into the distance of motion. The shorter the wavelength, the higher the signal frequency: Light of one-half the wavelength gives twice the frequency. If the incident light contains several wavelengths, the output of the detector is no longer sinusoidal; the ac signals that correspond to the individual wavelengths will superimpose.

The motion of the mirror, or *modulation,* can be realized in several ways. Mechanically driven mirrors, tuning forks with a mirror cemented to one of the tines, and piezoelec-

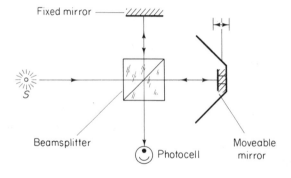

Fixed mirror

Beamsplitter

Photocell

Moveable mirror

S

Figure 4.4-3   Michelson interferometer as used for Fourier transform spectroscopy.

Figure 4.4-4 Compact Michelson-type interferometer spectrometer.

tric crystals of various shapes have all been used for this purpose. In the prototype model shown in Figure 4.4-4, the beamsplitter is an *Abbe cube* (a partially reflecting 45° boundary inside a solid glass cube). Light enters from the left. To the right of the cube is a loud-speaker, with a small first-surface mirror glued to its membrane. To the rear of the cube, not visible in the illustration, is a larger, stationary first-surface mirror. The recombined beams emerge from the side of the cube facing the reader and are picked up there through a fiber guide, which in turn leads to a photodetector (not shown).

Data reduction. Call x the path difference introduced by the motion of the mirror. If the light were monochromatic, the *amplitude* distribution in the interferogram would follow Equation [3.1-24]

$$y = \mathbf{A} \sin\left(2\pi \frac{x}{\lambda}\right)$$

Therefore, if the mirror is moving at a velocity $v = x/t$, the *intensity* distribution is

$$I = I_0 \sin^2\left(2\pi \frac{vt}{\lambda}\right) \qquad [4.4\text{-}5]$$

Using the identity $\sin^2 \alpha = \frac{1}{2} - \frac{1}{2} \cos 2\alpha$, this is equal to

$$I = \tfrac{1}{2} I_0 \left[1 - \cos\left(4\pi \frac{vt}{\lambda}\right)\right] \qquad [4.4\text{-}6]$$

Thus the signal obtained from the detector is modulated sinusoidally at a frequency $2v/\lambda$. In practice, it is best to let the mirror oscillate at a rate that causes the signal frequency to fall into the audio range—for which amplifiers and wave analyzers are readily available. If, from the variation of the photocurrent, we recover the power distribution (of the light), we have the spectrum of the source. To do this, we use *Fourier transformation.** Its principle is explained as follows.

A point source S is located in space at infinity, as illustrated in Figure 4.4-5. Light coming from S is incident at right angles on a line AB. An observer, standing at O anywhere along AB, will find that the illuminance is the same at all points. Furthermore, if the phase of the light at points O and O' were measured, the observer would find that the phase of the light intersecting AB is the same at both points.

But then let the source be at S', off the normal constructed on AB. Now the observer, moving from O to O', will find that there is a continual change of phase. This phase change has a certain periodicity, a function of both the angle θ and the wavelength.

Conversely, assume line AB to be an *emitter* of radiation. With a uniform phase along AB, the beam will proceed toward S. But if the phase along AB were to change sinusoidally, with a certain space frequency, the beam will proceed toward S'. An example is the *phase-array radar.* These radars are composed of sometimes more than a thousand individual, stationary antennas; in contrast to the rotating dish of conventional radar, there are no moving parts. Instead, the beam is steered electronically, inducing phase changes between any two adjacent antennas. This causes the beam to be deflected in any direction, away from the array's surface normal ("boresight").

Obviously, a reciprocal relationship exists between the spatial distribution of the phase along a line (in terms of x) and the angular distribution (in terms of θ) and one can be *transformed* into the other:

$$f(x) \rightleftharpoons F(\theta) \qquad\qquad [4.4\text{-}7]$$

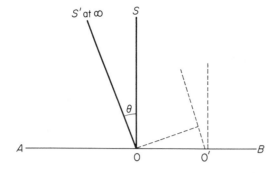

Figure 4.4-5 Deriving the concept of Fourier transformation. [Reprinted with permission from R. C. Jennison, *Fourier Transforms and Convolutions for the Experimentalist* (Elmsford, NY: Pergamon Press, Inc., 1961).]

---

*Jean Baptiste Joseph Baron de Fourier (1768–1830), French physicist and mathematician. Because of his mathematical skills, Fourier became an artillery officer, served as Napoleon's science adviser to Egypt, later became prefect of Isère, near Grenoble, Secretary of the French Academy of Sciences, and head of the École Polytechnique. After his return from Egypt, Fourier wrote a 21-volume treatise on Egyptology. His *Analytical Theory of Heat* became the foundation of thermodynamics. In 1815 Fourier showed that any periodic oscillation, no matter how irregular, can be broken into a series of sine functions, an insight, first rejected, but today considered a classic of mathematical physics.

The transformation, however, need not be confined to *spatial* distribution versus *angular* distribution. Frequently, we have two spatial distributions, especially in the case of two-dimensional transforms (to be discussed in Chapters 4.5 and 4.6). In other cases, the transform may involve the frequency domain, as with the oscillating mirror just described. Thus, to be as general as possible, we replace the variable θ by another variable, ξ, and write

$$f(x) \rightleftharpoons F(\xi) \tag{4.4-8}$$

where we follow the convention that the capital F represents the *Fourier transform* of the function designated by the lowercase f.

Fourier transformation.   In Chapter 3.2 we have seen that if two or more sinusoidal waves of equal frequency superimpose, the result is another sinusoidal wave. If the waves are of different frequencies, the result is a *complex* wave. Now we examine how a complex wave can be decomposed to find its contributions.

According to *Fourier's theorem,* any periodic function can be represented by a sum of sine and cosine terms. This is an extension of Equation [3.1-1] which describes the height of a point on a sinusoidal wave as a function of time,

$$y = \mathbf{A} \sin(\omega t) \tag{4.4-9}$$

The result is a *Fourier series,*

$$y = \tfrac{1}{2} A_0 + \sum_{m=1}^{\infty} A_m \cos(\omega t) + \sum_{m=1}^{\infty} B_m \sin(\omega t) \tag{4.4-10}$$

where the amplitude coefficients, A_m and B_m, are given by the integral expressions

$$A_m = \frac{2}{\lambda} \int_0^\lambda f(x)\cos(\omega t)dx \tag{4.4-11}$$

and

$$B_m = \frac{2}{\lambda} \int_0^\lambda f(x)\sin(\omega t)dx \tag{4.4-12}$$

A Fourier series represents the composite periodic wave, extending from infinity to infinity. By including a theoretically infinite number of terms, Fourier's theorem makes it possible to synthesize any waveform, even a "square" wave.

But Fourier analysis is not limited to periodic events, or to infinitely long waves. A *wavetrain*, that is, a group of waves of limited length, can, and must, be represented by a *Fourier integral.* A property of these integrals is that the component waves differ only by infinitesimal increments of wavelength, rather than by multiples of wavelengths.

Consider now Figure 4.4-6, which is an extension of Figure 4.4-5. As before, the light comes from infinity and proceeds along the axis; thus we have plane wavefronts incident on the aperture (left). A lens placed close to the aperture focuses the light into the F_2 plane (right).

For simplicity, consider only one dimension, normal to the direction of propagation and lying *in* the plane of the diagram. The height of the object is called x (not shown), the conjugate height of the image x', and the height in the transform ξ. We now determine the amplitude of the light that is *diffracted* toward an arbitrary point, P, in the image. This

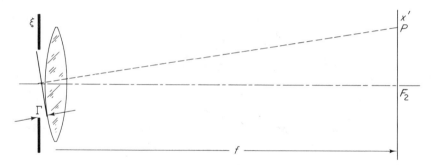

Figure 4.4-6 Deriving reciprocal relationship between light distribution within aperture (*left*) and light distribution in the image plane (*right*).

amplitude is the sum of the contributions from all elements of width $d\xi$ in the aperture. Inside the aperture we construct a line on which the optical path length is constant from any point to P. But P is off-axis; hence at any given point on the line there must be a path *difference*, Γ, between the incident light and the diffracted light. If the path difference is measured in wavelengths, then from Equation [3.3-9], the *phase* difference is

$$\delta = 2\pi\Gamma \qquad\qquad [4.4\text{-}13]$$

Each element of width $d\xi$ thus contributes an amplitude at P that may be represented by

$$d\mathbf{A} = d\xi \cos 2\pi\Gamma \qquad\qquad [4.4\text{-}14]$$

Now, since the cosine term represents the real part of a complex function, we have, in analogy with Equation [3.1-11],

$$d\mathbf{A} = d\xi e^{2\pi i \Gamma} \qquad\qquad [4.4\text{-}15]$$

From similar triangles in Figure 4.4-6,

$$\frac{\Gamma}{\xi} = \frac{x'}{f} \qquad\qquad [4.4\text{-}16]$$

We solve this equation for Γ, substitute it in Equation [4.4-13] and for convenience set $f = 1$. This gives

$$\delta = 2\pi x'\xi \qquad\qquad [4.4\text{-}17]$$

The total amplitude distribution in the image, $f(x')$, is then found by integration:

$$f(x') = \int_{-\infty}^{+\infty} F(\xi)e^{2\pi i x'\xi}d\xi \qquad\qquad [4.4\text{-}18]$$

where $F(\xi)$ is the amplitude distribution inside the aperture.

In the absence of absorption, the propagation of light is reversible. Therefore, we can reverse the process; from the amplitude in the image, we can find the amplitude in the aperture. Therefore, the *Fourier transform*, $F(\xi)$, of a function $f(x)$ is

$$\boxed{F(\xi) = \int_{-\infty}^{+\infty} f(x)e^{-2\pi i x\xi}\,dx} \qquad\qquad [4.4\text{-}19]$$

and the *inverse Fourier transform, $f(x)$, of $F(\xi)$* is

$$\boxed{f(x) = \int_{-\infty}^{+\infty} F(\xi)e^{2\pi ix\xi} \, d\xi}$$ [4.4-20]

The two functions $f(x)$ and $F(\xi)$ are called a *Fourier transform pair.* The exponential terms, as in Euler's formula, page 200, represent trigonometric functions,

$$e^{\pm 2\pi ix\xi} = \cos(2\pi x\xi) \pm i \sin(2\pi x\xi)$$ [4.4-21]

There are several, slightly different ways of writing these integrals. For example, we may reverse the signs in the exponents of Equations [4.4-19] and [4.4-20] but there must always be $+i$ in one equation and $-i$ in the other. Furthermore, we could take the 2π terms outside the integral:

$$F(\xi) = \frac{1}{2\pi} \int_{-\infty}^{+\infty} f(x)e^{-ix\xi} dx$$ [4.4-22]

but then

$$f(x) = \int_{-\infty}^{+\infty} F(\xi)e^{ix\xi} d\xi$$ [4.4-23]

Or we could split the $1/(2\pi)$ term between the two functions:

$$F(\xi) = \frac{1}{\sqrt{2\pi}} \int \cdots \quad \text{and} \quad f(x) = \frac{1}{\sqrt{2\pi}} \int \cdots$$ [4.4-24]

Note the symmetry in all these expressions. Except for the sign in the exponent, x and ξ can be interchanged at will to convert one of the equations into the other. This illustrates the purpose of transformations in optics: We have seen a spectrum produce an interferogram, and now we see an interferogram converted into a spectrogram.

Point sampling and results.   Now that we have a recording of the interference fringes, the irradiance distribution across the fringes must be processed further. Michelson merely used the visibility curve, that is, the envelope over the fringes. Today we sample the irradiance at preselected, equally spaced *data points* (irrespective of whether they are located in a maximum or between maxima). The irradiance received at each datum point need not be recorded; it is simply received by a photodetector and fed directly into a computer, which is programmed to perform the transformation.

Sometimes as many as 1000 data points are used (which increases the resolution), but usually 40 to 100 points are sufficient. Again, if the light were monochromatic, the irradiance distribution would follow Equation [4.4-6]. But if the light contains more than one frequency, each frequency contributes its own set of fringes and integration is needed to add these contributions. The integral can be evaluated, that is, the transformation can be carried out, in real time. This means that as soon as the interferogram is received, its transform, the spectrogram, is ready for inspection.

Figure 4.4-7 shows several typical examples. As we have seen, if the fringes extend out to infinity, the light would be perfectly monochromatic and the spectrum line causing the fringes would be of infinitesimal width (neither of which is possible). If there are simple

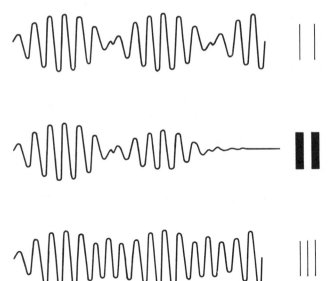

Figure 4.4-7 Interferograms obtained by scanning (*left*) and resulting spectrograms obtained by Fourier transformation (*right*).

harmonic variations of contrast, out to infinity, we would have two lines, close together and of infinitesimal width (top), which also is impossible. In reality we may have such variations and a gradual loss of contrast farther away from the axis. This then means two lines, each of finite width, like the sodium D example referred to earlier (center). Three lines result from an even more complex interferogram (bottom).

ADVANTAGES OF FOURIER TRANSFORM SPECTROSCOPY

1. Most important, a Fourier transform spectrometer can have a *wide aperture.* A grating or prism spectrometer must have a narrow slit. Expressed another way, a grating or prism scanning spectrometer looks at one wavelength at a time, rejecting all others. A Fourier transform instrument looks at all wavelengths all of the time. This has the advantage that spectra can be obtained even of very faint light sources or within a very *short interval of time.* High-resolution spectra can be obtained in much less time than with a grating or prism spectrometer.

2. In theory, the resolvance of a Fourier spectrometer is unlimited if the path difference is also unlimited. In practice, that is not possible. The mirror in a Michelson interferometer, for example, can move through a limited distance only, limiting the resolution. In analytical terms, the interferogram cannot be integrated from $-\infty$ to $+\infty$; the integral is *truncated.*

3. In any spectrometer, throughput and resolvance are inversely proportional. Thus some dimension of the spectroscope, for example the slit width, must be made smaller to gain higher resolvance. But the product of throughput $\times$ resolvance is not the same in every case: For a grating spectrometer it is slightly higher than for a prism, and for a Michelson-type Fourier spectrometer it is at least two orders of magnitude higher than for a grating.

Fourier transform spectroscopy is the superior method. Even more important, Fourier spectroscopy is not simply the application of another little invention; rather, it marks a turning point in philosophy, *away* from high-precision delicate optics, *toward* a simple, rugged sensor coupled with sophisticated *electronic data processing.*

SUGGESTIONS FOR FURTHER READING

L. MERTZ, *Transformations in Optics* (New York: John Wiley & Sons, Inc., 1965).

K. IIZUKA, *Engineering Optics,* 2nd edition (New York: Springer-Verlag New York, Inc., 1987).

R. N. BRACEWELL, *The Fourier Transform and Its Applications,* 2nd edition (New York: McGraw-Hill Book Company, 1978).

D. R. MATTHYS and F. L. PEDROTTI, "Fourier Transforms and the Use of a Microcomputer in the Advanced Undergraduate Laboratory," *Am. J. Phys.* **50** (1982), 990–95.

PROBLEMS

4.4-1. Using a Michelson interferometer with light of 500 nm wavelength, maximum contrast occurs every 250th fringe. What is the difference of the two wavelengths emitted by the source?

4.4-2. The spectrum of calcium contains two closely spaced lines centered around 395.1 nm. If interference fringes produced by these lines reach maximum contrast at every 116th fringe, what are the two wavelengths?

4.4-3. If we assume that the wavetrains from the 546-nm green mercury line are 11 mm long, what is the hypothetical bandwidth? (Hypothetical because of *Doppler broadening,* to be discussed in a later chapter.)

4.4-4. What is the bandwidth of the orange krypton line, 606 nm, assuming that the coherence length is 80 cm?

4.4-5. If the mirror in a Michelson-type Fourier spectrometer moves at a velocity of 4 mm s^{-1} and the light comes from a helium-neon laser (633 nm):
(a) What is the frequency of modulation of the photocurrent?
(b) If this within the audio range?

4.4-6. If a Michelson spectrometer is used with sodium light (589 nm), what should be the velocity of the mirror to modulate the photocurrent at a frequency of 5 kHz?

4.4-7. Write the equation of a complex wave that results from the superposition of two simple harmonic waves which have amplitudes in the ratio of $2:1$, frequencies in the ratio of $1:3$, and which are in phase, meaning that at time $t = 0$ their displacement is $y = 0$.

4.4-8. Continue with Problem 4.4-7 and draw plots of the two contributions and of the resultant wave.

4.5

Optical
Data Processing

EVEN A SIMPLE LENS is an optical data processor. It transforms a two-dimensional array of data, the object, into another two-dimensional array of data, the image, and it does so without scanning. *One-dimensional* data can very well be processed electronically. But humans, since time immemorial, have become accustomed to *two-dimensional* data, from hieroglyphics to letters of the alphabet to maps and paintings. Such data can be processed best by two-dimensional transformation; that is the domain of *optical data processing*.

ABBE'S THEORY OF IMAGE FORMATION

Optical data processing, for all its high technology, is a direct descendant of Abbe's venerable *theory of image formation*.* Abbe developed it more than a century ago to account for image formation in the microscope.

As shown in Figure 4.5-1, light from a point source (bottom) enters a collimating lens

---

*Ernst Abbe (1840–1905), German physicist, professor at the University of Jena, chief designer and partner of Carl Zeiss at the Carl Zeiss Optical Company in Jena. Abbe is known also for the theory of pupils, the sine condition, refractometer, and blood counting chamber, all named after him. In the course of his work, Abbe discovered that microscope objectives of low numerical aperture, however well they were made, produced images inferior to those produced by simpler lenses of high aperture, and on this basis formulated his theory of image formation. E. Abbe, "Beiträge zur Theorie des Mikroskops und der mikroskopischen Wahrnehmung," *Arch. mikrosk. Anat.* **9** (1873), 413–68.

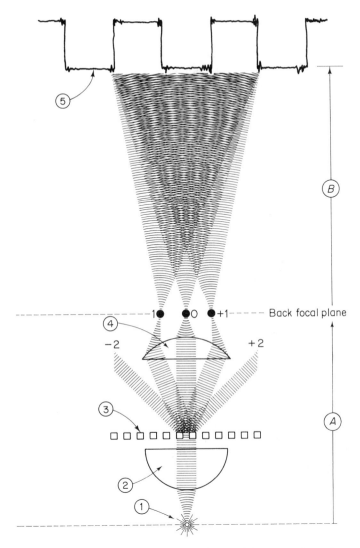

Figure 4.5-1 Image formation in the microscope. 1, Light source; 2, condenser; 3, object; 4, objective lens; 5, image; A, B, Fourier transformations.

(the condenser of the microscope) and is incident on the object. The object is a grid with a series of closely spaced parallel lines.

Much of the light passes through the grid and forms the zeroth-order maximum. Other light is diffracted by the grid; that light forms the higher orders. Since the light is collimated, the diffraction pattern is of the *Fraunhofer type* or, in terms of optical data processing, it is the *Fourier transform* of the light distribution across the grid. Another Fourier transformation then converts that transform into the (intermediate) image which is seen through the eyepiece. According to Abbe's theory, therefore, *an image is the result of two successive Fourier transformations.*

Also note that an image requires two contributions: undiffracted light (the zeroth order), which provides the overall *illumination,* and diffracted light (the higher-order maxima), which provides the transfer of *information.* If more higher orders could be admitted and contribute to image formation, the image would contain more information, more detail. This is why a microscope objective of high numerical aperture yields a "better" image; "magnification" as such does not mean much.

Since, by necessity, any lens is limited in diameter, only the zeroth order and the first few higher orders enter the objective and contribute to the image. This is why an image contains less information than the object, a fact that is most obvious if the object has details that are very small or very close together. If the details are so small, or the lens is so narrow, that no higher orders can enter its front surface, the object structure cannot be resolved, no matter what the magnification.

At first, Abbe's theory was thought to apply only to periodic objects, to the microscope, and to coherent light. Today we know that it applies to any object, to any optical system, and to any type of light. But why should we be concerned with the transformations that occur in image formation? Because these transformations make it possible to modify the image, using the process of *spatial filtering.*

Lab Experiment Illustrating Abbe's Theory

1. Use as the object a grid that has about 50 lines to the millimeter. Place the grid on an optical bench and illuminate it, from the left, with collimated white light. Orient the grid so that its lines are horizontal.

2. Assemble a compound microscope, using a $+10$-diopter lens as the objective and a $10\times$ Huygens eyepiece placed 30 cm to the right of the objective. Looking through the eyepiece, focus on the grid by slightly moving the objective back and forth.

3. Place a sheet of ground glass 10 cm to the right of the objective and slowly move the ground glass back and forth. In one position the light forms a distinct pattern (Figure 4.5-2, left). This is the Fourier transform; it appears in the back focal plane of the objective.

The bright spot in the center is the zeroth-order maximum. The less intense spots above and below are the higher-order maxima. The zeroth order is white. The higher orders are colored, red on the outside, blue on the inside.

4. Without moving anything else, replace the ground glass with an adjustable slit. Align the slit with the transform. Block out all maxima except the zeroth order (Figure 4.5-2, center). A small piece of paper, held closely behind the slit, will help you do this. Look through the eyepiece. Sufficient light passes through, but *no image is seen:* the zeroth-order maximum alone does not produce an image.

5. Open the upper and lower bracket of the slit so that the two first-order maxima are admitted

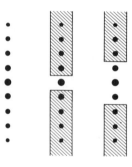

Figure 4.5-2   Fourier transform of a grid object, using point light source (*left*). Same transform, but the zeroth-order maximum only is admitted to the image space (*center*). Zeroth- and two first-order maxima admitted (*right*).

also (Figure 4.5-2, right). The lines in the grid become visible. Fully open the slit, admitting all maxima. The grid is seen even better: Higher-order maxima are needed for seeing more detail.

A series of spatial-filtering experiments, well suited for the more advanced optics lab, has been described by A. Eisenkraft, "A closer look at diffraction: Experiments in spatial filtering," *The Phys. Teacher* **15** (1977), 199–211.

TWO-DIMENSIONAL TRANSFORMS

Spatial filtering.   Now assume that the object is a *cross-grating,* placed in the left-hand focal plane of a converging lens (Figure 4.5-3). The Fourier transform of the cross-grating is a two-dimensional array of maxima. Then another lens is placed somewhere to the right of the first lens and a screen is placed in the right-hand focus of the second lens. On that screen we see the transform of the transform, that is, we see an image of the original cross-grating.

To modify the image, we place a *mask* in the transform plane. A mask is essentially a stop with one or several holes in it that will admit to the image plane only some of the light. First we use a mask that has merely one hole, coincident with the axis and so small that only the zeroth order can pass through. The higher orders are blocked. The result is that, while the screen receives enough light (from the zeroth order), *no image* will form (Figure 4.5-4, top).

Next we use a mask with a vertical *slit.* Such a mask will admit maxima that extend in the vertical direction. These maxima are caused by diffraction at the horizontal lines of the object; consequently, the image will show horizontal lines (bottom). If the mask were wide

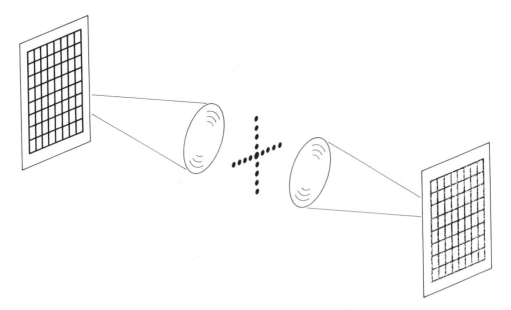

Figure 4.5-3   Image formation showing object (*left*), image (*right*), and Fourier transform (*between the two lenses*).

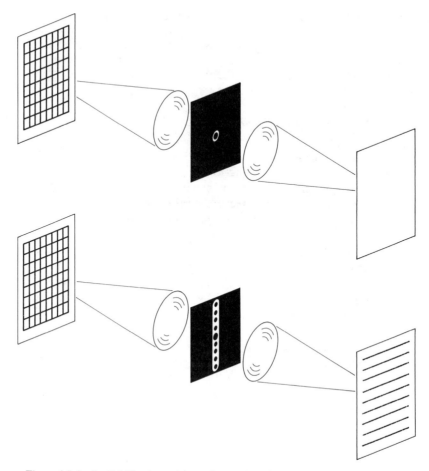

Figure 4.5-4   Spatial filtering using as the mask a pinhole (*top*) and a vertical slit (*bottom*).

open, or if we would use no mask, all of the grid, horizontal and vertical lines, would be visible.

But we can also use a mask that *blocks* the maxima. This time, the object may be a flower pot placed behind a grid. The grid consists of a regular series of lines. Its Fourier transform is a regular, predictable system of maxima. A mask blocking these maxima is easy to make.

But consider first that the mask has holes which coincide with the maxima. The result is that the grid (which causes these maxima) is visible. The flowers produce a more diffuse transform; this transform will for the most part be blocked by the mask and the flowers will not be visible (Figure 4.5-5, top). On the other hand, if the mask blocks the maxima caused by the grid, the grid cannot be seen but the flowers, with their more diffuse transform, can be (bottom).

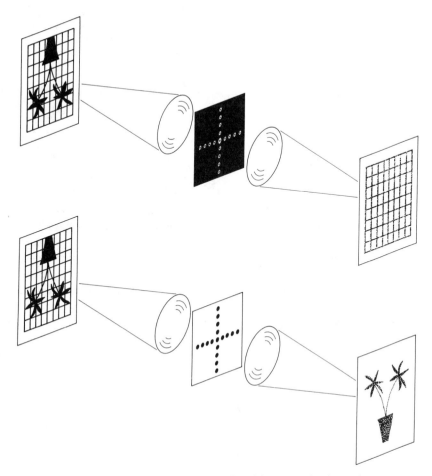

Figure 4.5-5 Spatial filtering using a mask that either transmits the maxima generated by a cross-grid (*top*) or that blocks them (*bottom*).

Theta modulation.   *Theta modulation,** another form of spatial filtering, also derives from Abbe's theory. Consider a scene that contains a house, a lawn, and sky (Figure 4.5-6, left). These elements are made out of small pieces of plastic diffraction grating oriented at different angles theta, θ—hence the term "theta modulation" (center).

The object is projected by a lens onto a screen. Focus the image. Cover the lens with a sheet of aluminum foil, the dull side facing the light source. On the foil you will see three pairs of spectra that correspond to the elements in the object scene, red being diffracted most, blue least. Cut out part of the foil (such as hole A in Figure 4.5-6, right); let only the red pass through in the two spectra that belong to the house, green for the lawn, blue for the sky. The mask thus made transmits only part of the light and the image appears in these colors.

*J. D. Armitage and A. W. Lohmann, "Theta Modulation in Optics," *Appl. Opt.* **4** (1965), 399–403.

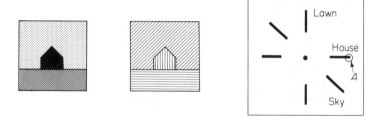

Figure 4.5-6   Theta modulation. Landscape scene in half-tone representation (*left*), theta-modulated (*center*), and Fourier transformed (*right*).

Input and output.   Consider now in more detail the general form of an optical data processor. Again we use a sequence of two lenses. The object is placed in the left-hand focal plane of the first lens, the *input plane.* Light emerging from that plane is made parallel by the (first) lens, hence the distance between the two lenses is immaterial.

The two lenses together form an image in the right-hand focal plane of the second lens, the *output plane.* For reasons of simplicity, we use two lenses of the same focal length and separate the lenses by twice that focal length. The *transform plane,* therefore, is halfway between the lenses. The system, shown in Figure 4.5-7, has three planes and two lenses, each separated from the next by one focal length ($1f$), with the total length from input to output equal to $4f$: that is called the $4f$ *configuration.*

In order to describe the size and shape of the transform, we need to know the coordi-

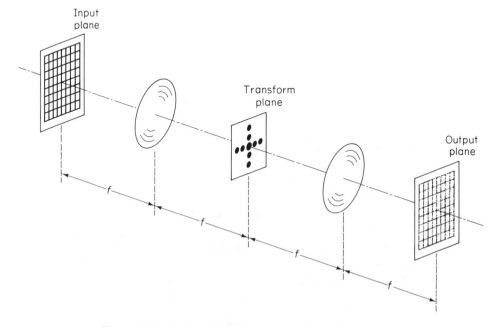

Figure 4.5-7   The $4f$ configuration of an optical data processor.

nates in the various planes. In the input plane, any point is defined by two variables, x and y. In the transform plane, we need other coordinates, following the notation of Fourier transformation. These are the coordinates ξ and η. In the output plane the coordinates are again x' and y'. The coordinates x and ξ, and y and η, are inversely proportional to each other: the closer the lines in a grating, for example, the wider apart the maxima.

In Chapter 4.4 we have seen that the (one-dimensional) Fourier transform, $F(\xi)$, of a function $f(x)$ is

$$F(\xi) = \int_{-\infty}^{+\infty} f(x)e^{-2\pi i x \xi}\, dx \qquad [4.5\text{-}1]$$

Now we go to two dimensions. For an input we use function $f(x, y)$; that produces the Fourier transform

$$F(\xi, \eta) = \int\int_{-\infty}^{+\infty} f(x, y)e^{-2\pi i(\xi x + \eta y)}\, dx\, dy \qquad [4.5\text{-}2]$$

This equation describes the *spatial-frequency spectrum.*

The second lens provides another Fourier transformation,

$$f(-x', -y') = \int\int_{-\infty}^{+\infty} F(\xi, \eta)e^{2\pi i(x\xi + y\eta)}\, d\xi\, d\eta \qquad [4.5\text{-}3]$$

the minus signs in the left-hand term indicating that the image, compared to the object, is upside down.

The output (image), as we have mentioned before, may be modified in various ways, placing a mask in the transform plane. The simplest type of a mask is one which eliminates the high spatial frequencies, that is, a *low-pass filter* (Figure 4.5-8a). Such filters are used to make a laser beam more uniform, eliminating its granularity, which may be due to inhomogeneities in the laser windows and because of dust.

A filter that eliminates the low spatial frequencies is a *high-pass filter* (b). Such a filter enhances the details of an image (the high frequencies) and emphasizes its edges (*edge enhancement*). A *band-pass filter* is a ring-shaped, annular aperture (c); it enhances details of a certain, predetermined size.

For example, consider a cytological specimen such as a Pap smear. Without spatial filtering, we see large epithelial cells, medium-sized leukocytes, and small particles of undetermined origin, a wide range of microscopic detail. Using a band-pass filter, large structures and very small detail can be suppressed. The various cell nuclei, however, that in a Pap

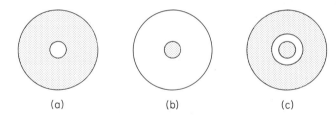

(a)   (b)   (c)

Figure 4.5-8 Various forms of simple spatial filters: (a) low-pass filter; (b) high-pass filter; (c) band-pass filter.

smear are the objects of interest, will stand out, a virtual necessity in any computer-aided screening of cytological samples.*

Phase contrast is another, particularly useful application of spatial filtering. Abbe had always assumed that the object is an *amplitude grating*. Its diffraction maxima all have the same phase. But the object may instead be a *phase grating* (see page 277).

In Figure 4.5-9 we represent the various transforms by vectors. An amplitude grating, left, has an intense zeroth-order maximum and less intense higher orders, all of the same phase. A phase grating has an even more intense zeroth-order maximum (longer vector!), but the higher orders differ in phase by $\pi/2$ radian or 90° (center). The transform of a phase grating, therefore, is distinctly different from that of an amplitude grating.

It then occurred to Zernike† that if the transform were modified by attenuating the amplitude and delaying the phase of the zeroth order, a *phase object could be made to look like an amplitude object* (Figure 4.5-9, right). That is of particular interest in the microscopic examination of living tissue where the details often differ from their surroundings only by their refractive index and, without staining, are hardly visible at all. Zernike called his process the "phase-strip method for observing phase objects in good contrast" or *phase contrast,* for short.

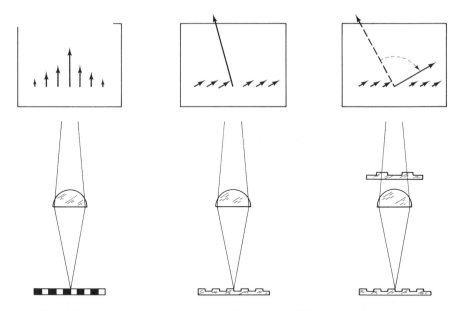

Figure 4.5-9   Vector representation of amplitude object (*left*), phase object (*center*), phase contrast (*right*).

*See P. J. S. Hutzler, "Spatial frequency filtering and its application to microscopy," *Appl. Opt.* **16** (1977), 2264–72.

†Frits Zernike (1888–1966), Dutch physicist, professor of physics at the University of Groningen, received the Nobel Prize in physics in 1953 for the discovery of phase contrast. F. Zernike, "Beugungstheorie des Schneidenverfahrens und seiner verbesserten Form, der Phasenkontrastmethode," *Physica* **1** (1934), 689–704; and "How I Discovered Phase Contrast," *Science* **121** (1955), 345–49.

Correlation and convolution.   Finally, we consider the important case that somehow *two* patterns interact with each other. That is called *convolution*. Again, we discuss first an experimental example and then go to the more theoretical aspects.

Assume that two transparencies are mounted, at some distance from one another, on an optical bench. The first transparency contains a circle (Figure 4.5-10, top left). The light, before it reaches the circle, is passing through a sheet of ground glass. The second transparency contains a triangle, actually a series of six pinholes. The system now acts as a *multiple-pinhole camera* that forms as many images of the circle as there are pinholes in the triangle (top right).

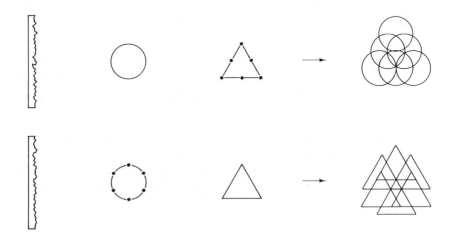

Figure 4.5-10   Circle and triangle: examples of cross-correlation.

If, instead of the triangle, the circle is made out of pinholes, we would have a *pinhole projector* that produces a series of triangles in a circle (bottom row). In either case, the result is an extended and, depending on the size of the pinholes, slightly blurred, image.

Now assume that the first transparency has a transmittance function $M(x, y)$. The transparency is illuminated with light of intensity I_0; that produces an intensity distribution passing through the transparency of

$$I(x, y) = I_0 M(x, y) \qquad [4.5\text{-}4]$$

The second transparency has a transmittance function $N(x, y)$. The intensity distribution that emerges from the second transparency is the product of both,

$$P(x, y) = I_0 M(x, y) N(x, y) \qquad [4.5\text{-}5]$$

This is the convolution of the two transparencies, a measure of the *correlation* of the two functions.

The difficulty is that when one of the two transparencies is moved in the *x-y* plane, either $M(x, y)$ or $N(x, y)$ changes while the pattern as such does not change. Hence, the function must be invariant under translation and the pattern have "internal coordinates" that move with it. The output, now called $P(\xi, \eta)$, is the integral of the product of $M(x, y)$

and $N(x - \xi, y - \eta)$ for all values of x and y over the extent of the pattern,

$$P(\xi, \eta) = \iint_{-\infty}^{+\infty} M(x, y)N(x - \xi, y - \eta)dx\, dy \qquad [4.5\text{-}6]$$

which is the *convolution integral.* The direction of translation, either $+$ or $-\xi$, and $+$ or $-\eta$, is immaterial. This is an example of a two-dimensional convolution between *different* functions, an example of *cross-correlation.*

Mathematically, as well as experimentally, determining the convolution is not easy, not even for a simple function. However, when we carry out the convolution of two Fourier transforms, rather than of the functions themselves, the operation is much simpler; in that case the result is merely the *product* of the Fourier transforms of the components.

The situation is different if a pattern is convolved with an identical pattern, that is, if it is convolved with itself. In that case, $\xi = \eta = 0$, and Equation [4.5-6] becomes

$$P(\xi, \eta) = \iint_{-\infty}^{+\infty} M(x, y)M(x, y)dx\, dy = \iint_{-\infty}^{+\infty} M^2(x, y)dx\, dy \qquad [4.5\text{-}7]$$

This is an example of *autocorrelation.* In that case the product in the integrand is as large as it can be for any given function. That means that in the center of the pattern detail the output reaches a maximum, forming a bright spot of light that is easy to see, an important feature in any *pattern recognition.* That subject I will discuss more fully in the next chapter.

OPTICAL VERSUS ELECTRONIC DATA PROCESSING

Virtually all electronic computers process information, in the form of *bits,* along single channels, one pulse after another. The only independent variable available to them is time.

Optical computers, by contrast, have the inherent property that they possess two degrees of freedom, represented by the two variables which define a point in a plane. Therefore, optical computers process information *always in parallel,* numerous channels side by side and simultaneously. That is an important advantage, considering that a well-corrected lens can easily resolve a two-dimensional array of 1000×1000 pixels. Consequently, optical computers have become well suited to test different architectures of parallel data processing.

Moreover, the information in an optical computer is carried by pulses of *light* rather than by electric pulses. Optical pulses do not interact with one another as electrons do. That means that different optical pathways can cross through each other without interference: Light traveling in one channel does not affect the light traveling in another channel. And since nothing travels faster than light, optical pulses provide an efficient means of communication. Indeed, the speed of an all-optical computer is extraordinarily high.

Many components used in an optical computer are elements familiar from other optical systems: lenses, fibers and integrated waveguides, lasers, light modulators, detector arrays, and other optoelectronic devices that provide the necessary input/output interfaces. Even a simple lens, as we have seen, can perform as difficult a task as Fourier transformation, and it does so with the greatest of ease. It can also speedily correlate any two-dimensional arrays of data. It can even serve as a fan-out device that spreads data to hundreds of workstations with a degree of synchronization far better than what is needed today (or tomorrow). Given the right interface components, optical computers are likely to make significant contributions to the science and art of high-speed computation.

SUGGESTIONS FOR FURTHER READING

J. W. GOODMAN, *Introduction to Fourier Optics* (New York: McGraw-Hill Book Company, 1968).

H. STARK, *Applications of Optical Fourier Transforms* (New York: Academic Press, Inc., 1982).

G. L. ROGERS, *Noncoherent Optical Processing* (New York: John Wiley & Sons, Inc., 1977).

W. K. PRATT, *Digital Image Processing* (New York: John Wiley & Sons, Inc., 1978).

G. BAXES, *Digital Image Processing* (Englewood Cliffs, NJ: Prentice-Hall, Inc., 1984).

R. A. SCHOWENGERDT, *Techniques for Image Processing and Classification in Remote Sensing* (New York: Academic Press, Inc., 1983).

PROBLEMS

4.5-1. **(a)** If the slit in Figure 4.5-11 were moved up by a distance twice its width, how would the transform change?

   (b) If the slit were rotated clockwise, how would the transform rotate?

   (c) If the slit were made narrower, what would you see?

Figure 4.5-11

4.5-2. If the maxima in a transform pattern form a straight line extending between the numbers 8 and 2 on a clock, how are the lines in a grating oriented that cause these maxima?

4.5-3. The object illustrated in Figure 4.5-12 is known as an *Abbe diffraction plate.*
  (a) If such an object is placed in a microscope of high numerical aperture, what do you see in the Fourier transform plane?
  (b) How would the transform, and the image, change if the objective were of low numerical aperture?

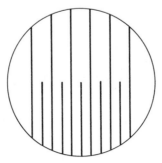

Figure 4.5-12

4.5-4. What does the Fourier transform, seen in the back focal plane of a microscope objective, look like if the object is a linear grating and the aperture below the condenser, instead of being circular, is an equilateral *triangle?*

4.5-5. The image sent back from space by a vehicle exploring Jupiter is marred by many closely spaced, slightly oblique but parallel scan lines. What kind of spatial filter will suppress these lines?

4.5-6. How can the grainy structure of a newspaper picture be smoothed out by a suitably shaped spatial filter?

4.5-7. An image contains a great many closely spaced concentric circles. If a long, narrow, horizontal slit is then placed in the transform plane, how will the image change?

4.5-8. A wild beast is kept in a cage made with many vertical bars. When in a photograph the bars are to disappear so that the beast seems to be out in the open, what kind of a spatial filter is needed to remove the bars while having the least effect on the rest of the picture?

4.5-9. The scene illustrated in Figure 4.5-13 is theta-encoded. How will the spectra, seen with white light, be oriented? What should a space filter look like that is to make the sky blue, the water green, the boat white, and the sail red?

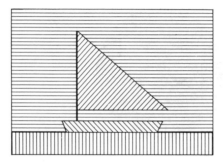

Figure 4.5-13

4.5-10. If the mask in theta modulation is made so that the barn is red, the sky blue, and the grass green, and if a red filter is then placed in the path, how will it change the image?

4.5-11. A wire mesh containing 25 wires per centimeter each way serves as the object in an optical data processor. If two lenses, one on each side of the mesh, have focal lengths of 60 cm each and if the light has a wavelength of 600 nm, what is the distance between adjacent maxima in the transform plane?

4.5-12. If only the first-order maxima are used for image formation, what percentage of the theoretically available information is being utilized? [*Hint:* Add the intensities in all the maxima (excluding the zeroth order), using the figures given on page 252, and estimate the percentage of light in the first orders.]

4.5-13. A camera with a spatial filter is aimed at a group of high-rise buildings. How much of the buildings, their contours and windows, and of clouds, birds, and other detail can be seen if the spatial filter transmits:

(a) Only the low space frequencies?

(b) All frequencies?

(c) Only the high space frequencies?

4.5-14. Determine the convolution of two transparencies, each containing three holes as shown in Figure 4.5-14.

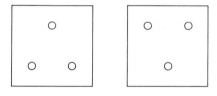

Figure 4.5-14

4.6

Holography

DISCOVERIES ARE OFTEN MADE because the time is right. Some discoveries have been made almost simultaneously yet independently. This is not so with *holography*. Holography is the discovery of one man, Dennis Gabor, at that time, 1948, working in a British industrial research laboratory. His discovery came at a time when the technology was not yet ready to take full advantage of it. It was several years later, with the invention of the laser, that Gabor's discovery reached its full potential.

Introductory example. The essence of holography is that *the process of image formation is being interrupted and split into two:* In a first step the object is transformed into a photographic record, called the *hologram,* and in a second step, called *reconstruction,* the hologram is transformed into the image. No lens is needed in either step.*

---

*Gabor had originally conceived of holography, or "wavefront reconstruction" as it was called then, as a means of correcting for the spherical aberration of the electron microscope. Ironically, that goal has still not been reached. The term *holography* comes from the Greek ὅλος = all, whole and γραφεῖν = to write.

Dennis Gabor (1900–1979), Hungarian-born physicist, inventor, and philosopher. At age 14 Gabor read advanced physics texts, at 17 he wondered what happens to light as it travels from object to image. Gabor studied at the universities of Budapest and Berlin, became research engineer at the British Thomson-Houston Co. Ltd. in Bristol, professor of applied electron physics at the Imperial College of Science and Technology in London, and later staff scientist at the CBS Laboratories in Stamford, Connecticut. His first publication on the subject was "A New Microscopic Principle," *Nature* (*London*) **161** (1948), 777–78. Shortly thereafter he wrote "Microscopy by reconstructed wave-fronts," *Proc. Roy. Soc. London* **A197** (1949), 454–87, and "Microscopy by Reconstructed Wave Fronts: II," *Proc. Physical Soc.* **B64** (1951), 449–69, which both became classics in holography. In 1971, Gabor received the Nobel Prize in physics for his discovery.

This basic concept of holography can readily be understood from a simple example. Light, as shown in Figure 4.6-1, top left, is incident on a point object and is diffracted by it into a series of concentric rings, maxima and minima. The rings are recorded photographically as a transparency, a *transmission-type hologram.* In the second step (top right), light is incident on the hologram and, the same as with a zone plate, focused by it into a point.

Next (center row), assume that the object consists of two points. The diffraction pattern now contains two sets of rings, in essence two overlapping zone plates. Each of these will focus the light; the image, thus, consists of two points. Finally (bottom row), the object may be an aggregate of many points, such as the letters ABC. These letters produce a multiplicity

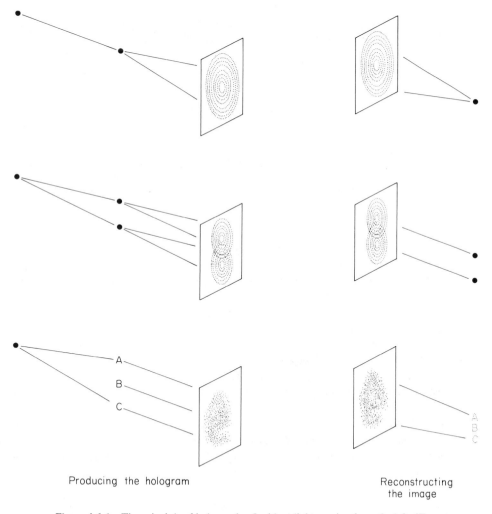

Producing the hologram   Reconstructing
                                                          the image

Figure 4.6-1   The principle of holography. Incident light coming from the left. (*Top row*) Point object forms concentric rings, essentially a zone plate; reconstruction of zone plate gives point image. (*Center row*) The same transformations occur with two points and, (*bottom row*) with a more complex object.

of lines and rings. But, as before, each set of rings, which came from one point in the object, focuses the light into one point. The sum of these points is again an ABC.

PRODUCING THE HOLOGRAM

A hologram is the result of interference. Interference occurs between two contributions, a *signal,* which is the light diffracted by, or scattered off, the object, and a "coherent background" or *reference,* which is light reaching the photographic film directly. But photographic film can record only intensities (irradiances); it cannot record the *phase* of the light. These phases, in other words, are lost.

However, if the signal is combined with a significantly stronger background, the resultant (phase) will become similar to the phase of the background alone (Figure 4.6-2). The loss of phase, in that case, will not be essential anymore.

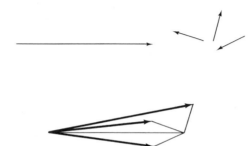

Figure 4.6-2 Adding strong coherent background (*top left*) to random signals (*top right*) gives similar resultant vectors (*bottom*).

To provide enough of the background radiation, Gabor used objects such as a wire mesh or slender letters with relatively wide spaces between them. He then took light from a mercury arc (lasers had not been invented as yet), passed the light through a pinhole and a filter to isolate one of the Hg lines, and then let it pass through the object. The signal beam and the reference, in other words, were *coaxial* (Figure 4.6-3, top).

The next major advance was made by Leith and Upatnieks, who added the reference at an *offset angle* (Figure 4.6-3, bottom). That made possible the recording of holograms of *solid, three-dimensional objects.* *

In practice, holograms are often recorded as shown in Figure 4.6-4. Part of the incident light goes to a mirror and from there is reflected toward the film. This is the reference beam. The signal beam is scattered off the object and from there reaches the film. Both contributions combine and form fringes. Ordinarily, these fringes are very closely spaced and cannot be seen by the unaided eye, hence the typical hologram appears to be uniformly gray. Under the microscope, however, a hologram is seen to consist of a myriad of tiny domains, each containing a series of fringes of various lengths and spacing.

---

*E. N. Leith and J. Upatnieks, "Wavefront Reconstruction with Continuous-Tone Objects," *J. Opt. Soc. Am.* **53** (1963), 1377–81; "Wavefront Reconstruction with Diffused Illumination and Three-Dimensional Objects," *J. Opt. Soc. Am.* **54** (1964), 1295–1301; and *Holograms,* U.S. Patent 3,894,787, July 15, 1975.

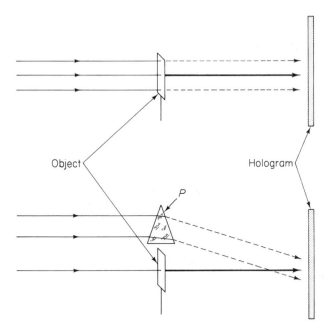

Figure 4.6-3 Coaxial recording of signal and reference (*top*) and off-axis recording (*bottom*). The light is incident from the left; the signal (heavy line) and the reference (dashed line) proceed to the right. *P*, prism.

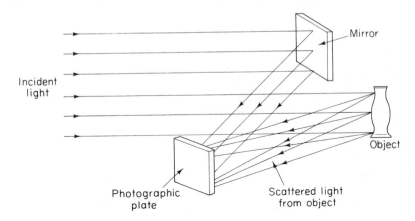

Figure 4.6-4 Recording the hologram.

A laser is not really needed for holography; it is merely the use of solid, three-dimensional objects that calls for light whose coherence length exceeds the path differences due to the unevenness of such objects.

RECONSTRUCTION

A hologram, as we have seen, results from interference between two beams, the signal, A_1, and the reference, A_2. These are amplitudes that can readily be added to one another, and then be squared, but since a photographic film responds to intensities, we find that for each point (in the hologram) the intensity is

$$I(x, y) = (\mathbf{A}_1 + \mathbf{A}_2)^2$$
$$= (\mathbf{A}_1 + \mathbf{A}_2)(\mathbf{A}_1 + \mathbf{A}_2)*$$
$$= |\mathbf{A}_1|^2 + |\mathbf{A}_2|^2 + \mathbf{A}_1\mathbf{A}_2^* + \mathbf{A}_1^*\mathbf{A}_2 \qquad [4.6\text{-}1]$$

where the asterisk * indicates the complex conjugate.

In the reconstruction process, light of amplitude $\mathbf{A}_3$ is used to illuminate the hologram. The hologram has a transmittance function $T(x, y)$ and, since it had been produced by the amplitudes $\mathbf{A}_1$ and $\mathbf{A}_2$,

$$T(x, y) = \mathbf{A}_1\mathbf{A}_2^* + \mathbf{A}_1^*\mathbf{A}_2 \qquad [4.6\text{-}2]$$

The $\mathbf{A}_3$ light, therefore, is diffracted (*modulated*) by the hologram,

$$\mathbf{A}_4 = \mathbf{A}_3 T(x, y) \qquad [4.6\text{-}3]$$

Note that in Equation [4.6-1] it is only the two right-hand terms that carry the information because only *they* are the result of any interference between signal and reference. Hence, if we consider merely these terms, then

$$\mathbf{A}_4 \propto \mathbf{A}_3\mathbf{A}_1\mathbf{A}_2^* + \mathbf{A}_3\mathbf{A}_1^*\mathbf{A}_2 \qquad [4.6\text{-}4]$$

If the reconstruction amplitude $\mathbf{A}_3$ is equal, or at least proportional, to the reference amplitude $\mathbf{A}_2$, the amplitudes $\mathbf{A}_2$ and $\mathbf{A}_3$ in Equation [4.6-4] cancel and

$$\mathbf{A}_4 \propto \mathbf{A}_1 \qquad [4.6\text{-}5]$$

This means that the amplitude diffracted by the hologram is proportional to the amplitude initially diffracted by the object: *The image is a reconstruction of the object.*

At this point I emphasize again the fundamental difference between a hologram and a conventional photograph. In a photograph, the information is stored in an *orderly* fashion: each point in the object relates to a *conjugate* point in the image. In a hologram there is no such relationship; light from every object point goes to the entire hologram. This has two consequences. If the hologram were shattered, or cut into small pieces, each fragment would still reconstruct the whole scene, not just part of the scene (although the resolution would be less).

In addition, since the hologram as such is often larger in size than the object, it receives light not only from the side facing the film but also from the adjoining sides (which could not be seen in conventional photography). Thus, as the observer's head moves sideways in viewing the hologram, the image is seen in three dimensions.

Holograms give both real and virtual images (Figure 4.6-5). A real image can be projected and focused on a screen. A virtual image can be seen but it cannot be projected. If both the $\mathbf{A}_2$ and the $\mathbf{A}_3$ waves are plane, the two images have the same magnification and are symmetric, with the hologram in the middle. This is an example of a *Fraunhofer hologram*. If the hologram has been recorded with divergent light, the real image is magnified and the virtual image is reduced in size. That is an example of a *Fresnel hologram*.

In either case, the observer looking into the $\mathbf{A}_4$ beam will find that beam hard to distinguish from the $\mathbf{A}_1$ beam. Indeed, the image seen in holography looks startlingly similar

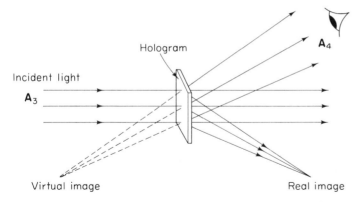

Figure 4.6-5 Reconstruction. The hologram diffracts the incident light A_3, producing a real image and a virtual image.

to the object, including its three-dimensionality and the parallax typical of the real world. It is this realism that has made holography so intriguing a subject to both scientists and casual observers.

Practical considerations.   Holograms must be recorded on photographic film of high resolvance.* Look again at Figure 4.6-4 and note that the reference, the light reflected by the mirror, and the signal, the light scattered by the object, subtend at the film a certain angle. If this angle is too large, the fringes formed between signal and reference are so closely spaced that even the best emulsions cannot resolve them. The experimental setup must also be stable and free from vibration to prevent the fringes from becoming obliterated during exposure.

Reconstruction always yields a photographic positive. If a contact print were made of a hologram and, hence, the blacks and whites reversed, the image would still look the same, just as the "negative" of a simple diffraction grating would show the same spectrum lines as the original "positive."

In an emulsion that is sufficiently *thick,* wavefronts traveling in one direction can be made to interfere with wavefronts traveling in the opposite direction. The wavefronts then form a three-dimensional standing-wave pattern, rather than the two-dimensional pattern of a conventional hologram. If viewed in white light, such *volume holograms* give reconstructions in full color.

APPLICATIONS OF HOLOGRAPHY

Holography has a great many fascinating applications. Its influence on precision measurements, pattern recognition, and even art has only begun to bear fruit. Most important of all is that holography has provided us with *conclusive proof that any image formation is a two-*

---

*Photographic emulsions suitable for holography are made by several manufacturers such as the Eastman Kodak Company and Agfa-Gevaert. Depending on the resolvance required, and the speed desired, various types of film of different designations are available from these manufacturers.

step process. Whereas in conventional image formation these two steps follow each other without delay, in holography they can be separated, conceptually as well as experimentally.

Holographic interferometry.   Testing for stresses, strains, and surface deformations is one of the most useful practical applications so far. In the *double-exposure technique* two exposures are made of the object, one before loading and the other after. The original object and the object after deformation are recorded holographically on the same film. When the hologram is reconstructed by illuminating it with the reference wave, both images are viewed simultaneously and, since they are slightly different, the two images interfere. Thus any distortions of the object will show in the form of fringes. In the *continuous-exposure* or *real-time technique* the object keeps moving during exposure (such as the vibrating body of a violin). This produces interference by light coming from different positions of the object during motion and, likewise, generates fringes. These fringes represent contours of constant amplitude of the vibration.

Particle size determination.   Particles such as aerosols and atmospheric pollutants in suspension are in constant motion; thus, with conventional photography it is difficult to focus even on a single particle, and it is next to impossible to focus on all of them at the same time.

A hologram, though, can be recorded without focusing. Using a pulsed laser, for example, freezes the motion. Later, when viewing the image, the observer can then examine at leisure any particles that were present at the instant the holographic recording was made.

Information storage.   Information can be stored and retrieved more efficiently in the form of holograms than in the form of real images. Perhaps this is how information, then called *engrams,* is stored in the brain; at least it would help explain why attempts to locate certain "centers" in the brain have never met with much success and why major brain injury often does not lead to predictable, circumscribed defects.

Acoustic holography.   In our daily life there is no evidence for images formed by sound. But actually, such images should exist and, indeed, holography has made them possible because it requires no lenses.

There are different ways to realize acoustic holography. In one of them, two sound generators, as in Figure 4.6-6, emit the reference and the signal, respectively. On a calm surface of water, these two contributions produce ripples. The ripple pattern *is* the hologram. The pattern may be photographed and then reconstructed, or reconstruction may be made in real time by coherent light reflected off the surface as shown.

Holographic optical elements.   Holographic optical elements are holograms of typical optical components such as lenses or mirrors. HOEs perform precisely the same type of operation as the lens or mirror of which they were made, but they do it by diffraction, rather than by refraction or reflection. That seems to be an interesting alternative to more conventional elements.

Computer-generated holograms.   We have seen earlier that a zone plate can easily be drawn using a suitable computer program (such as the program on page 266). The printout, preferably made with a laser printer, is reduced in size and reconstructed; that

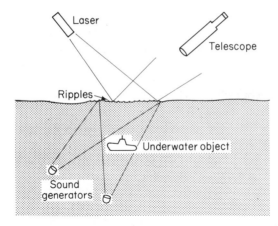

Figure 4.6-6   Acoustic holography.

results in a single point. But much more complex holograms can be synthesized as well. Their reconstructions produce surprisingly beautiful, three-dimensional images of objects, which, oddly enough, have never existed in the first place.

Pattern recognition.   One of the most exciting applications of holography is *pattern recognition,* also called *character recognition* in reference to alphanumeric characters. Early pattern recognition systems were based on geometric optics. Consider, for example, that we want to read the letter A (Figure 4.6-7). A set of characters, A, B, C, . . . , are printed as negatives on a strip of film and this film is moved through the image plane. If the character to be read matches the character on the film, the output from a photodetector is zero, triggering a printer. But in reality this does not work: the character and the negative must be aligned perfectly, both in position and size, an unrealistic requirement.

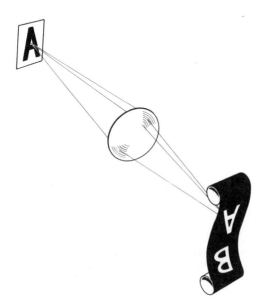

Figure 4.6-7   Pattern recognition based on geometrical optics.

Instead, modern pattern recognition systems are based on holography. In place of a mask containing a real image of the letter A, we now use the *Fourier transform* of the A.

As in holography in general, the Fourier transform hologram of the A is the result of a superposition of two sets of wavefronts, the signal and the reference. The signal is diffracted by an original A and the reference is a beam of collimated light. Subsequently, when the hologram of the A is illuminated by collimated light, an A is reconstructed. When the transform is illuminated with light from another A, plane wavefronts result that can be focused *into a bright spot* (Figure 4.6-8, top). The spot is easily recognized by the eye, by photography, or photoelectrically. On the other hand, if the wavefronts are coming from a B, or from some other character, they do *not* transform into perfectly plane wavefronts and do *not* produce a focused spot but rather a diffuse patch of light (center). Hence, we can search a given matrix of characters and determine whether a particular character is present (bottom).

The holograms shown in Figure 4.6-8 appear to be amplitude filters. But because they are generated by interference between signal and reference, they represent in fact both the amplitude and the phase of the light. They are called "complex," "matched," or *Vander Lugt filters.* *

Plenty of difficulties still lie ahead before true *reading machines* can become a reality. Some letters and words are "inside" others. For example, F is inside E, P is inside R and B, T and L have the same horizontal and vertical lines, and *arc* is inside *search*. Clearly, the more alike the two characters, the less the power of discrimination. A major problem, also,

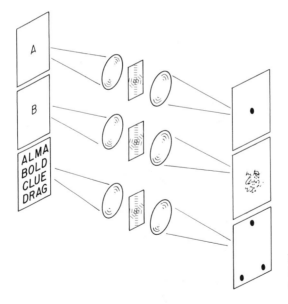

Figure 4.6-8 Pattern recognition by holographic *Vander Lugt filter* (the filters are seen between the lenses).

---

*The term "complex" is used, not because of the complexity of making them, but because of the analogy to complex numbers, which also carry both amplitude and phase information. The term "matched" comes from radar technology but now refers also to light coming from an identical, "matched" character. Filters of this type were first made by A. Vander Lugt, "Signal Detection by Complex Spatial Filtering," *IEEE Trans. Inf. Theory* **IT-10** (1964), 139–45.

will be to teach the machine to recognize the "meaning" of a letter set in different typeface; the letter A, for example, can be printed A, **A**, 𝔄, 𝒜, a, **a**, ɑ, α, and an almost infinite number of variations is possible when it comes to *handwriting*.

SUGGESTIONS FOR FURTHER READING

E. N. Leith, "White-Light Holograms," *Sci. Am.* **235** (Oct. 1976), 80–95.

H. M. Smith, *Principles of Holography,* 2nd edition (New York: John Wiley & Sons, Inc., 1975).

P. Hariharan, *Optical Holography—Principles, Techniques and Applications* (New York: Cambridge University Press, 1986).

H. J. Caulfield, editor, *Handbook of Optical Holography* (New York: Academic Press, Inc., 1979).

Yu. N. Denisyuk, *Fundamentals of Holography* (Moscow: Mir Publishers, 1984).

C. M. Vest, *Holographic Interferometry* (New York: John Wiley & Sons, Inc., 1979).

PROBLEMS

4.6-1.   If the angle subtended at the hologram by the signal and the reference is 15°, what is the spacing of the fringes provided the wavelength is 492 nm?

4.6-2.   Holograms are taken with light of 632.8 nm wavelength. What is the limiting angle between the signal and the reference if the space frequency in the hologram is not to exceed 200 lines/mm?

4.6-3.   Following an example by Gabor, assume that the amplitudes of the signal and the reference are related as 1 : 10. Since the two beams when they combine may be completely *in* phase, or completely *out of* phase, what is the maximum ratio of their intensities?

4.6-4.   Using the lengths of the amplitude vectors drawn in Figure 4.6-2, determine the highest *ratio* of the intensities possible.

4.6-5.   Reconstruction is made of a hologram that is part of a zone plate, *not* including its center. If the hologram is viewed in collimated light, show where the real and the virtual image will be formed.

4.6-6.   A projectile is fired through a wind tunnel placed in one arm of a Mach-Zehnder interferometer. If the fringe pattern were "reconstructed," what would it show?

4.6-7.   If a hologram is recorded with light of 400 nm wavelength and reconstructed with light of 800 nm, how does the image change as compared to a reconstruction at 600 nm?

4.6-8.   Consider a three-dimensional object and image and discuss the essential difference between *stereoscopic* and *holographic* image formation.

4.6-9.   The theoretical storage capacity of a hologram is one bit of information per λ^3 of volume and the storage *density* n^3/λ^3. Therefore, how many bits can be stored in 1 mm^3 if the index is $n = 1.483$ and the wavelength $\lambda = 546$ nm?

4.6-10. A pattern recognition system is built as a reading-aid for the blind, using a hand-held probe as the reading head. How critical is it to follow printed material exactly line by line?

5.1

Light Sources and Detectors

ELECTROMAGNETIC ENERGY IS GENERATED, emitted, transmitted, and received in the form of elemental units of energy, each of them called a *quantum.* The quantum aspect of light, just as rays and waves, is an integral part of optics. Neither the photoelectric effect nor the process of laser emission could be explained without it. And, more important still is the completeness of the system of optics which we discover through phenomena such as blackbody radiation, atomic spectra, absorption, and lasers, all of which require us to know about the concepts of *quantum optics.*

LIGHT SOURCES

Many light sources, from the sun to wax candles to electric light bulbs, are *thermal sources;* they produce light and other electromagnetic radiation as a consequence of their temperature. Other light sources are nonthermal sources; they produce light by "cold" emission as it occurs in luminescence and in stimulated (laser) emission. These will be discussed in another chapter. Similarly, light *detectors* may be either thermal detectors or nonthermal, quantum detectors. Quantum detectors are of particular interest, both theoretical and practical; some of them are so sensitive they respond to individual quanta.

    Thermal light sources come in types as diverse as kerosene lanterns, where carbon, freed by the combustion process, is heated to *incandescence,* and electric light bulbs, where a filament is heated. The first electric light bulbs had carbon filaments. Then came metal filaments, in particular those made of refractory metals that have a high melting point such as *tungsten.**

Tungsten melts at 3410°C, but before that happens it slowly evaporates and forms deposits on the inner surface of the glass envelope, reducing the output. Some halogens, in particular iodine, retard this process. On the inside of the hot (quartz) envelope, the iodine combines with the tungsten deposits to form tungsten iodide. Near the even hotter filament the molecule dissociates again, depositing the tungsten back onto the filament and letting the iodine reenter the cycle. The filament can then be heated close to the melting point, which accounts for the high efficacy of such *quartz iodine lamps.*

Fluorescent lamps, in contrast to incandescent lamps, do not produce much heat. They are often filled with argon, neon, and/or mercury. The inside of such lamps is coated with a variety of phosphors, often $3Ca_3(PO_4) \cdot Ca(FCl)_2$, and other additives such as manganese to enhance the reds. *High-pressure mercury lamps* may contain traces of dysprosium, holmium, gallium, or indium iodide, which tend to broaden the Hg lines and generate new ones. *High-pressure xenon lamps* can be built for power outputs in excess of 120 kW and power densities higher than that of the sun.

Blackbody radiator. The laws basic to thermal light sources can be illustrated best by the example of the *blackbody radiator.* A blackbody can be approximated by a hollow block of cast iron with a small hole in it. Radiant energy entering the hole becomes trapped inside; hence, the cavity is a near-perfect absorber. On the other hand, when the cavity is heated, like an oven, it radiates energy out of the hole. The spectral distribution of that radiation is a function of temperature alone; the material as such plays no role.

The radiation is produced by oscillations of the molecules of the cavity's wall and, according to classical theory, is proportional to the number of standing waves generated by these oscillators. This assumption agrees well with experimental evidence at least in the infrared, but it does not agree in the ultraviolet. As the wavelength approaches zero, the number of waves theoretically would become infinite but obviously this cannot be and the amount of energy need to become zero as well, a dilemma then called the "ultraviolet catastrophe."

The problem was solved by Max Planck.† Planck assumed that the oscillators emit

---

*The first electric light bulbs were made in 1879 by Thomas Alva Edison (1847–1931), prolific American inventor. Out of a total of 1100 patents, in one four-year period alone he obtained 300, one patent every five days. Among them, besides the electric-lamp patent (No. 223,898, dated Jan. 27, 1880), were the phonograph (his own favorite) and the motion picture camera and projector. When Edison announced he would try producing light from electricity to compete with natural gas then used for illumination, gas stocks at the New York and London stock exchanges tumbled; so much faith had the public in his ability. Edison first tried various metals. It didn't work; so he turned to cotton thread, which, when "carbonized" (scorched), conducted electricity without melting. The second light bulb Edison made lasted for 40 hours. "I think we've got it," he exulted, "if it can burn 40 hours, I can make it last a hundred."

†Max Karl Ernst Ludwig von Planck (1858–1947), German physicist, professor of theoretical physics at the University of Berlin, and president of the then Kaiser Wilhelm Gesellschaft, now Max Planck Gesellschaft. Planck introduced his concept of quantum units of light at a meeting of the Berlin Physical Society on 19 October 1900. Acceptance of his idea at first was slow; Planck himself resisted it for several years. "My futile attempts to fit the elementary quantum of action somehow into the classical theory continued for a number of years," he wrote, "and they cost me a great deal of effort." M. Planck, 'Ueber irreversible Strahlungsvorgänge," *Ann. Phys.* (4) **1** (1900), 69–122. In 1918 Planck received the Nobel Prize for physics.

energy, not in the form of a smooth continuous flow, but only as discrete, elemental units of energy called *quanta* or *photons*.

The least amount of energy, E, that can be emitted according to Planck is given by

$$E = h\nu \qquad [5.1\text{-}1]$$

where h is a factor of proportionality, since called *Planck's constant*, and ν is the frequency of the light. The currently accepted best value of Planck's constant is

$$h = 6.626\ 075\ 5 \times 10^{-34}\ \text{J s}$$

where the unit, joule-second, can also be written joule/hertz. Planck's constant must be determined experimentally; it cannot be predicted from theory. It has been measured in a variety of ways, and all of these measurements agree.

The spectral distribution of the radiation emitted by a blackbody can be described by *Planck's radiation law*. This law is often written in terms of radiant power emitted per unit surface area, that is, in terms of *exitance, M*, as a function of wavelength, λ,

$$M(\lambda) = \frac{C_1}{\lambda^5} \frac{1}{e^{C_2/\lambda T} - 1} \qquad [5.1\text{-}2]$$

where T is the absolute temperature, in degrees kelvin (K), and C_1 and C_2 are two *radiation constants*. The first radiation constant, C_1, is defined as, and has a numerical value of,

$$C_1 = 2\pi hc^2 = 3.742 \ldots \times 10^{-16}\ \text{W m}^2 \qquad [5.1\text{-}3]$$

and the second radiation constant, C_2, is defined as, and has a value of,

$$C_2 = \frac{hc}{k} = 1.4388 \times 10^{-2}\ \text{m K} \qquad [5.1\text{-}4]$$

where h is again Planck's constant, c the velocity of light, and k *Boltzmann's constant*, $1.3806 \ldots \times 10^{-23}\ \text{J K}^{-1}$.

Wien's displacement law.   There are two sequels to Planck's radiation law. First, if we plot Planck's law for different temperatures, we find that with increasing temperature (a) *more energy* is emitted and (b) *the peak emission shifts toward the shorter wavelengths*. That is illustrated in Figure 5.1-1.

The shift to shorter wavelengths agrees with common experience. For example, when a block of iron is gradually heated, it emits at first only infrared. As the temperature is raised, it begins to glow dark red. As the temperature is raised further, the block becomes bright red ("red hot"), and finally "white hot."

The peak wavelength, λ_{max}, is given by

$$\lambda_{\text{max}} = \frac{2.8978 \times 10^{-3}}{T}\ \text{m K} \qquad [5.1\text{-}5]$$

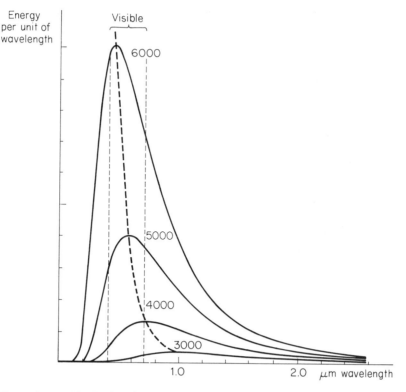

Figure 5.1-1   Distribution of radiant energy as a function of wavelength. At higher temperatures the peak emission shifts to shorter wavelengths.

where T is again the absolute temperature ($0°C = 273.16$ K). This is *Wien's displacement law.**

The *color temperature* of a light source refers to the temperature of a blackbody that emits radiation of the same color as the source. The color temperature of the sun, for example, is 5600 K, and that of a 200-W tungsten-filament lamp is about 3200 K. Accordingly, daylight color film is balanced for 5600 K, and indoor color film for 3200 K.

Stefan–Boltzmann's law.   Second, the total radiation output of a blackbody is found by integration over all wavelengths,

$$M = \int_0^\infty M(\lambda)d\lambda = \sigma T^4 \qquad [5.1\text{-}6]$$

---

*Wilhelm Carl Werner Otto Fritz Franz Wien (1864–1928), German physicist. Wien studied under Helmholtz, became his assistant, later successor to Röntgen at both the universities of Würzburg and Munich. Wien contributed to thermodynamics and electric discharges but is best known for his displacement law. He received the Nobel Prize in physics in 1911. W. Wien, "Temperatur und Entropie der Strahlung," *Ann. Phys.* (Neue Folge) **52** (1894), 132–65.

This is *Stefan–Boltzmann's law.* * It shows that the exitance increases as the fourth power of the absolute temperature. The constant of proportionality, σ, is called *Stefan's constant;* it has a numerical value of 5.67 . . . $\times 10^{-8}$ W m^{-2} K^{-4}.

Kirchhoff's law.   A blackbody is a perfect radiator and absorber. Transparent and reflective materials are poor radiators and absorbers. Consequently, an object that is a good radiator at a given wavelength is also a good absorber at the same wavelength, a relationship known as *Kirchhoff's law.* †

All of these radiation laws apply to blackbody radiators only. If the body is not black ("graybodies"), the amount of energy radiated is less by a factor ϵ called the *emissivity* of the surface. Stefan–Boltzmann's law, for example, then becomes

$$M = \epsilon \sigma T^4 \qquad [5.1\text{-}7]$$

Only for a true blackbody, in other words, will ϵ be unity. For other bodies, the emissivity usually lies between 0.2 and 0.9, and for highly polished metals it may be as low as 0.01.

Example 1

If the peak emission from the sun is at 475 nm, what is its surface temperature, in degrees Celsius?

Solution.   From *Wien's displacement law,*

$$T = \frac{2.8978 \times 10^{-3}}{475 \times 10^{-9}} = 6100 \text{ K}$$

$$6100 - 273 = \boxed{5827°\text{C}}$$

Comment.   The sun, however, is not a perfect blackbody. If it were, we could apply *Stefan–Boltzmann's law;* this would result in 5750 K. The discrepancy is due to absorption at shorter wavelengths in the outer layers of the photosphere of the sun, which does not affect the position of the 475-nm maximum.

Example 2

What is hotter, heaven or hell?

Solution.   Heaven, according to *Isaiah 30:26,* receives from the moon as much radiation as the earth does from the sun and in addition seven times seven (forty-nine) times as much as the earth

---

*Josef Stefan (1835–1893), Austrian, professor of physics at the University of Vienna, found the relationship by experiment; Ludwig Edward Boltzmann (1844–1906), Stefan's one-time student and later successor as professor of physics at the same university up to his death by suicide, derived it independently from theory. Their discoveries were published in J. Stefan, "Über die Beziehung zwischen der Wärmestrahlung und der Temperatur," *Sitzungsber. Math.-Naturwiss. Classe kais. Akad. Wiss., Wien* **79**, II. Abtheilung (1879), 391–428; L. Boltzmann, "Ableitung des Stefan'schen Gesetzes, betreffend die Abhängigkeit der Wärmestrahlung von der Temperatur aus der electromagnetischen Lichttheorie," *Ann. Phys.* (Neue Folge) **22** (1884), 291–94.

†G. Kirchhoff, "Ueber das Verhältniss zwischen dem Emissionsvermögen und dem Absorptionsvermögen der Körper für Wärme und Licht," *Ann. Phys.* (4) **19** (1860), 275–301.

from the sun, or fifty times as much radiation in all. If we assume that the average temperature of the earth is 27°C = 300 K, then from Stefan–Boltzmann's law,

$$\left(\frac{\text{heaven}}{300}\right)^4 = 50$$

which means that the temperature of heaven is

$$(300)(50^{1/4}) = 798 \text{ K} = \boxed{525°C}$$

Hell, according to *Revelation 21:8,* has as its main topographic feature a lake of molten brimstone (sulfur). Since the boiling point of sulfur is 444.7°C, the temperature of hell must be less than that: Above 444.7° the lake would evaporate, and below it, it would harden. Thus, *heaven is hotter than hell.* [From *Appl. Opt.* **11** (1972), 8, A14.]

DETECTORS

Detectors of electromagnetic radiation, likewise, fall into two groups, *thermal detectors* and *quantum detectors.* Thermal detectors are based on absorption and heating; if the absorbing material is black, they are independent of wavelength. Quantum detectors are based on the *photoelectric effect.*

Thermal detectors.   Thermal detectors are relatively slow to respond; it may take them almost a second to reach equilibrium. A good example of a thermal detector is the *Golay cell,* a thin black membrane placed over a small, gas-filled chamber. Heat absorbed by the membrane causes the gas to expand, which in turn can be measured, either optically (by a movable mirror) or electrically (by a change in capacitance). Golay cells are used in the infrared.

A *thermocouple,* in essence, is a junction between two dissimilar metals. As the junction is heated, the potential difference changes. In practice, two junctions are used in series, a *hot* junction exposed to the radiation, and a *cold* junction shielded from it. The two voltages are opposite to each other; thus the detector, which without this precaution would show the absolute temperature, now measures the temperature differential. A *thermopile* contains several thermocouples and, therefore, is more sensitive. A *bolometer* contains a metal element whose electrical resistance changes as a function of temperature; if instead of the metal a semiconductor is used, it is called a *thermistor.* Unlike a thermocouple, a bolometer or thermistor does not generate a voltage; they must be connected to a voltage source.

Quantum detectors.   A beam of intense light, it seems, should cause more of an effect than a beam of dim light. With a quantum detector, though, that is not necessarily so. The wavelength of the light plays an important role; in fact, there is a certain threshold above which there is no effect at all, no matter what the intensity. The classical example where this occurs is the *photoelectric effect.*

Assume the radiation is incident on a plate M mounted in an evacuated tube as shown in Figure 5.1-2. The plate is made of a material that, when irradiated, releases electrons, then called *photoelectrons.* Opposite M is another plate, the collector plate C. If C is made

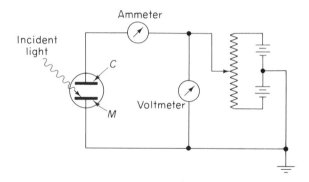

Figure 5.1-2 Photoelectric effect.

positive with respect to M, the photoelectrons released by M are attracted by, and travel to, C. As the potential V, read on an high-impedance voltmeter, is increased, the current, i, read on an ammeter, increases too, but only up to a given *saturation level,* because then all of the electrons emitted by M are collected by C.

On the other hand, if C is made *negative,* some photocurrent will still exist, provided the electrons ejected from M have enough kinetic energy to overcome the repulsive field at C. But as C is made more negative, a point is reached where *no* electrons reach C and the current drops to zero. This occurs at the *stopping potential, V_0* (Figure 5.1-3). In short: A significant amount of photocurrent is present only if the collector, C, is made positive, that is, if it is the *anode.* The light-sensitive surface M, then, is the photo*cathode.*

Now, if more intense light falls on the photocathode, it will release more electrons but their energies, and their velocities, will remain the same. Instead, the energy of the photo-electrons depends on the *frequency* of the light: blue light produces more energetic photo-electrons than red light. In addition, the response of a quantum detector is all but instantaneous: there is *no time lag,* at least not more than 10^{-8} s, between the receipt of the irradiation and the resulting current.

These facts are hard to reconcile. Conceivably, the energy imparted to the electrons could come either from the light (but that would mean higher energies with more intense light and perhaps some time delay for very dim light), or it could come from heat energy stored in the material, the light merely acting as a trigger (but we do not find higher energies as the surface is heated).

The answer again follows from the fact that the light is received in the form of discrete quanta. But part of the energy contained in a quantum is needed to make the electron

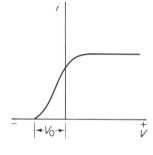

Figure 5.1-3 Dependence of photocurrent on potential applied to collector plate.

escape from the surface; that part is called the *work function, W*. It is only the excess energy, beyond the work function, that appears as kinetic energy (of the electron). The maximum kinetic energy with which the electron can escape, therefore, is

$$\boxed{KE_{max} = h\nu - W}$$

[5.1-8]

which is *Einstein's photoelectric-effect equation.**

It is often convenient to measure energies on an atomic scale not in joule but in *electron volt*, eV. By definition, 1 eV is the amount of energy acquired, or lost, by an electron moving through a potential difference of 1 volt:

$$1 \text{ eV} = (1e)(1V)$$

Since the charge on the electron, e, is very nearly 1.60×10^{-19} C,

$$1 \text{ eV} = 1.60 \times 10^{-19} \text{ J}$$

[5.1-9]

If we then use $c = \lambda\nu$, solve for λ, and substitute $E = h\nu$, we find that

$$\lambda = \frac{hc}{E}$$

Inserting the actual values for h and c and substituting Equation [5.1-9] leads to

$$\lambda = \frac{(6.63 \times 10^{-34} \text{ J s})(3 \times 10^8 \text{ m/s})}{1.6 \times 10^{-19} \text{ J/eV}} = \frac{1240 \text{ nm eV}}{E}$$

[5.1-10]

which connects the wavelength, in nanometers, with the energy, in electron volts.

Example

The work function determines the longest wavelength to which a detector can respond: The lower the work function, the longer the wavelength. The lowest work functions are found among the alkali metals. Potassium, for example, as shown in Table 5.1-1, has a work function of 2.25

TABLE 5.1-1 PHOTOELECTRIC PROPERTIES OF SOME ALKALI METALS

Alkali	Work function (eV)	Threshold (nm)
Sodium	2.28	543
Potassium	2.25	551
Rubidium	2.13	582
Cesium	1.94	639

---

*In 1905, Albert Einstein applied Planck's theory to the photoelectric effect, postulating that light is not only *emitted* but also *received* in the form of quanta. A. Einstein, "Über einen die Erzeugung und Verwandlung des Lichtes betreffenden heuristischen Gesichtspunkt," *Ann. Phys.* (4) **17** (1905), 132–48. This paper by Einstein on the photoelectric effect is much more radical than the one he wrote in the same year on the theory of relativity. In fact, it was this paper that earned him the Nobel Prize in physics.

eV. At the threshold, *no* photoelectrons are released and Equation [5.1-8] becomes zero:

$$KE_{max} = h\nu - W = 0$$

so that

$$W = h\nu = E$$

Then, from Equation [5.1-10],

$$\lambda = \frac{1240 \text{ nm eV}}{W} = \frac{1240}{2.25} = \boxed{551 \text{ nm}}$$

which is the longest wavelength that can be detected using a potassium surface.

PRACTICAL QUANTUM DETECTORS

In contrast to thermal detectors, quantum detectors respond to the *number* of quanta, rather than to the energy contained in them. The simplest type is probably the *vacuum phototube,* shown earlier in Figure 5.1-2; such a diode is an example of a *photoemissive detector.*

The ratio of the number of photoelectrons released to the number of photons received is called the *quantum efficiency.* Ordinarily, this efficiency is no higher than a few percent. But if several diodes are combined in series to form a *photomultiplier,* the efficiency becomes much higher.

A *photocell* is the solid-state equivalent of the vacuum photodiode; most often it is a *semiconductor.* As the name implies, a semiconductor conducts electricity better than an insulator but not as well as a conductor (that is, a metal). In an insulator, the electrons are tightly bound to their respective atoms. In a metal, the electrons can move freely; hence, even a small voltage applied to the conductor will cause a current.

A semiconductor, such as silicon or germanium, lies somewhere in the middle. Interestingly enough, some semiconductors, such as cadmium sulfide CdS, gallium arsenide, and silicon, conduct electricity poorly only in the dark; when exposed to light, they conduct very well. They are *photoconductive detectors.*

Still another type of quantum detector is made from two semiconductors, one of them transparent to light, for instance a layer of CdS deposited on selenium. When light is incident on the junction, the electrons start moving, but only in one direction producing a current; the junction, in other words, converts light energy into electrical energy. Such *photovoltaic detectors* are used as solar cells and as exposure meters in photographic cameras.

An *image tube* not only detects light but also preserves the spatial characteristics of an image. Some image tubes contain an array of photoconductors, one for each pixel. When exposed to light, the elements form a latent image that can be read by an electron beam scanning across them or, as shown in Figure 5.1-4, the photoelectrons emitted by the cathode can be focused by an electron lens and made visible on a phosphor screen mounted in the same tube.

If the image is merely amplified, we have an *image intensifier.* If the image is formed in the IR, the UV or the X-ray range and converted into the visible, we have an *image*

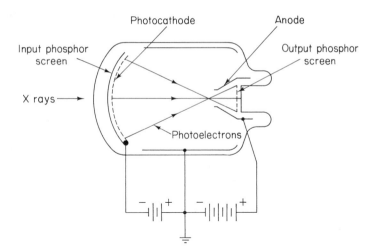

Figure 5.1-4   X-ray converter. (Reproduced by permission, Hamamatsu Corp.)

converter; and if the system is built around an array of many short fibers or capillaries, fused into a wafer, we have a *microchannel image intensifier.* *

SUGGESTIONS FOR FURTHER READING

P. L. KNIGHT and L. ALLEN, *Concepts of Quantum Optics* (Elmsford, NY: Pergamon Press, Inc., 1985).

E. GOLDIN, *Waves and Photons* (New York: John Wiley & Sons, Inc., 1982).

G. J. ZISSIS and A. J. LAROCCA, "Optical Radiators and Sources," in W. G. Driscoll and W. Vaughan, editors, *Handbook of Optics,* Sec. 3, pp. 1–83 (New York: McGraw-Hill Book Company, 1978).

E. L. DERENIAK and D. G. CROWE, *Optical Radiation Detectors* (New York: John Wiley & Sons, Inc., 1984).

PROBLEMS

5.1-1. A carbon *arc,* in contrast to a filament lamp, usually requires direct current. Why? Hence, what is the advantage of using a carbon arc, and using dc?

5.1-2. A *zirconium arc* contains a metal-disk anode with a small aperture and a needle-shaped cathode centered inside the aperture. What are the advantages of such a light source as compared to a filament lamp?

---

*See M. Laughton, "The Microchannel Image Intensifier," *Sci. Am.* **245** (Nov. 1981), 62–71.

5.1-3. Planck's constant can be written in units of either J Hz^{-1} or J s. Use J s and verify the *units* of the two radiation constants.

5.1-4. Knowing that $h = 6.626 \times 10^{-34}$ J Hz^{-1}, verify the *numerical values* of the two radiation constants.

5.1-5. A carbon filament, which may be considered a blackbody, is heated first to 500°C and then to 1000°C. What are the peak wavelengths emitted at these temperatures?

5.1-6. Can daylight (peak wavelength 555 nm) be obtained from a tungsten filament lamp without using an additional filter?

5.1-7. What is the peak wavelength emitted by the human body (normal temperature = 37°C)?

5.1-8. If a certain material is heated to 1000°C and then emits infrared energy of 2.3 μm wavelength, what must be its temperature to emit yellow light of 575 nm?

5.1-9. Will the *color temperature* of light change and, if so, how will it change, when a blue filter is placed in front of the light source?

5.1-10. What is the temperature, in °C, of a surface 1 mm^2 in size that has an emissivity of 0.15 and an exitance of 2.27×10^5 W m^{-2}?

5.1-11. When monochromatic light coming from a pinhole is collimated and focused into a small aperture in front of a photocell, as shown in Figure 5.1-5, the ammeter reads 10 mA. Now a glass plate A is inserted into one half of the beam (neglect plate B). Plate A is of a thickness such as to retard the light by one-half of a wavelength. How much light will reach the detector? How much photocurrent will result? What happens to the energy?

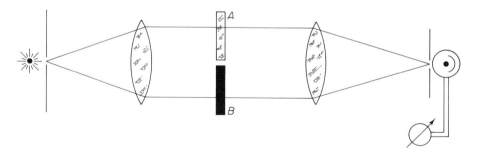

Figure 5.1-5

5.1-12. Continue with Problem 5.1-11. Leave plate A in place and insert in addition an opaque plate B, blocking one-half of the beam. What happens? How much current will result?

5.1-13. Determine the wavelength of a quantum whose energy is 2.48 eV.

5.1-14. Express Planck's constant in terms of electron volts.

5.1-15. What is the energy, in eV, of a quantum of:
   (a) Infrared of 2.0 μm?
   (b) Ultraviolet of 250 nm wavelength?

5.1-16. If a ruby laser ($\lambda = 694$ nm) delivers a pulse of 0.29 J energy, how many photons are contained in that pulse?

5.1-17. If a certain material has a work function of 3.1 eV, what is the longest wavelength (cutoff wavelength) that could produce any photoelectrons?

5.1-18. Light of 400 nm wavelength hits a surface that has a work function of 2.48 eV. What is:
   (a) The highest kinetic energy, in joules, of the electrons ejected from the surface?
   (b) The cutoff wavelength of the light?

Radiometry/ Photometry

RADIOMETRY IS THE SCIENCE OF MEASURING radiant quantities; it applies throughout the electromagnetic spectrum. *Photometry* is part of radiometry; it applies only to that part of the spectrum that is perceived by the human eye as the sensation of *light*. Most radiometric quantities have the adjective *radiant* and all of their symbols carry the subscript e, for "electromagnetic." Most photometric quantities have the adjective *luminous* and all of their symbols carry the subscript v, for "visual."

Radiometry and photometry have long been confounded by ambiguity. Sometimes, different terms are used for the same quantities. Terms like "candlepower" are ill conceived. Still others, such as nox, phot, glim, skot and scot, bril and brill, helios, lumerg, pharos, stilb, and blondel, merely delight the historian.

In recent years much progress toward clarity has been made, especially since international agreement has been reached to adopt simple, logical, and easily convertible units, based on the International System of Units, SI. These terms and units are defined and used in this chapter.

TERMS AND UNITS

Power. *Power, ϕ, is the term central to all of radiometry and photometry.* It is the amount of energy, Q, that is generated, transmitted, or received, per unit of time, t:

$$\phi = \frac{Q}{t}$$
[5.2-1]

The unit of *radiant power*, ϕ_e, is the same as that of power in general, *watt*, W:

$$1 \text{ watt} = \frac{1 \text{ joule}}{1 \text{ second}}; \qquad W = J \, s^{-1}$$

Luminous power, ϕ_v, is that part of the radiant power that appears as light. Its unit is the *lumen*, lm.*

Radiation geometry.   A quantity such as power, however, does not take into account the specific "radiation geometry," that is, the specific configuration of the source, the beam, or the receiver. To include these configurations, we need another set of terms: *exitance, radiance/luminance, intensity,* and *irradiance/illuminance,* shown schematically in Figure 5.2-1. These terms and the associated symbols and units are summarized in Table 5.2-1.

Exitance.   *Exitance, M,* is the amount of power, ϕ, that "exits" (emerges) from a source, per unit area, A:

$$M = \frac{\phi}{A}$$
[5.2-2]

The unit of *radiant exitance, M_e,* is watts per square meter, $W \, m^{-2}$. The unit of *luminous exitance, M_v,* is lumens per square meter, lm m^{-2}.

Intensity.   The term *intensity* means many things. In the theory of waves it means that energy is proportional to the square of the amplitude, in electrostatics it means force per unit charge, and in acoustics it means power per unit area. In optics, intensity, I, is the amount of power, ϕ, that is emitted by a *point* source and proceeds within a cone of a given *solid angle*, ω:

$$I = \frac{\phi}{\omega}$$
[5.2-3]

---

*The unit of radiant power is named after James Watt (1736–1819), Scottish instrument maker and civil engineer ("civil" in contrast to military engineers). When Watt repaired a Newcomen steam engine, used at that time to pump water out of coal mines, he hit upon the idea of keeping the cylinder with the reciprocating piston hot at all times, greatly increasing the engine's efficiency. By 1800 some 500 Watt engines were working throughout England, ushering in the Industrial Revolution.

The unit of energy is named after James Prescott Joule (1818–1889), British experimental physicist. The son of a wealthy brewer, Joule had the means to devote his life to research, mainly precision measurements of the heat produced by electric current, friction, compression, combustion, and irradiation. Even on his honeymoon, he took time out to measure the temperature of the water at the top of a waterfall and at its bottom, an idea that led to the concept of the mechanical equivalent of heat, now set at 1 cal = 4.184 J.

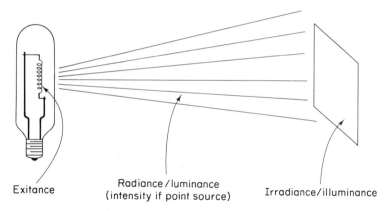

Exitance

Radiance/luminance
(intensity if point source)

Irradiance/illuminance

Figure 5.2-1   Geometrical concepts used in radiometry and photometry.

TABLE 5.2-1   RADIOMETRIC AND PHOTOMETRIC TERMS, SYMBOLS, AND UNITS

Term	Symbol	Unit
Power	ϕ	
Radiant power	ϕ_e watt	W
Luminous power	ϕ_v lumen	lm
Exitance	M	
Radiant exitance	M_e watt/meter^2	W m^{-2}
Luminous exitance	M_v lumen/meter^2	lm m^{-2}
Intensity	I	
Radiant intensity	I_e watt/steradian	W sr^{-1}
Luminous intensity	I_v candela	cd
Radiance	L_e watt/(meter^2 steradian)	W m^{-2} sr^{-1}
Luminance	L_v candela/meter^2	cd m^{-2}
Irradiance	E_e watt/meter^2	W m^{-2}
Illuminance	E_v lux	lx

A solid angle is three-dimensional. Its unit is the ratio of area to distance squared, a *steradian*. One steradian is the solid angle subtended at the center of a sphere of 1 m radius by an area on its surface 1 m^2 in size, regardless of the shape of the area. Since a sphere has a (total) surface area of $4\pi R^2$, the total solid angle about a point is

$$\omega = \frac{4\pi R^2}{R^2} = 4\pi \text{ sr}$$

Substituting in Equation [5.2-3] then shows that a point source, emitting uniformly into all space, has an intensity of

$$I = \frac{\phi}{4\pi} \qquad\qquad [5.2\text{-}4]$$

Radiant intensity, I_e, is radiant power per solid angle. Its unit is watts per steradian, W sr^{-1}. In *luminous intensity*, I_v, watts are replaced by lumens; its unit, therefore, is lumens per steradian, lm sr^{-1}, a unit called *candela*, cd.

RADIANT-LUMINOUS CONVERSION

The candela is the link that connects radiometric and photometric quantities. Its definition permits us to convert one into the other. In earlier years, the unit of luminous intensity was the flame of a candle. Then came the "new candle," a blackbody radiator heated to the temperature of solidification of platinum (1772°C). Today, *the candela is* [defined as] *the luminous intensity, in a given direction, of a source that emits monochromatic radiation of frequency* 540 × 10^{12} *hertz and that has a radiant intensity in that direction of* (1/683) *watt per steradian.**

Why 540 × 10^{12} Hz (which corresponds to a wavelength of 555 nm)? Because that is the center of the bright-yellow part of the spectrum to which the human eye is most sensitive. At other wavelengths the eye's *luminous efficiency*† is less. A blue lamp, for example, even if it emits the same amount of radiant power as a green lamp, appears dimmer because the eye is less sensitive to blue light than to green. The fact that light of 555 nm appears brightest to the human eye, however, holds only for daylight or *photopic vision*, which is a function of the *cones* of the retina. When the eye is adapted to night or *scotopic vision*, which is a function of the *rods*, the peak sensitivity shifts to 510 nm (Figure 5.2-2). Luminous efficiency has no units; it is given as a percentage (see Table 5.2-2).

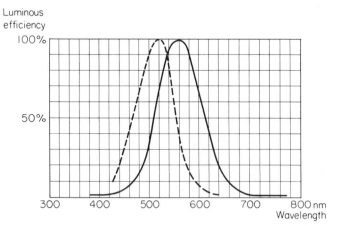

Figure 5.2-2 Sensitivity of the human eye. Photopic vision shown by solid line, scotopic vision by dashed line. Both curves are normalized to their own maxima.

---

*Resolution adopted by the 16th General Conference on Weights and Measures, Paris, 1979.

†Not to be confused with the *efficacy* of a light source, which is the ratio of light emitted to power consumed, in units of lumens/watt, lm W^{-1}. In the case of an electric light source, the power consumed is the product of voltage and current, V × A.

TABLE 5.2-2   SPECTRAL DISTRIBUTION OF PHOTOPIC LUMINOUS
EFFICIENCY

Wavelength	Luminous efficiency	Wavelength	Luminous efficiency
420	0.004	570	0.952
430	0.012	580	0.870
440	0.023	590	0.757
450	0.038	600	0.631
460	0.060	610	0.503
470	0.091	620	0.381
480	0.139	630	0.265
490	0.208	640	0.175
500	0.323	650	0.107
510	0.503	660	0.061
520	0.710	670	0.032
530	0.862	680	0.017
540	0.954	690	0.008
550	0.995	700	0.004
560	0.995		

Example 1

A high-intensity lamp which, because of its small bulb, can be considered a "point light source"
is rated at 25 W and has a luminous intensity of 60 cd. What is:
(a) Its luminous output?
(b) Its luminous efficacy?

Solution.   (a) From Equation [5.2-4] we know that a 1-cd point source emits a total of 4π
lumens. Thus, the luminous output is

$$\phi = (4\pi)(60) = \boxed{754 \text{ lm}}$$

(b) The luminous efficacy is

$$\frac{754}{25} = \boxed{30 \text{ lm W}^{-1}}$$

Example 2

If at 555 nm 1 lumen is equivalent to $1/683$ watt, how many watts are equivalent to 1 lumen of red
light of 630 nm?

Solution.   From Table 5.2-2, the luminous efficiency at 630 nm is 0.265. Thus,

$$1 \text{ W} = (683)(0.265) = 181 \text{ lm}$$

and

$$1 \text{ lm} = \frac{1}{(683)(0.265)} = \boxed{5.525 \text{ mW}}$$

Radiance/luminance.   The terms *radiance* and *luminance* refer to the amount
of radiation that comes from an *extended* source, rather than from a point source. It is the
amount of power emitted from, transmitted through, or reflected or scattered off a surface

per unit area of that surface, per unit solid angle. A little shorter, it is intensity per unit area. Actually, it is not the area as such that counts but the *projected* area, $A \cos \theta$,

$$ L = \frac{\phi}{(A \cos \theta)(\omega)} = \frac{I}{A \cos \theta} \qquad [5.2\text{-}5] $$

where θ is the angle subtended by the normal to the (emitting) surface and the direction of the radiation. Thus, the amount of radiation that leaves a surface element ΔA in a given direction is proportional to $\cos \theta$:

$$ I_\theta = I_\perp \cos \theta \qquad [5.2\text{-}6] $$

which is *Lambert's law.* Substituting Equation [5.2-6] in [5.2-5] gives

$$ L = \frac{I_\perp \cos \theta}{A \cos \theta} = \text{constant} \qquad [5.2\text{-}7] $$

which shows that the subscript θ can be dropped: *The radiance of a Lambertian surface is the same in all directions.*

Example

> This definition of a Lambertian surface makes perhaps more sense when we consider a flat metal disk with a rough surface, heated to incandescence. If the disk has an area of 1 cm^2 and a radiance of 1 W cm^{-2} sr^{-1}, it radiates 1 W sr^{-1} in a direction normal to its surface. In a direction 45° to the normal, it radiates only (1 W sr^{-1}) (cos 45°) = 0.707 W sr^{-1}. However, when seen from the 45° direction, the disk appears compressed into an ellipse, and hence its projected area is less also. The two reductions cancel and the radiance remains the same.

The unit of *radiance, L_e,* is watts per square meter per steradian, W m^{-2} sr^{-1}. The unit of *luminance, L_v,* is lumens per square meter per steradian. But since a lumen per steradian is a candela, a more convenient unit of luminance is candela per square meter, cd m^{-2}.

The *lambert, L,* and other units of luminance, are deprecated:

1 L = 1000 millilambert, mL

1 mL = 10/π cd m^{-2}

footlambert, fL: 1 fL = 1/π cd/ft^2 = 3.426 cd m^{-2}

A Lambertian surface of 1 lm/m^2 exitance produces 1/π cd/m^2 luminance. Similarly, a surface of 1 lm/ft^2 exitance produces 1 fL luminance. Some examples are shown in Table 5.2-3.

TABLE 5.2-3   LUMINANCE OF VARIOUS OBJECTS

Object	Luminance (cd m^{-2})
Atomic bomb	2×10^{12}
Sun	2.3×10^{9}
Xenon arc	1×10^{9}
60-W light bulb, inside frosted	1.2×10^{5}
Snow in bright sunlight	4×10^{4}
Average landscape in summer	8000
Average landscape under overcast sky	2000
Luminance recommended for comfortable viewing	1400

Irradiance/illuminance.   Now let the radiation be *incident* on a surface. This means that the irradiance/illuminance, E, is equal to the power, ϕ, incident per unit area, A:

$$E = \frac{\phi}{A}$$

[5.2-8]

the same relationship that defines the exitance of radiation from a surface.

Since the total power emitted from a point source into all space is $\phi = 4\pi I$ and since the surface area of a sphere is $A = 4\pi R^2$, a source of intensity I produces a total irradiance/illuminance of

$$E = \frac{4\pi I}{4\pi R^2}$$

Canceling 4π and changing R into d, for distance, gives

$$E = \frac{I}{d^2}$$

[5.2-9]

which is the *inverse-square law*. It shows that the irradiance/illuminance, E, at distance d, is directly proportional to the intensity, I, of the source and inversely proportional to the square of the distance (Figure 5.2-3).

There is an interesting similarity worth noting that exists between irradiance/illuminance and magnetic induction. Look again at Figure 4.1-2 (page 287) and consider Biot–Savart's law,

$$\mathbf{B} = \mu_0 \frac{1}{4\pi} i \int \frac{1}{R^2} d\mathbf{W} \times \hat{\mathbf{R}}$$

[5.2-10]

Note that the induction, B, is inversely proportional to the *square* of the distance.

But instead of integrating over the length of the wire, W, from $-\infty$ to $+\infty$, we could as well change variables and replace W by the angle θ subtended by the wire and the direction to the test point. That means that now the integration is performed from $\theta = 0$ (at infinity on one end) to $\theta = 180° = \pi$ rad (at infinity on the other end). Then

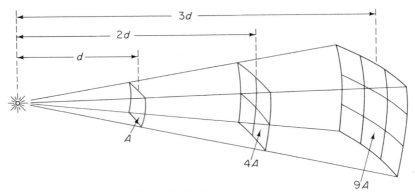

Figure 5.2-3 Inverse-square law.

$$B = \mu_0 \frac{1}{4\pi} i \int_0^\pi \frac{1}{R^2} \sin \theta \, d\theta$$

[5.2-11]

$$= \mu_0 \frac{1}{4\pi} i \frac{1}{d} \int_0^\pi \cos \theta \, d\theta$$

Solving the integral yields

$$B = \mu_0 \frac{1}{4\pi} i \frac{1}{d} (-\cos \pi + \cos 0)$$

[5.2-12]

and thus

$$B = \mu_0 \frac{1}{2\pi} i \frac{1}{d}$$

[5.2-13]

which shows that instead of the inverse-*square* relationship, we now have $1/d$.

This is exactly the same change that occurs when we proceed from a point source to an extended source. Of course, there are no true point sources, nor are there infinitely large sources. For a realistic light source, in short, the exponent in Equation [5.2-9] must lie somewhere between the limits 1 and 2, never reaching either one.

Example

Consider an illuminated slit 7 cm long and 25 cm away from a (small) photodetector. Determine the exponent that should replace the 2 in the inverse-square law.

Solution. If we examine Equations [5.2-12] and [5.2-13] we notice that they differ by a factor $(\cos \theta_1 - \cos \theta_2)/2$. To find to which power the distance d needs to be raised, we set

$$\frac{1}{d^x} = \frac{1}{d} \left(\frac{\cos \theta_1 - \cos \theta_2}{2} \right)$$

$$d^x = d \left(\frac{2}{\cos \theta_1 - \cos \theta_2} \right)$$

and, since by definition $\theta_2 = 180° - \theta_1$,

$$d^x = d\left(\frac{1}{\cos\theta}\right) \qquad\qquad [5.2\text{-}14]$$

Inserting the actual figures, $d = 25$ cm, $\theta_1 = \arctan[25/(7/2)] = 82°$ and $\theta_2 = 180° - 82° = 98°$, and using Equation [5.2-14] yields

$$25^x = 25\left(\frac{1}{\cos 82°}\right) = 180$$

$$x = \frac{\log 180}{\log 25} = \boxed{1.613}$$

Indeed, with an extended light source, the exponent in the inverse-"square" law will always be *less than 2.*

So far we have assumed that the radiation reaches the surface at normal incidence. If it does not, as shown in Figure 5.2-4, the irradiance/illuminance is *less* by a cosine factor,

$$E' = \frac{I}{(d')^2}\cos\theta \qquad\qquad [5.2\text{-}15]$$

In addition, distance d' is *longer* than d:

$$d' = \frac{d}{\cos\theta}$$

Substituting in Equation [5.2-15] gives

$$E' = \frac{I}{(d/\cos\theta)^2}\cos\theta = \frac{I}{d^2}\cos^3\theta \qquad\qquad [5.2\text{-}16]$$

which shows that farther away from point P the irradiance/illuminance falls off as $\cos^3\theta$.

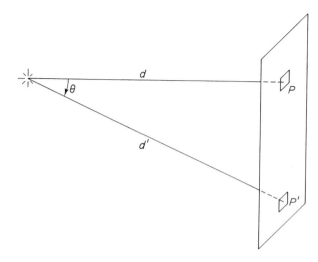

Figure 5.2-4 Light incident on an extended surface.

Finally, assume that the light comes from an *extended* source. If the source has an area S, its radiance/luminance, following Equation [5.2-5], is $L = I/(S \cos \theta)$. Solving for I and substituting in Equation [5.2-16] gives

$$E' = \frac{LS \cos \theta}{d^2} \cos^3 \theta = \frac{LS}{d^2} \cos^4 \theta \qquad [5.2\text{-}17]$$

which shows that now the irradiance/illuminance falls off as $\cos^4 \theta$, a relationship known as the *cosine-fourth law*.

The ratio S/d^2 is the solid angle Ω subtended by the source at point P. Therefore, setting $\theta = 0$ gives

$$E = L\Omega \qquad [5.2\text{-}18]$$

This shows that the irradiance produced by an extended source depends only on the radiance of the source and on the solid angle it subtends; it does not depend on the distance between source and receiver.

The unit of *irradiance*, E_e, is watts per square meter, W m^{-2}, the same as that of radiant exitance. The unit of *illuminance*, E_v, is lumens per square meter, lm m^{-2}, or *lux*, lx, for short.

The relationship between candela and lux is illustrated in Figure 5.2-5. The source (A) has an intensity of 1 cd. Such a source, from Equation [5.2-4], has a *total* output, in all directions, of $\phi = 4\pi I = 12.57$ lm. But the candela is defined as lumens per steradian, and 1 steradian is the solid angle subtended by an area 1 m^2 in size at a distance of 1 m. That is

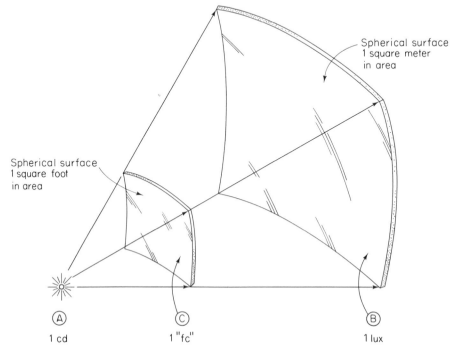

Figure 5.2-5   Diagram showing relationship between candela, footcandle, and lux.

just the area marked B in Figure 5.2-5; this area receives an illuminance $E = 12.57$ lm/ 12.57 m^2, which is 1 lux.

The unit of "*footcandle*," fc, is not a good choice. It seems to suggest that the luminous intensity, in units of "candles," is to be multiplied by a distance. Instead, it means lumens per square foot, lm/ft^2. If the area in Figure 5.2-5 were 1 ft^2 in size and 1 ft away from the source, it would have an illuminance of 1 footcandle (C).

Example

A 4-cd light source is suspended 80 cm above the center of a table 1 m in diameter. What is the illuminance:
(a) In the center of the table?
(b) Near the edge of the table?

Solution. (a) For a point light source, from Equation [5.2-9],

$$E = \frac{I}{d^2} = \frac{4}{(0.8)^2} = \boxed{6.25 \text{ lx}}$$

(b) With the dimensions given, the direction to the edge of the table, and therefore the angle of incidence at the edge, is

$$\theta = \arctan \frac{0.5}{0.8} = 32°$$

The illuminance, therefore, is

$$E' = \frac{I}{d^2} \cos^3 \theta = \frac{4}{(0.8)^2} \cos^3 32° = \boxed{3.8 \text{ lx}}$$

Brightness is not the same as intensity or illuminance. Brightness is a function of illuminance and reflectivity. The term "*illumination*" should not be used either; illumination is the process of exposing an object to light, rather than a property of the object.

Throughput.   In our discussion of magnification we found that the Lagrange invariant, *nyu*, is constant throughout any optical system (page 43). Now we extend this (two-dimensional) convention to three dimensions. Then the height of the object, y, becomes its area, A, and the plane angle u becomes a solid angle, ω. This leads to a new quantity, $n^2 A\omega$, called *throughput, T*. Throughput is a geometric measure of *how much light can be transmitted* through a system of index n and cross section A and that accepts a solid angle ω:

$$T = n^2 A\omega \qquad\qquad [5.2\text{-}19]$$

The terms A and ω occur in the definition of radiance, $L = \phi/(A\omega)$. Therefore, solving for $A\omega$, substituting in Equation [5.2-19], and setting $n = 1$ shows that throughput is also equal to the ratio of power to radiance, $T = \phi/L$. From this we conclude that the throughput of a given system *cannot be increased*, neither by focusing nor by "forcing" the light through a "funnel." In fact, the throughput within a given system becomes gradually *less*, due to inevitable losses on surfaces and in matter. The unit of throughput* is square meter steradian, m^{2}sr.

---

*There is as yet no agreement on whether *throughput* is the best term. The French call it étendue, from which comes "extent," either *geometric extent, Aω*, or *optical extent, n^{2}Aω*. And there is more opposition. W. H. Steel of

RADIOMETERS AND PHOTOMETERS

Comparison photometers.   Systems for measuring radiation in general are called *radiometers*. Systems for measuring light are called *photometers*. The classical example of a comparison photometer is the *Bunsen grease-spot photometer*. It contains a sheet of white paper with a great spot in the center which, by way of two mirrors, can be viewed from both sides at the same time. When the grease spot receives light from only one side, it appears, seen from that side, dark on a bright background; seen from the other side, it appears bright on a dark background. The distances, d_1 and d_2, from the two sources are adjusted so that the spot appears at least contrast. Then, from the inverse-square law,

$$E_1 = \frac{I_1}{d_1^2} = E_2 = \frac{I_2}{d_2^2}$$

and thus

$$\frac{I_1}{I_2} = \left(\frac{d_1}{d_2}\right)^2 \qquad\qquad [5.2\text{-}20]$$

which could be called the *photometric comparison law*.

Photoelectric radiometers.   Basically, a photoelectric *radio*meter measures irradiance, although it may be calibrated instead in terms of power, intensity, or radiance. A photoelectric *photo*meter must have a detector whose spectral response, with or without additional filters, matches that of the eye. Various quantities are determined as follows.

1. *Power*. The total flux emitted by a light source is found by placing the source inside a hollow sphere with a diffusely reflecting inner surface, called an *integrating sphere*. The photodetector is placed near a small window in the sphere's surface.

2. *Irradiance*. Measuring irradiance is easy. All that is needed is a radiation detector. The surface of the detector must be completely filled with the radiation, and the surface must be oriented normal to the direction of the radiation.

3. *Intensity and radiance*. Measuring intensity and radiance is of much practical interest. Radiance, because it includes intensity, is the more general case. Measuring radiance requires certain precautions: (1) *all* of the radiation to be measured must reach the detector and (2) extraneous radiation must be excluded. With a weak source, it may not be enough to let the radiation simply fall on the detector. Instead, the source may need to be *imaged* on the detector's surface, using a lens, a suitably shaped stop, or even a telescope or a microscope.

The simplest luminance meters are the small light meters customary on photographic cameras. Aim the meter at a large, uniformly illuminated wall: You will get a certain reading. Then move back. The reading will be the same—as long as the solid angle subtended at the meter by the size of the wall does not become less than the meter's acceptance angle.

If the target is small and has an odd shape, the meter should look only at the sampling

---

the Australian National Standards Laboratory, for one, asks rhetorically: "If the input is doubled, what happens to the throughput?" *Ans.* Nothing. Input and output are actual quantities of light entering and leaving a system; throughput describes the *capacity* of the system to convey light, rather than the quantity being conveyed.

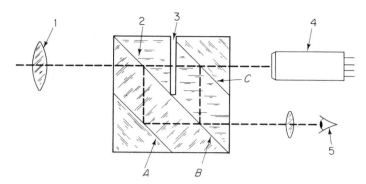

Figure 5.2-6   Schematic diagram of compact luminance meter. Light from the target proceeds to lens 1 and glass cube 2, which contains a slot 3 for a field-limiting mask, to photodetector 4. A series of partially coated interfaces *A*, *B*, and *C* direct light to observer 5, who can see the field both inside and outside the area limited by the mask. [Redrawn from D. A. Schreuder, "Optical System for a Universal Luminance Meter," *Appl. Opt.* **5** (1966), 1965–66, by permission.]

field. This is accomplished best by a see-through mask, such as that illustrated in Figure 5.2-6.

SUGGESTIONS FOR FURTHER READING

F. GRUM and C. J. BARTLESON, *Optical Radiation Measurements* (New York: Academic Press, Inc., 1979).

R. W. BOYD, *Radiometry and the Detection of Optical Radiation* (New York: Wiley-Interscience, 1983).

C. L. WYATT, *Radiometric System Design* (New York: Macmillan Publishing Company, 1987).

E. HELBIG, *Grundlagen der Lichtmesstechnik* (Leipzig: Akademische Verlagsgesellschaft Geest & Portig K.-G., 1977).

IES Lighting Handbook (New York: Illuminating Engineering Society, 1981).

PROBLEMS

5.2-1.   The total power emitted by the sun is estimated to be 3.9×10^{26} J s^{-1}. What is the power incident per square meter just outside the earth's atmosphere, assuming a distance of 150 million km?

5.2-2.   How much energy does a 40-W light bulb consume in 15 min? How much is this in *kilocalories?*

5.2-3.   What is the intensity of a light source that projects 50.3 mW into a hemisphere?

5.2-4.   How much power is produced by a light source that emits 83 cd into all space?

5.2-5.   What is the solid angle subtended at the apex of a cone of light that emerges from a point source and, 14 cm away, fills an aperture 50 mm in diameter?

5.2-6. Light originating at a point source enters a lens 5 cm in diameter. If the light at the source subtends an angle of 0.17 sr, what is its vergence on entering the lens?

5.2-7. A white screen receives 1940 lm of green light of 520 nm wavelength. How much radiant power is incident on the screen?

5.2-8. How much more power must a red light source have to elicit the same response as a green light source? Consult page 7 for peak wavelengths and Table 5.2-1 for luminous efficiencies.

5.2-9. A light bulb rated at 25 W has a luminous intensity of 151.2 cd. Determine the output of the bulb (in lumens) and its efficacy.

5.2-10. If a 60-cd light source consumes 0.68 A at 110 V and if it emits light uniformly in all directions, what is its efficacy?

5.2-11. If the luminance of a large, flat, diffusely reflecting surface, determined at right angles and from a distance of 40 cm, is 3.6 cd m^{-2}, how does the luminance change when determined:
(a) From an angle of 60°?
(b) From a distance of 2 m?

5.2-12. The luminance of an area 10 cm in diameter is measured from a distance of 40 cm, using a calibrated photometer with an acceptance angle of 20°. By what factor must the reading of the photometer be corrected?

5.2-13. If light from a 40-cd source falls on a screen 2.5 m away, what is the illuminance?

5.2-14. A 5.5-cd source is placed at the center of curvature of a mirror of 8 cm radius of curvature and 8 cm diameter. How much light is incident on the mirror?

5.2-15. If a book is held 3 m from a high-intensity lamp and then is moved to 60 cm from the lamp, by how much has the illuminance increased?

5.2-16. A light bulb is suspended 4 m above the floor. To what distance should it be lowered in order to increase the illuminance to 1.77 of its initial value?

5.2-17. A 7-cd light bulb is suspended 75 cm above the center of a table 1.5 m in diameter. What is the illuminance on the table's top near the edge of the table?

5.2-18. A 200-"candle power" incandescent lamp is mounted 5 ft above a drafting table, tilted 60° with respect to its horizontal position. What is the illuminance, in "footcandles"?

5.2-19. A 45-cd lamp is located 1.2 m away from a screen on which it produces the same illuminance as another lamp 40 cm away. What is the intensity of the other lamp?

5.2-20. Two high-intensity lamps of equal efficacy, one rated 25 W and the other 100 W, are mounted 90 cm apart from each other. How far from the 25-W lamp must the screen of a Bunsen grease-spot photometer be placed in order to have equal illuminance on both sides?

5.2-21. The flat center of a street receives light from two lamps of 2000 cd each, mounted on poles 14 m tall and 60 m apart from each other. What is the illuminance at that point?

5.2-22. A point light source, as shown in Figure 5.2-7, is located directly above a point M on a horizontal plane. At which distance, h, above the plane must the light source be placed in order to produce maximum illuminance at a point P on the plane located 20 cm away from M?

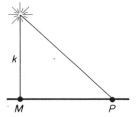

Figure 5.2-7

Atomic Spectra

"IT IS IMPOSSIBLE EVER TO KNOW the chemical composition of the sun!" said French philosopher Auguste Comte, before Bunsen and Kirchhoff established *emission spectroscopy.* Although it was known before, they wrote, that "certain substances when heated in a flame impart definite bright lines to their spectra, . . . differences in temperature have no effect upon the position of these lines . . . which are characteristic of each substance. . . . From these facts it appears certain that the appearance of the bright lines may be regarded as . . . proof of the presence of the particular substance." In one of their early experiments, Bunsen and Kirchhoff were able to detect less than 1/3,000,000 mg of sodium chlorate, convincing proof of the sensitivity of their method.

LINE EMISSION SPECTRA

We have discussed before (in Chapter 5.1) how a solid material, when heated, produces a spectrum of electromagnetic radiation and how this spectrum changes as a function of temperature. These spectra were invariably *continuous spectra.* Now we come to *line spectra* and to the specific realm of Bunsen and Kirchhoff's *emission spectroscopy.** In contrast to

---

*Robert Wilhelm Bunsen (1811–1899), German chemist and professor of chemistry at the University of Breslau, later at Heidelberg. Bunsen invented the gas burner and the grease-spot photometer, both named after him, the carbon zinc battery and the prism spectroscope, at that time containing a hollow prism filled with carbon disulfide. Bunsen, a careful and patient experimentalist, and Gustav Robert Kirchhoff (1824–1887), a theoretical physicist,

continuous spectra which issue from solids and from gases at high pressure, line emission spectra come from vapors and gases at low pressure, where the atoms or molecules are far apart and do not significantly interact.

Each chemical element has unique spectrum lines that are characteristic of the element (like a fingerprint is of a person) and therefore permit a positive identification.* Some elements, such as hydrogen and neon, have relatively few lines; others, such as iron, have many thousand.

An example of a comparatively simple spectrum is that of hydrogen (Figure 5.3-1). In this example the lines follow each other in an orderly fashion. A group of such lines is called a *series*. The most prominent, brightest line (of atomic hydrogen) lies to the right; it has the longest wavelength, and is designated α. The next line to the left is called β, the third line γ, and so on. At the short-wavelength end, on the far left, the lines crowd together and form the *series limit*.

The Rydberg equation.   The position of the lines (in the spectrum of hydrogen) follows a simple relationship. This relationship fits all known lines (of hydrogen) and has led to the prediction of new ones. It can be expressed in the form of *Rydberg's equation*,†

$$\frac{1}{\lambda} = \Re\left(\frac{1}{m^2} - \frac{1}{n^2}\right) \qquad\qquad [5.3\text{-}1]$$

where λ is the wavelength, in nm; $\Re$ the *Rydberg constant*‡ for hydrogen, approximately 0.011 nm^{-1}; m an integer, 1, 2, 3, . . . , which is constant within any one series, and n the

---

proved to be an ideal combination of talents. At age 20, Kirchhoff calculated the distribution of currents in an arbitrary network, establishing what we now call *Kirchhoff's law*. In 1850 he became professor at the University of Breslau, where he met Bunsen. They became friends, and when Bunsen moved to Heidelberg, he managed to obtain an appointment for Kirchhoff at the same university. In the course of their work they found that the residue of some mineral water produced spectrum lines not seen before; they attributed the lines to a new alkali metal that, because of the bluish color of one of the lines, they called *cesium*. A few months later, Bunsen discovered in the mineral lepidolite another alkali metal which, because of its prominent red line, he called *rubidium*. Their fundamental paper is G. Kirchhoff and R. Bunsen, "Chemische Analyse durch Spectralbeobachtungen," *Ann. Phys.* (4) **20** (1860), 161–89.

*The spectrum *lines* are actually images of the entrance slit of the spectrograph. If the spectrograph had a triangular aperture, the "line" spectrum would be a series of triangles.

†Named after Johannes Robert Rydberg (1854–1919), Swedish mathematician and physicist. Rydberg, unlike most others, used wavenumbers, $\sigma = 1/\lambda$, in trying to predict the position of the lines. When he heard that Johann Jakob Balmer (1825–1898), a Swiss secondary-school teacher, had found an empirical formula, $\lambda = 364.56$ $m^2/(m^2 - 4)$ nm; $m = 3, 4, 5, . . .$, he changed Balmer's formula to wavenumbers, recognized that Balmer's is a special case and that $\Re$ is a universal constant. Being interested all his life in the periodic system, Rydberg maintained that between hydrogen and helium there must be two more elements, which he called "nebulium" and "coronium," their alleged lines later found to be due to ionized oxygen/nitrogen and iron. Balmer's contribution is found in "Notiz über die Spectrallinien des Wasserstoffs," *Ann. Phys.* (Neue Folge) **25** (1885), 80–87; Rydberg's in "Recherches sur la constitution des spectres d'émission des éléments chimiques," *K. Sven. vetenskaps-akad. handl.* (4) **23** (1890), No. 11, 1–155.

‡The most precise figure found to date is 0.010 973 731 476 (32) nm⁻¹. See J. E. M. Goldsmith, E. W. Weber, and T. W. Hänsch, "New Measurement of the Rydberg Constant Using Polarization Spectroscopy of Balmer-α," *J. Opt. Soc. Am.* **68** (1978), 1411.

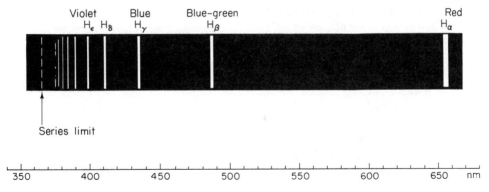

Figure 5.3-1   The Balmer series in the spectrum of hydrogen.

quantum number or "running figure," another integer which runs as a whole number 2, 3, 4, . . . through the series.

The first or α line in any one series is found when we set

$$n = m + 1 \qquad [5.3\text{-}2]$$

the second, β, when $n = m + 2$; the third, γ, when $n = m + 3$; and so on. At the series limit $n \to \infty$.

After Rydberg's discovery, Lyman applied Rydberg's equation using $m = 1$, and found a series of lines in the ultraviolet. Paschen then set $m = 3$ and found another series, in the infrared. As time went on, and more sensitive detectors became available, especially in the IR, more and more series were discovered. A summary of what is known today is shown in Table 5-3.1.

A particularly strong hydrogen line is the Lyman α line, $\lambda = 121.5$ nm. It is by far the most powerful line emitted by the sun. The Lyman α line is completely absorbed by air; it is

TABLE 5.3-1   SUMMARY OF HYDROGEN SERIES*

Series	Year of discovery	Integer, m	Quantum number, n	Spectral region
Lyman	1904	1	2, 3, 4, . . . , ∞	UV
Balmer	1885	2	3, 4, 5, . . . , ∞	Visible
Paschen	1908	3	4, 5, 6, . . . , ∞	IR
Brackett	1922	4	5, 6, 7, . . . , ∞	IR
Pfund	1924	5	6, 7, 8, . . . , ∞	IR
Humphreys	1953	6	7, 8, 9, . . . , ∞	IR
Hansen and Strong	1973	7	8, 9, 10, . . . , ∞	IR

*Some of the original references are T. Lyman, "Preliminary Measurement of the Short Wave-Lengths Discovered by Schumann," *Astrophys. J.* **19** (1904), 263–67; F. Paschen, "Zur Kenntnis ultraroter Linienspektra. I. Normalwellenlängen bis 27 000 Å.-E.," *Ann. Phys.* (4) **27** (1908), 537–70; A. H. Pfund, "The Emission of Nitrogen and Hydrogen in the Infrared," *J. Opt. Soc. Am.* **9** (1924), 193–96; P. Hansen and J. Strong, "Seventh Series of Atomic Hydrogen," *Appl. Opt.* **12** (1973), 429–30.

also absorbed by deoxyribonucleic acid, DNA, the building block of all living matter on Earth. On other planets which have no protective atmosphere, therefore, life either cannot exist or must be based on genetic material other than DNA.

Atomic structure of matter.   To understand why spectrum lines are arranged in such an orderly fashion, we need to review the *atomic structure of matter*. Early theories, such as *Rutherford's model,** held that an atom consists of a massive, positively charged nucleus and of one or several negatively charged electrons revolving around the nucleus.

For simplicity, assume that a single electron revolves in a circular orbit of radius R. Following the laws of classical mechanics, the centripetal force, F, needed to keep a body of mass m and velocity v in a circular orbit is

$$F = \frac{mv^2}{R} \qquad [5.3\text{-}3]$$

In the case of an atom, the centripetal force is supplied by electrostatic attraction, which, from *Coulomb's law*, is

$$F = \frac{1}{4\pi\epsilon_0}\frac{q_1 q_2}{R^2} \qquad [5.3\text{-}4]$$

where ϵ_0 is the electric permittivity of free space. Combining Equations [5.3-3] and [5.3-4] by eliminating F gives

$$\frac{mv^2}{R} = \frac{1}{4\pi\epsilon_0}\frac{e^2}{R^2} \qquad [5.3\text{-}5]$$

where m is the mass of the electron, approximately 9.1×10^{-31} kg, and e is the charge on the electron, approximately 1.6×10^{-19} C.

The kinetic energy of the electron is

$$\mathrm{KE} = \frac{1}{2}mv^2 = \frac{1}{2}\frac{1}{4\pi\epsilon_0}\frac{e^2}{R} = \frac{e^2}{8\pi\epsilon_0 R} \qquad [5.3\text{-}6]$$

and the potential energy, by integrating Equation [5.3-4], is

$$\mathrm{PE} = -\frac{e^2}{4\pi\epsilon_0 R} \qquad [5.3\text{-}7]$$

---

*Ernest Rutherford (1871-1937), New Zealand-born physicist. In 1895 Rutherford received a scholarship to go to Cambridge; from there he went to McGill University in Montreal, to Manchester, England, and, succeeding his early mentor J. J. Thomson, again to Cambridge. Interested in newly discovered radioactivity, Rutherford found that the radiation given off by radioactive material were α particles which are the nuclei of helium (atoms without the electrons), and β particles which are electrons. Their emission accounts for the transformation of the chemical elements, a discovery for which he received the Nobel Prize in physics in 1908. In an even more famous experiment he discovered, by α particle scattering on a thin gold foil, that the atom has a small, massive nucleus accompanied by electrons revolving around it, not unlike planets revolving around the sun; in this way he became the first to elucidate the structure of the atom. E. Rutherford, "The Scattering of α and β Particles by Matter and the Structure of the Atom," *Philos. Mag.* (6) **21** (1911), 669–88.

The total energy, E, of the electron is the sum of its kinetic and potential energies,

$$E = \frac{e^2}{8\pi\epsilon_0 R} - \frac{e^2}{4\pi\epsilon_0 R} = -\frac{e^2}{8\pi\epsilon_0 R} \qquad [5.3\text{-}8]$$

The minus signs in Equations [5.3-7] and [5.3-8] mean that, since the electron is bound to the atom, energy must be supplied to remove it from the nucleus. The closer the electron is to the nucleus, the greater the energy that is needed.

The angular momentum, L, again from classical mechanics, is

$$\mathbf{L} = m\mathbf{v} \times \mathbf{R} \qquad [5.3\text{-}9]$$

Such a system is mechanically stable, because the Coulomb force provides the centripetal force necessary to keep the electron in orbit. Rutherford's model, however, does not account for the discrete frequencies that occur in the emission process. This dilemma is resolved in *Bohr's model*.

Bohr's model of the atom. Two years after Rutherford, Bohr *postulated* (assumed) that electrons move only in certain orbits, called *stationary states*. In these orbits, the angular momentum of an electron can have only certain values which are integral multiples of $h/(2\pi)$.* Thus instead of Equation [5.3-9] we now have

$$mvR = n\frac{h}{2\pi} \qquad [5.3\text{-}10]$$

where n is the (principal) *quantum number,* an integer, 1, 2, 3, . . . , and h is *Planck's constant.*

Eliminating v between Equations [5.3-5] and [5.3-10] then gives for the radii of the *quantized* orbits

$$R = n^2 \frac{h^2\epsilon_0}{\pi e^2 m} \qquad [5.3\text{-}11]$$

For example, hydrogen has but one electron. With the atom in its lowest energy state ($n = 1$), the electron revolves in an orbit of radius

$$R = (1)^2 \frac{(6.63 \times 10^{-34})^2(8.85 \times 10^{-12})}{(\pi)(1.6 \times 10^{-19})^2(9.1 \times 10^{-31})} = 5.3 \times 10^{-11} \text{ m} \qquad [5.3\text{-}12]$$

---

*Niels Henrik David Bohr (1885–1962), Danish physicist, professor of theoretical physics at the University of Copenhagen, founder of the Institute for Theoretical Physics. Bohr's scientific work extended through 57 years, a period of fundamental changes. He began when the structure of the atom was still unknown; he ended when atomic physics had reached maturity. Bohr established the quantum orbits of the hydrogen atom, discovered element 72, hafnium (after the Latin name for Copenhagen), laid the groundwork for quantum field theory, and in 1922 received the Nobel Prize in physics. During World War II he escaped from Denmark aboard a small fishing vessel and came to the United States. Introduced at the Los Alamos Laboratories as a fictitious "Mr. Nicholas Baker," he helped see the Manhattan Project through to completion. Later in his life, Bohr's writings became more philosophical, expounding over and over again the concept of complementarity, which, he thought, "would afford people the guidance they need." N. Bohr, "On the Constitution of Atoms and Molecules," *Philos. Mag.* (6) **26** (1913), 1–25, 476–502, 857–75.

which means the hydrogen atom is approximately 1 Å (10^{-10} m) in diameter.*

To find the total (quantized) energies in the various states, we substitute Equation [5.3-11] in [5.3-8] and obtain

$$E = -\frac{e^4 m}{8\epsilon_0^2 n^2 h^2} \qquad [5.3\text{-}13]$$

Again for hydrogen and $n = 1$,

$$E = -\frac{(1.6 \times 10^{-19})^4 (9.1 \times 10^{-31})}{(8)(8.85 \times 10^{-12})^2 (1^2)(6.63 \times 10^{-34})^2} = -2.17 \times 10^{-18} \text{ J}$$

Converting J to eV gives

$$E = -\frac{21.7 \times 10^{-19}}{1.6 \times 10^{-19}} = -13.6 \text{ eV} \qquad [5.3\text{-}14]$$

a figure that will soon come up again.

ATOMIC TRANSITIONS

It is easier to represent the energy states of an atom in the form of *energy levels,* rather than in the form of orbits. An energy-level diagram of hydrogen, as an example, is shown in Figure 5.3-2. Each horizontal line represents an allowed energy state. The vertical arrows between levels represent various discrete transitions. The lowest energy level is the level at which the electron revolves in the innermost orbit (where $n = 1$); that level is called the *ground state.* For hydrogen, the ground state is at -13.6 eV, as derived. When the hydrogen gas is heated, or subject to an electric discharge, the atoms are raised to higher, *excited states.* For example, at the next higher level (where $n = 2$), following Equation [5.3-13], the electron has reached an energy level of -3.4 eV, at $n = 3$, it has reached -1.5 eV, and so on; but quantum theory does not allow the electrons to have energies between -13.6 and -3.4 eV, or between -3.4 and -1.5 eV.

The *transition* from one energy state to another is accomplished by a transfer of energy. If energy, in whatever form, is supplied to the system, the system is raised from a lower energy state, E_1, to a higher, excited state, E_2. Such a transition, $E_1 \rightarrow E_2$, is called *absorption* (see Chapter 5.4). The reverse process, the downward transition $E_2 \rightarrow E_1$, is called *emission.* During the emission process, as the atom *returns from a higher energy state, E_2, to a lower energy state, E_1,* it emits a *quantum of energy, $h\nu$.* The amount of energy being emitted is the difference between the energies in the two states,

$$E_2 - E_1 = h\nu \qquad [5.3\text{-}15]$$

which is *Bohr's frequency condition.*

We solve $c = \lambda\nu$ for λ and set, following Equation [5.3-14], $h\nu = 13.6$ eV. That shows

---

*Here I have made several simplifying assumptions, neglecting the facts that the orbits are elliptical, rather than circular, that the proton is not stationary (which would require it to have infinite mass), and that the mass of the electron changes relativistically as a function of velocity.

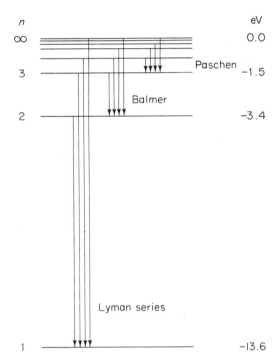

Figure 5.3-2 Energy-level diagram for hydrogen showing some of the transitions in the Lyman, Balmer, and Paschen series. In the ground state ($n = 1$) the electron has an energy of -13.6 eV, in the $n = 2$ state it has -3.4 eV, in the $n = 3$ state -1.5 eV, and so on.

that the wavelength of radiation emitted through a transition of hydrogen from the highest to the ground state is

$$\lambda = \frac{c}{\nu} = \frac{hc}{h\nu} = \frac{(6.63 \times 10^{-34} \text{ J s})(3 \times 10^8 \text{ m/s})}{(13.6 \text{ eV})(1.6 \times 10^{-19} \text{ J/eV})} = 91.4 \text{ nm} \qquad [5.3\text{-}16]$$

which is the limit of the *Lyman series*. For the other series of hydrogen we use the same reasoning, noting that in Equation [5.3-13] the *square* of the quantum number occurs in the denominator.

Finally, if the quantum number, n, is used to describe the higher energy state and another constant, m, is introduced to describe the lower state, then

$$\nu = \frac{me^4}{8\epsilon_0^2 h^3} \left(\frac{1}{m^2} - \frac{1}{n^2} \right) \qquad [5.3\text{-}17]$$

which is again the *Rydberg equation*. In short, we have arrived at the same result, once by experimental observation and then by theoretical reasoning.

Other line spectra.   The spectrum of *helium* can be described by an equation very similar to Rydberg's equation for hydrogen. The helium spectrum has four series, a first and second Lyman series ($m = 1$ and $m = 2$), both in the vacuum ultraviolet, the Fowler series ($m = 3$), and the Pickering series ($m = 4$).

The spectra of other atoms are more complicated. Only a few are understood as thoroughly as the spectra of hydrogen and helium; these include the spectra of the alkali metals, lithium, sodium, potassium, rubidium, and cesium. The spectra of most other elements are

TABLE 5.3-2 WAVELENGTHS (nm) OF
SELECTED SPECTRUM LINES OF VARIOUS
ELEMENTS, LISTED IN ORDER OF
INCREASING ATOMIC NUMBER

Hydrogen	1[H]	656.285
Helium	2[He]	587.562
Lithium	3[Li]	610.364
		670.784
Sodium	11[Na]	588.995
		589.592
Potassium	19[K]	766.491
		769.898
Cesium	55[Cs]	852.110
Mercury	80[Hg]	435.835
		546.074
Thallium	81[Tl]	535.05

still more complex and a discussion of them is not needed for an understanding of atomic spectra in general. The most prominent lines of certain elements are listed in Table 5.3-2.

Conclusions. What does it all mean, in terms of both theoretical insight and practical application?

1. Surely, since Bunsen and Kirchhoff's early work emission spectroscopy has come a long way. Today we have a data base that provides computer access to some 80,000 spectra of nearly 70,000 different compounds. That helps us identify unknown substances by comparing their spectra to those of other, verified substances.

2. Emission spectroscopy has greatly clarified the structure of matter, revealing at the same time some of the most fundamental laws of physics, from Bohr's model of the hydrogen atom to electron spin to quantum mechanics.

3. On the more mundane level, new and better light sources have been produced. Replacing pure sodium vapor, for example, by a mixture of sodium and xenon causes the narrow sodium lines to widen into the green and red of the spectrum, providing a more pleasant and more efficient means of illumination.

4. Understanding atomic emission greatly helps understanding *stimulated* emission. This is the basis of laser action which we will discuss in Chapter 5.5.

SUGGESTIONS FOR FURTHER READING

H. G. KUHN, *Atomic Spectra,* 2nd edition (New York: Academic Press, Inc., 1969).

R. CHANG, *Basic Principles of Spectroscopy* (New York: McGraw-Hill Book Company, 1971).

R. D. COWAN, *The Theory of Atomic Structure and Spectra* (Berkeley: University of California Press, 1981).

J. M. HOLLAS, *High Resolution Spectroscopy* (Woburn, MA: Butterworth Publishers, 1982).

PROBLEMS

5.3-1. What is the wavelength of the tenth line in the Lyman series of hydrogen?

5.3-2. Using the Rydberg equation, find the wavelength and color of the Balmer α line in the spectrum of hydrogen.

5.3-3. What is the difference, in nm, between the wavelengths of the tenth line and the series limit of the Balmer series of hydrogen?

5.3-4. What is the series number, the quantum number, and the wavelength of:
   (a) The Lyman β line?
   (b) The Balmer γ line?
   (c) The Paschen α line?

5.3-5. Calculate the velocity of the electron in the first Bohr orbit of the hydrogen atom. (Use Equation [5.3-5] and $\epsilon_0 = 8.85 \times 10^{-12}$ $C^2 N^{-1} m^2$.)

5.3-6. Determine the magnitude of the Coulomb attraction between the nucleus and the electron in the ground state of the hydrogen atom, using:
   (a) The centripetal force equation.
   (b) The electrostatic force equation.
   Take the velocity of the electron from Problem 5.3-5.

5.3-7. With the ionization energy of hydrogen being 13.6 eV, what is its energy in the $n = 4$ state? (*Hint:* Use the statement following Equation [5.3-16].)

5.3-8. The Balmer series limit occurs as the hydrogen atom drops from the $n = \infty$ state to the $n = 2$ state. What is the wavelength of this limit? (*Hint:* Read the energy difference from Figure 5.3-2 and use Bohr's frequency condition and Equation [5.3-16] to find the wavelength.) Compare the result with the series limit obtained in Problem 5.3-3.

5.4

Absorption

IN GENERAL, THE PROPAGATION OF LIGHT is reversible. If a ray of light, passing through an optical system in a given direction, is sent through the system in the opposite direction, it will exactly retrace its path. But absorption is not reversible. It is an *irreversible thermodynamic process.*

Sometimes, the terms absorption and extinction are used synonymously. That is not correct. Extinction is the broader term; it includes attenuation by both absorption and scattering.

ABSORPTION SPECTRA

We have seen earlier (in Chapter 5.3) that when a downward transition occurs, from a higher energy level to a lower energy level, a quantum of light is *emitted.* But now assume that a quantum of energy is incident on an atom. If the quantum has the right amount of energy, the quantum will *raise* the atom from the lower energy state to a higher, excited state, causing an upward transition. That, in essence, is the atomic basis of *absorption.*

Absorption spectra, the same as emission spectra, can be classified as either continuous spectra or line spectra. Continuous absorption spectra occur with dyes and other colored liquids and solids. Line absorption spectra result when (white) light passes through a gas at low pressure and at a temperature lower than that of the source. These lines have the same wavelengths as the lines which the gas would emit if it were hot; they identify an element just as emission spectra do. Absorption lines, superimposed on the continuous spectrum of

TABLE 5.4-1 FRAUNHOFER LINES

Letter designation	Element	Position in spectrum	Wavelength (nm)
F'	Cd	Blue	480.0
F	H	Blue	486.1
e	Hg	Green	546.1
d	He	Yellow	587.6
D_2	Na	Yellow	589.0
D_1	Na	Yellow	589.6
C	H	Red	656.3

the sun, are known as *Fraunhofer lines.** Some representative examples are listed in Table 5.4-1.

The exponential law.   Let a collimated beam of light travel in the $+x$ direction and let it pass through a thin slice of material of thickness Δx (Figure 5.4-1). We call ϕ_0 the power of the light incident on the material and ϕ' the power that emerges from it.

On passing through the material, a certain fraction of the light, $\Delta\phi$, is lost:

$$\phi_0 - \phi' = \Delta\phi \qquad [5.4\text{-}1]$$

The magnitude of this loss is proportional to ϕ_0, to the thickness Δx, and to a constant of proportionality called *absorptivity,* α:

$$\phi' - \phi_0 = -\Delta\phi = \phi_0\alpha\,\Delta x \qquad [5.4\text{-}2]$$

The absorptivity (the old "absorption coefficient") is characteristic of the material, and also a function of wavelength.

Then assume that the medium is made into infinitesimally thin slices, each of thickness dx. Again, in each slice a constant fraction of light, $d\phi$, is lost and Equation [5.4-2] becomes

$$\frac{d\phi}{\phi_0} = -\alpha\,dx \qquad [5.4\text{-}3]$$

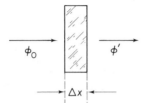

Figure 5.4-1 Absorption of light in a thin slice of matter of thickness Δx.

*By his own count, Fraunhofer saw 574 dark lines in the spectrum of the sun but, as we know today, some of these are due to absorption in the atmosphere of Earth, rather than the photosphere of the sun. J. Fraunhofer, "Bestimmung des Brechungs- und des Farbenzerstreuungs-Vermögens verschiedener Glasarten, in Bezug auf die Vervollkommnung achromatischer Fernröhre," *Ann. Phys.* (2) **26** (1817), 264–313.

In order to find the total loss within the medium (of thickness x), we integrate Equation [5.4-3] between the limits of ϕ and x:

$$\int_{\phi_0}^{\phi'} \frac{d\phi}{\phi} = -\alpha \int_0^x dx \qquad [5.4\text{-}4]$$

Since

$$\int_a^b \frac{d}{dx} f(x)\, dx = f(b) - f(a) \qquad \text{and} \qquad \int \frac{dx}{x} = \ln x$$

$$\ln\left(\frac{\phi'}{\phi_0}\right) = -\alpha x \qquad [5.4\text{-}5]$$

and

$$\frac{\phi'}{\phi_0} = e^{-\alpha x} \qquad [5.4\text{-}6]$$

If the absorbing medium is a solution, the concentration of the solution, c, usually given in grams or moles per liter, must be included also and Equation [5.4-6] becomes

$$\phi' = \phi_0\, e^{-\alpha x c} \qquad [5.4\text{-}7]$$

which is the *exponential law of absorption.* * Note that in deriving this law, integration leads to *natural,* Napierian logarithms (to base e). In contrast, as we will see shortly, when it comes to practical applications, we use *common* logarithms (to base 10).

Transmittance, T, is the ratio of the radiant power transmitted through the sample to the power incident on it, both measured at the same wavelength:

$$T = \frac{\phi'}{\phi_0} \qquad [5.4\text{-}8]$$

Absorbance, A, is the logarithm to base 10 of the inverse of the transmittance:

$$A = \log_{10}\left(\frac{1}{T}\right) = \log_{10}\left(\frac{\phi_0}{\phi'}\right) \qquad [5.4\text{-}9]$$

(The term "optical density" is deprecated because of possible confusion with "density," which is mass per unit volume. The term "absorption" is a process rather than a property.)

---

*The exponential law as presented has evolved through many efforts. Most notable among its originators were the following: Pierre Bouguer (1698–1758), French oceanographer, was the first to find a relationship similar to the exponential law. He published it in his book *Essai d'optique sur la gradation de la lumière* (Paris: Jombert, 1729). Johann Heinrich Lambert (1728–1777), a largely self-taught German physicist and mathematician, contributed to thermodynamics, photometry (cosine law), absorption, planetary motion, and cosmology, liked to reduce his thoughts on logic and philosophy to mathematical models. He described the law in *Lamberts Photometrie (Photometria, sive de mensura et gradibus luminis, colorum et umbrae)* (Augsburg, 1760), the same year in which a second, greatly enlarged edition of Bouguer's book, renamed *Traité d'optique sur la gradation de la lumière* (Paris: H. L. Guerin & L. F. Delatour, 1760) was published (posthumously) that contained virtually the same relationship. August Beer (1825–1863), German physicist, professor of mathematics at the University of Bonn, recognized the role of the concentration, as he reported in his book *Einleitung in die höhere Optik* (Braunschweig: F. Vieweg und Sohn, 1853).

TABLE 5.4-2 TRANSMITTANCE AND ABSORBANCE

Percent transmittance, T	Absorbance, A
100	0.0000
90	0.0458
80	0.0969
70	0.1549
60	0.2218
50	0.3010
40	0.3979
30	0.5229
20	0.6990
10	1.0000
1	2.0000
0	∞ (opaque)

Representative figures that show the relationship between transmittance and absorbance are listed in Table 5.4-2.

Absorbance is an important term. In principle we could measure the *transmittance* of a solution at different concentrations and plot the curve thus obtained. But if we use *absorbances,* rather than transmittances, plotting is easier because then the relationship is linear and only a few points are needed to establish a straight line. Furthermore, absorbances are additive, whereas transmittances are multiplicative. Figure 5.4-2 illustrates the connection.

Absorptivity, α, as we saw in Equation [5.4-7], appears in the exponential law as a *natural* logarithm:

$$\ln\left(\frac{\phi'}{\phi_0}\right) = -\alpha x c \qquad [5.4\text{-}10]$$

But absorbance, A, is based on *common* logarithms:

$$A = \log_{10}\left(\frac{\phi_0}{\phi'}\right) \qquad [5.4\text{-}11]$$

To convert one into the other, we use the identities $\ln x = 2.3026 \ldots \log_{10} x$ or $\log_{10} x = 0.4343 \ldots \ln x$, as needed. This gives us a set of equations that we will use frequently:

$$\boxed{\begin{aligned} A &= \log_{10}\left(\frac{\phi_0}{\phi'}\right) \\[4pt] A &= 0.4343\,\alpha x c \\[4pt] T &= 10^{-A} = \frac{1}{10^A} \\[4pt] \alpha &= 2.3026\,\frac{A}{xc} \end{aligned}} \qquad [5.4\text{-}12]$$

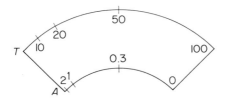

Figure 5.4-2 Scale of photometer showing logarithmic relationship between transmittance, T, and absorbance, A.

From the second equation in [5.4-12] we see that the absorbance, A, of a given sample is directly proportional to its thickness, x:

$$A = (k)(x)$$

If we have two samples of the same material (same α and same c), we divide one equation by the other:

$$\frac{A_1}{A_2} = \frac{(k)(x_1)}{(k)(x_2)}$$

cancel the constant, k, and rearrange:

$$A_2 = A_1 \frac{x_2}{x_1} \qquad \text{[5.4-13]}$$

Multiplying by -1 and taking the logarithm of 10 to the Ath power gives

$$\log 10^{-A_2} = \log 10^{-A_1(x_2/x_1)}$$

and, substituting twice the third equation in [5.4-12],

$$\boxed{T_2 = T_1^{x_2/x_1}} \qquad \text{[5.4-14]}$$

This is a very useful equation which we will need whenever we compare the transmittance of a sample of thickness x_1 with the transmittance of another sample, of the same material but of a different thickness, x_2.

Example 1

A neutral density filter of a given thickness absorbs 10% of the light incident on it. How much light will a filter absorb, made of the same material but 12 times as thick?

Solution. If 10% is being absorbed by the first filter, 90% is transmitted. Hence the transmittance is

$$T_1 = 90\% = 0.9$$

The transmittance of the second (thicker) filter, from Equation [5.4-14], is

$$T_2 = T_1^{x_2/x_1} = 0.9^{12} = 0.28$$

Therefore, if the second filter transmits 28% of the light, it absorbs

$$100 - 28 = \boxed{72\%}$$

Example 2

Another filter, 4 mm thick, absorbs 30% of the light. How thick a filter is needed to absorb 60%?

Solution.   Converting these percentages into transmittances, we have $T_1 = 70\%$ and $T_2 = 40\%$. Also, $x_1 = 4$ mm, while x_2 is to be found. Then again from Equation [5.4-14],

$$0.4 = 0.7^{x_2/4}$$

$$0.4^4 = 0.7^{x_2}$$

$$4 \log 0.4 = x_2 \log 0.7$$

(For this last step, either common or natural logarithms may be used.) The thickness of the second filter, then, is

$$x_2 = \frac{4 \log 0.4}{\log 0.7} = \boxed{10.3 \text{ mm}}$$

In general, and in accordance with the exponential law, a solution of a given thickness (path length) and concentration will absorb the same fraction of light as the same solution twice as thick but of half the concentration. There are situations, however, where α varies as the concentration is changed.

Selective absorption.   Some materials, such as metals, have high absorptivity from the ultraviolet through the infrared; even a thin metal membrane is virtually opaque to light. Other materials such as dyes block out only part of the spectrum. This is called *selective absorption*. In general, there are no substances that have no absorption at least somewhere in the electromagnetic spectrum: glass is opaque to UV, water absorbs below 190 nm and in the IR, and absorption by air extends from the region of soft X rays to and including the vacuum ultraviolet.

This makes somewhat obsolete the earlier distinction between *general absorption* and *selective absorption*. In general absorption, the absorptivity of the material is very nearly the same over a wide range of wavelengths. In selective absorption, attenuation occurs at certain wavelengths more than at others. Within the visible spectrum, this causes the sensation of *color.** A sample of green glass, for instance, will absorb mainly red, and since red is complementary to green, the transmitted light will lack red and the sample will appear green.

Any color has three physical characteristics: *hue, lightness,* and *saturation*. Hue gives the color its name, such as blue, green, or red. Lightness refers to whether a color is pale (light) or dark (deep). Saturation is described by terms such as reddish white, moderate red, strong red, and vivid red. If more and more white is added to red paint, the dominant hue, red, stays the same but a series of tints will result, ending in a pale pink.

---

*Early in 1672, Newton wrote to the Royal Society of London announcing that he wanted to report "a philosophical Discovery . . . in my judgment the oddest if not the most considerable detection wch hath hitherto beene made in the operations of Nature." He let sunlight fall on a "triangular glass-Prisme" and found that "Light it self is a Heterogeneous mixture of differently refrangible Rays." The colors obtained could not be divided further; they could be combined back into white. I. Newton, "A New Theory about Light and Colours," *Philos. Trans. Roy. Soc. London* **5**, No. 80 (Feb 19, 1672), 3075–87.

EXPERIMENTAL METHODS

In order to determine the absorption characteristics of a sample as a function of wavelength, we need a *spectrophotometer,* that is, a photometer or radiometer connected to a device for wavelength selection. If the sample is a solid such as glass or plastic, the sample is made into a plane-parallel, polished plate. If it is a liquid or a substance in solution, it is placed in a cuvette and a comparison made between the solution and the pure solvent. Modern spectrophotometers plot the transmittance, or the absorbance, or both, as a function of wavelength.

Types of spectrophotometers. Spectrophotometers can be built as one-beam-indicator, one-beam-substitution, and two-beam instruments. The latter may have either one or two detectors. In the *one-beam one-detector method,* the light proceeds in a single path through a monochromator and through the sample to the photodetector (Figure 5.4-3). The sample and the reference (blank) are alternately brought into the path and the transmittance through each is measured at each wavelength desired.

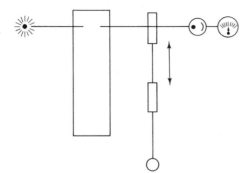

Figure 5.4-3 One-beam one-detector method.

In the *one-beam substitution* (zero) *method,* not shown, sample and reference again are alternately brought into the path. While measuring the reference, however, a neutral density filter or other attenuator is brought into the beam to make the two intensities equal. The advantage is that the response of the detector need not be linear.

In the *two-beam two-detector method,* the light coming from the monochromator is divided by a beamsplitter, passed through the sample and the blank, respectively, and is incident on two detectors (Figure 5.4-4). The potentiometer is adjusted so that the galvanometer reads zero. Two-beam instruments are independent of voltage fluctuations.

In the *two-beam one-detector method* (Figure 5.4-5), the light is split, passes through the sample and the blank, is recombined by another beamsplitter, and falls on *one* detector; thus there can be no mismatch between two photosurfaces. The reference beam is attenu-

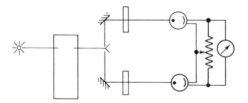

Figure 5.4-4 Two-beam two-detector method.

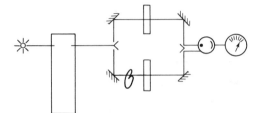

Figure 5.4-5   Two-beam one-detector method.

ated (chopped) and the degree of attenuation directly read in terms of transmittance or absorbance.

Photographic emulsions.   Absorption plays a key role in the process of *photography*. Photographic emulsions generally contain silver halides such as silver bromide, AgBr. If a quantum of light is incident on an AgBr crystal, it causes an electron to be freed from the bromine ion,

$$Ag^+Br^- \rightarrow Ag^+ + Br^0 + e^- \qquad [5.4\text{-}15]$$

The free electron becomes attached to a lattice defect in the silver ion, Ag^+, reducing the ion to a neutral, metallic silver atom,

$$e^- + Ag^+ \rightarrow Ag^0 \qquad [5.4\text{-}16]$$

That produces a *latent image*. The latent image as such is invisible, but it can be made visible by *development*. Developers often used are aminophenol derivatives and hydroquinone; they reduce the exposed crystal's other Ag^+ ions to grains of metallic silver. These grains render the exposed parts of the emulsion dark, forming a *negative*. To prevent the unexposed crystals from turning into silver, they are removed (dissolved) from the emulsion without affecting the developed image, using a "fixer" such as sodium thiosulfate, called "hypo," $Na_2S_2O_3$.

As the film is gradually exposed to more light, the (developed) negative becomes darker and its absorbance higher. A plot of *exposure* (which is the product of irradiance $\times$ time) *versus* absorbance produces a curve known as the *Hurter–Driffield curve.** The linear portion of the curve, *B–C* in Figure 5.4-6, has a slope called the *gamma* of the emulsion. The

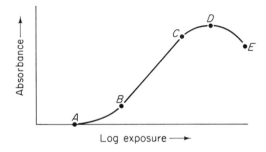

Figure 5.4-6   H-D curve of a typical photographic film.

---

*F. Hurter and V. C. Driffield, "Photo-Chemical Investigations and a New Method of Determination of the Sensitiveness of Photographic Plates," *J. Soc. Chem. Ind. (London),* **9** (1890), 455–69.

steeper the curve, the higher the gamma, and the higher the contrast that results. Parts *A–B* and *C–D* have less contrast.

Photochromic glass.   Materials are called *photochromic* if, under the influence of light, they become darker or change color. When the light is removed, they return to their initial state. The process of darkening, or *activation,* takes about a minute, until the absorbance reaches a certain saturation level. Light between 320 and 400 nm wavelength is most effective. The return process, or *clearing,* is a function of the incident IR radiation, that is, it depends on the ambient temperature (even at room temperature there is some *thermal fading*); clearing can be accelerated by light between 550 and 650 nm *(optical bleaching).* In the region between activation and bleaching, between 450 and 500 nm, light has no effect.

Photochromism is similar to photography: in both there is a dissociation of silver halides. But, in a photographic emulsion the halogen ion diffuses away while the silver remains in place. In photochromic glass, the glass matrix is so tight that the halogen cannot diffuse away; it remains close to the silver and recombines with it when activation ceases. Hence, the process of photochromism is *reversible,*

$$AgCl \overset{h\nu}{\rightleftharpoons} Ag^+ + Cl^- \qquad [5.4\text{-}17]$$

Samples have been exposed for half a million cycles of activation and bleaching, and have shown no fatigue. Typically, a small amount of copper, or other catalysts, is added to the glass; this increases the rate and depth of darkening by a factor of 100 or more.

Thermal fading, as the name implies, depends on the ambient temperature (on a hot day photochromic glass clears faster), but activation (darkening) does not. Thus the absorbance will reach a state of dynamic equilibrium that for a given irradiance decreases with increasing temperature. On a cold day, and with a given amount of light, photochromic glass becomes darker than on a hot day.

SUGGESTIONS FOR FURTHER READING

W. J. Price, *Analytical Atomic Absorption Spectroscopy* (London: Heyden & Son Ltd., 1972).

F. W. Billmeyer, Jr., and M. Saltzman, *Principles of Color Technology,* 2nd edition (New York: John Wiley & Sons, Inc., 1981).

D. L. MacAdam, *Color Measurement: Theme and Variations* (New York: Springer-Verlag New York, Inc., 1981).

C. E. K. Mees and T. H. James, *The Theory of the Photographic Process,* 4th edition (New York: Macmillan Publishing Company, 1977).

E. N. Mitchell, *Photographic Science* (New York: John Wiley & Sons, Inc., 1984).

PROBLEMS

5.4-1. If a certain filter transmits 38% of the incident light, what is its absorbance (optical density)?

5.4-2. If 37% of the incident light is absorbed in a sample, what is its absorbance?

5.4-3. If a neutral-density filter has an absorbance of 0.6, what is its transmittance?

5.4-4. A metallic coating absorbs 88% of the incident light. A thin film has one-half the absorbance of the coating. What is the transmittance of the film?

5.4-5. If a filter absorbs 20% of the light incident on it, how much light would the filter absorb if it were twice as thick?

5.4-6. A thin layer of a certain material absorbs 40% of the radiant energy passing through. How much energy will a layer absorb that is nine times as thick?

5.4-7. A 2.7-mm-thick neutral density filter causes 35% absorption. If the filter were 5 mm thick, how much would it transmit?

5.4-8. A 3-mm-thick color filter absorbs 74% of the light incident on it. What percentage of the light would pass through if the filter were 4.3 mm thick?

5.4-9. Twelve percent of the light is being absorbed in a filter 3.2 mm thick. How much light would be lost if the filter were 5.6 mm thick?

5.4-10. If 2.3-mm-thick sunglasses transmit 40%, how thick should the glasses be to transmit only 20%?

5.4-11. A certain type of protective glass has an absorptivity of $\alpha = 0.39$ mm^{-1}. What percentage of the light is transmitted through a plate 4 mm thick?

5.4-12. A pipe 5 m long contains a gas at normal atmospheric pressure. If the gas has an absorptivity of $\alpha = 0.08$ m^{-1}, how much light, in percent, is being absorbed?

5.4-13. A cylinder, 25 cm long and filled with a liquid, absorbs 62% of the light of a given wavelength. What is the absorptivity of the liquid?

5.4-14. The absorptivity of seawater at 570 nm is 0.22 m^{-1}. How long a path will reduce the transmittance to 80%?

5.4-15. Consider the Hurter–Driffield curve's section *D–E*, which represents severe *overexposure*. Therefore, what does the sun look like in the developed negative, compared to the surrounding sky?

5.4-16. If the memory unit in an optical computer is made of photochromic glass, which color of light should be used for:

(a) Storing the information?

(b) Reading the information?

(c) Erasing the information?

5.5

Lasers

THE WORD LASER IS AN ACRONYM for light amplification by stimulated emission of radiation. In 1916, Einstein predicted that the existence of equilibrium between matter and electromagnetic radiation required that besides emission and absorption there must be a third process, now called *stimulated emission*. This prediction attracted little attention until 1954, when Townes and coworkers developed a microwave amplifier (maser) using ammonia, NH_3. In 1958, Schawlow and Townes showed that the maser principle could be extended into the visible region and in 1960, Maiman built the first laser using ruby as the active medium. From then on, laser development was nothing short of miraculous, giving optics new impetus and wide publicity.

STIMULATED EMISSION OF RADIATION

Boltzmann distribution. So far we have discussed the transitions that occur between different energy states, the upward transition from a lower energy state to a higher state, $E_1 \rightarrow E_2$, which refers to absorption, and the downward transition, $E_2 \rightarrow E_1$, which refers to emission. Now we consider the number of atoms, per unit volume, that exist in a given state. This number, N, called the *population,* is given by *Boltzmann's equation,*

$$N = e^{-E/kT} \qquad [5.5\text{-}1]$$

where E is the energy level of the system, k is Boltzmann's constant, and T is the absolute temperature.

In any absorption or emission, at least two energy states are involved. The ratio of the populations in these two states, N_2/N_1, is called *Boltzmann's ratio* or *relative population*,

$$\frac{N_2}{N_1} = \frac{e^{-E_2/kT}}{e^{-E_1/kt}}$$

from which it follows that

$$N_2 = N_1 e^{-(E_2-E_1)/kT} \qquad [5.5\text{-}2]$$

Then we plot the energy in the higher state relative to that in the lower state, versus the populations in these states, E versus N in Equation [5.5-2]; the result is an exponential curve known as a *Boltzmann distribution*. When the distribution is *normal,* it means that the system is in thermal equilibrium, having more atoms in the lower state than in the higher state. If we substitute Bohr's frequency condition in Equation [5.5-2], we find that

$$N_2 = N_1 e^{-h\nu/kT} \qquad [5.5\text{-}3]$$

Einstein's prediction.   Perhaps, it seems, the transition between two energy states can either be up (as in absorption) or down (as in emission). But Einstein, in a talk given in 1916 and published in writing a year later, postulated that there must be a third process.*

Assume first that an ensemble of atoms is in thermal equilibrium and *not subject to* an external radiation field. At higher temperatures, a certain number of atoms is in the excited state; on return to the lower state, these atoms will emit radiation, in the form of quanta $h\nu$. That is called *spontaneous emission.*

The *rate* of the transition is the number of atoms in the higher state that make the transition to the lower state, per second. (The reciprocal of the rate is the *lifetime* of the transition.) If, as before, N_2 is the number of atoms (per unit volume) in the higher state, the rate of the spontaneous transition, P_{21}, is

$$P_{21} = N_2 A_{21} \qquad [5.5\text{-}4]$$

where A_{21} is a constant of proportionality.

Assume next that the system *is subject to* some external radiation field. In that case, one of two processes may occur, depending on the direction (the *phase*) of the field with respect to the phase of the oscillator. If the two phases coincide, a quantum of the field may cause the emission of another quantum. This process, anticipated by Einstein, is now called *stimulated emission.* Its rate is

$$P_{21} = N_2 B_{21} u(\nu) \qquad [5.5\text{-}5]$$

where B_{21} is another constant of proportionality and $u(\nu)$ is the energy density (in units of J m^{-3}), as a function of frequency, ν.

On the other hand, if the phase of the radiation field is opposite to that of the oscillator, the impulse transferred counteracts the oscillation, energy is consumed, and the system

---

*A. Einstein, "Zur Quantentheorie der Strahlung," *Mitt. Phys. Ges. Zürich,* **18** (1916), and *Phys. Z.* **XVIII** (1917), 121–28.

is raised to a higher state, as it occurs in *absorption*. Its rate is

$$P_{12} = N_1 B_{12} u(\nu) \qquad [5.5\text{-}6]$$

where B_{12} is still another constant of proportionality. These three constants, A_{21}, B_{21}, and B_{12}, are known as *Einstein's coefficients*. Before we proceed further, we summarize these findings in a schematic diagram (Figure 5.5-1).

With the system in thermal equilibrium, the net rate of downward transitions must equal the net rate of upward transitions,

$$N_2 A_{21} + N_2 B_{21} u(\nu) = N_1 B_{12} u(\nu) \qquad [5.5\text{-}7]$$

Dividing both sides by N_1 yields

$$\frac{N_2 A_{21}}{N_1} + \frac{N_2 B_{21} u(\nu)}{N_1} = B_{12} u(\nu)$$

$$\frac{N_2}{N_1} [A_{21} + B_{21} u(\nu)] = B_{12} u(\nu)$$

$$\frac{N_2}{N_1} = \frac{B_{12} u(\nu)}{A_{21} + B_{21} u(\nu)} \qquad [5.5\text{-}8]$$

If we then substitute Equation [5.5-3], we obtain

$$\frac{B_{12} u(\nu)}{A_{21} + B_{21} u(\nu)} = e^{-h\nu/kT} \qquad [5.5\text{-}9]$$

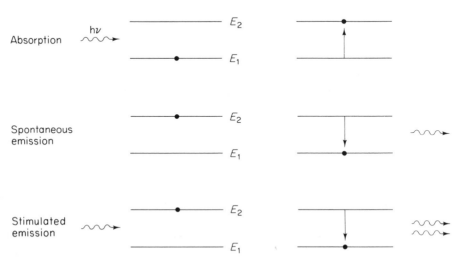

Figure 5.5-1 Transitions between energy states. If the atom is in a lower state, it may absorb a quantum of energy (*top*). If the atom is in the higher state, it may release a quantum spontaneously (*center*) or it may be stimulated to release a quantum (*bottom*).

Solving for $u(\nu)$ gives

$$u(\nu) = \frac{A_{21}}{B_{12}} \frac{1}{e^{h\nu/kT} - B_{21}/B_{12}}$$ [5.5-10]

To maintain thermal equilibrium, the system must release energy, probably in the form of electromagnetic radiation. The spectral distribution of this radiation follows Planck's radiation law,

$$M(\lambda) = \frac{C_1}{\lambda^5} \frac{1}{e^{C_2/\lambda T} - 1}$$

If Planck's law is written in terms of energy density, rather than exitance, $M(\lambda)$ changes to $u(\lambda)$ and, since u is related to M as

$$u = \frac{4}{c} M$$

the first radiation constant becomes

$$C_1 = 8\pi hc$$

and Planck's law

$$u(\lambda) = \frac{8\pi hc}{\lambda^5} \frac{1}{e^{-hc/\lambda kT} - 1}$$ [5.5-11]

The energy density must be consistent with Planck's law for any value of T. This is possible only when

$$B_{21} = B_{12}$$ [5.5-12]

and

$$\frac{A_{21}}{B_{21}} = \frac{8\pi h\nu^3}{c^3}$$ [5.5-13]

These two equations are called *Einstein's relations.* They show that (1) the coefficients for both stimulated emission and absorption are numerically equal and (2) the ratio of the coefficients of spontaneous *versus* stimulated emission is proportional to the third power of the frequency of the transition radiation. This explains why it is so difficult to achieve laser emission in the X-ray range, where ν is rather high.

Population inversion.   Ordinarily, when an atomic system is in thermal equilibrium, absorption and spontaneous emission take place side by side, but because $N_2 < N_1$, absorption dominates: an incident quantum is more likely to be absorbed than to cause emission. However, if we can find a material that could be *induced* to have a majority of atoms in the higher state, $N_2 > N_1$, then we have a condition called *population inversion* and, on return to the ground state, the system will probably lase.

Assume for a moment that the system has only two energy states that are involved in the process (a *two-level system*). Such a system represents the *ammonia maser,* the forerunner of the (optical) laser.

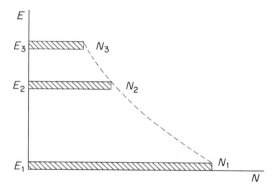

Figure 5.5-2 Population of a three-level system in equilibrium.

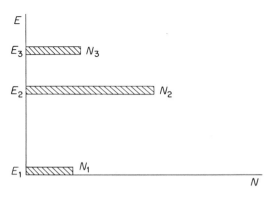

Figure 5.5-3 Three-level system showing population inversion of E_2 with respect to E_1.

A system that has three energy states is a *three-level system*. These levels may be designated E_1, E_2, and E_3. With the system in equilibrium, the uppermost level, E_3, is populated least, and the lowest level, E_1, is populated most (Figure 5.5-2).

The curve shown represents a *normal* Boltzmann distribution. Since the population in the various states is such that $N_3 < N_2 < N_1$, the system is absorptive rather than emissive. But assume that the system is being excited by supplying it with extraneous energy. This may cause either N_3 or N_2, or both, to exceed N_1; thus the system becomes top-heavy and reaches population inversion (Figure 5.5-3).

PRACTICAL REALIZATION

General construction.   First, a laser requires an *energy source* to supply the energy needed for raising the system to the excited state. That is called *pumping*, in analogy to pumping water up to a higher level of potential energy. It also requires an *active medium* which, when excited, reaches population inversion and lases. The medium may be a solid, liquid, or gas, and it may be one of thousands of materials that have been found to lase.

A *cavity* is optional. Some systems are built as laser *amplifiers;* they have no cavity. Most systems, however, are laser *oscillators;* here the medium is enclosed in a cavity that provides feedback and additional amplification. Generally, the cavity is formed by two mirrors facing each other. One of the mirrors is coated to full reflectance; the other mirror is partially transparent to let some of the radiation pass through to be used (Figure 5.5-4).

When the system is raised to the excited state, the first few photons will be emitted spontaneously. These photons trigger the release of more photons and become part of a continually growing wave, an *avalanche.* As the wave is reflected back and forth between the mirrors, the chance for triggering more transitions continually increases, thus a laser oscillator is much more efficient than a laser amplifier. These processes, excitation, population inversion, and avalanche action, are fundamental to stimulated emission; they have made lasers a reality.*

Excitation.   There are several ways of pumping a laser and producing the population inversion necessary for laser action. Most commonly used are the following.

In *optical pumping,* a light source is used to supply the energy. Short, intense flashes of light were already used in Maiman's ruby laser and still are customary today with solid-state lasers. In some applications, primarily in fusion experiments, pumping is accomplished by another laser.

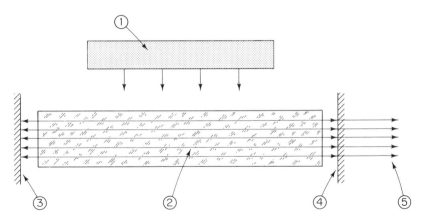

Figure 5.5-4 Basic components of a laser oscillator: Energy source (1) supplies energy to medium (2) contained between mirrors (3) and (4). Radiation (5) emerges through mirror (4).

*The laser is an example of the same idea having come to different people, in different parts of the world, at about the same time. The idea first occurred to Charles Hard Townes (1915–   ), then professor of physics at Columbia University, while, one day in 1951, sitting on a bench in Franklin Park in Washington, DC. Independently, Alexandr Mikhailovich Prokhorov (1916–   ), head of the Oscillation Laboratory of the P. N. Lebedev Institute of Physics at the University of Moscow, and his assistant Nicolai Gennadievich Basov (1922–   ), a 1943 graduate of the Kiev School of Military Medicine, now director of the Lebedev Institute, in 1952 presented a paper at an All-Union (SSSR) Conference on radio spectroscopy in which they discussed the possibility of building a "molecular generator." The pertinent publications are C. H. Townes, *Production of Electromagnetic Energy,* U.S. Patent 2,879,439, Mar. 24, 1959; N. G. Basov and A. M. Prokhorov, *Zh. Eksp. Teor. Fiz.* **28** (1955), 249–50, Engl. trans. "Possible Methods of Obtaining Active Molecules for a Molecular Generator," *Sov. Phy. JETP* **1** (1955), 184–85; and A. L. Schawlow and C. H. Townes, "Infrared and Optical Masers," *Phys. Rev.* **112** (1958), 1940–49. In 1964, Townes, Prokhorov, and Basov, and in 1981 Arthur Schawlow (1921–   ), earlier at Bell Laboratories, were awarded the Nobel Prize in physics.

Another way is by *electron excitation* as in an argon laser where the medium itself is the conductor. The electrons, emitted by the cathode, are accelerated toward the anode. Some of the electrons impinge on the atoms of the medium, ionize them, and raise them to the excited state, producing the population inversion needed. Similar, if less direct, is the excitation and subsequent *atom–atom collisions* that occur in a helium–neon laser.

A *direct conversion* of electric energy into radiation occurs in light-emitting diodes, LEDs, and in semiconductor lasers that derive from them.

Thermal excitation is typical of the gas-dynamic laser, where a hot gas is forced through a nozzle. An example is the CO_2 laser.

In a *chemical laser,* the energy comes from an exothermic chemical reaction. Hydrogen, for instance, can combine with fluorine,

$$H_2 + F_2 \rightarrow 2HF \qquad [5.5\text{-}14]$$

to pump the molecular upper level, excited HF, which lases.

Cavity configurations.   The simplest type of a resonant cavity is a combination of two *plane mirrors* (Figure 5.5-5, top). Such mirrors require precise alignment. This con-

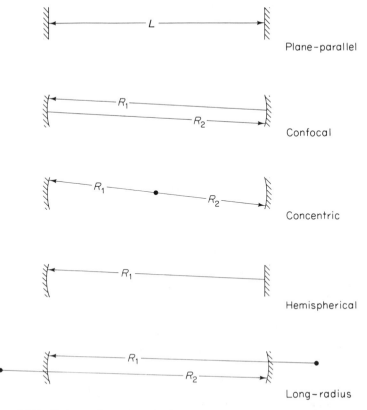

Figure 5.5-5 Cavity configurations. L, distance between mirrors; R, radii of curvature.

figuration is very efficient because of its good *filling* but, because of difficult alignment and low stability, is rarely used today.

A *confocal cavity* has two concave mirrors of the same radius of curvature, R, separated by a distance L that is equal to R, $L = R$. The focal length of the mirrors, therefore, is one-half of L and the focal points coincide in the center, hence the term "confocal." A confocal cavity is much easier to align than plane-parallel mirrors but the filling is poor; halfway between the mirrors, for instance, only a small fraction of the medium is utilized.

A *concentric cavity* (also called *spherical cavity*) has two concave mirrors of the same radius, one-half of the distance between them, $L = 2R$. Again, the filling is poor and alignment not easy. A *hemispherical cavity* is a hybrid between the plane-parallel and the confocal type. It has a concave mirror at one end and a plane mirror at the other. The plane mirror is placed at the center of curvature of the spherical mirror, thus $L = R_1$. Such a cavity is easier to align than those mentioned so far but again, the filling is poor and the output low. A *long-radius cavity* has two concave mirrors, their radii of curvature significantly longer than the distance between them, $R_1 = R_2 > L$. This is a good compromise between the plane-parallel and the confocal variety; it is the type of cavity used most often in today's commercial lasers.

Mode structure.   Assume for simplicity that the cavity is limited by two plane-parallel mirrors. Since the mirrors form a *closed* cavity, the standing-wave pattern inside the cavity has nodes, rather than loops, at both ends. If, as shown earlier in Figure 5.5-5, top, L is the length of the cavity, the longest wavelength possible is $\lambda_1 = 2L$. The next shorter wavelength is $\lambda_2 = L$, the next shorter wavelength $\lambda_3 = \frac{2}{3}L$, and so on; thus,

$$\lambda = \frac{2}{q} L \qquad [5.5\text{-}15]$$

where q is the number of half-wavelengths, or *axial modes,* that fit into the cavity.

It is often better to write Equation [5.5-15] in terms of frequency, ν. We take the relationship $v = \lambda\nu$, replace v by the velocity of light c, solve for ν, and substitute Equation [5.5-15]. That gives

$$\nu = q \frac{c}{2L} \qquad [5.5\text{-}16]$$

But a laser cavity must, by necessity, contain a medium (of index n), rather than free space. Hence, we replace the actual length of the cavity, L, by the optical path length, $S = Ln$, so that, instead of Equation [5.5-16], we have

$$\nu = q \frac{c}{2S} \qquad [5.5\text{-}17]$$

which is the *resonance condition for axial modes.* A wave of frequency ν that travels along the axis of the cavity, therefore, forms within the cavity a series of standing waves, called *stationary axial modes.*

Now consider two consecutive modes (which differ by $q = 1$). These modes have frequencies that, following Equation [5.5-17], are separated by a frequency difference $\Delta\nu$,

$$\Delta\nu = \frac{c}{2S} \qquad [5.5\text{-}18]$$

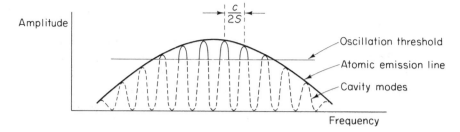

Figure 5.5-6   Output of laser containing several resonance frequencies.

Slightly different frequencies are closely, and evenly, spaced. Often several of them lie within the width of a single emission line. That means that the output of the laser then consists of a number of lines separated by $c/2S$, as shown in Figure 5.5-6.

In addition to the axial modes we also have *transverse modes,* called TEM, for transverse electromagnetic, modes. The TEM modes are generally few in number, and they are easy to see. Aim the laser at a distant screen and spread the beam out by a negative lens. Most often the light forms several bright patches, separated from one another by intervals called "nodal lines." Within each patch, the phase of the light is the same, but between patches the phase is reversed. Figure 5.5-7 shows several examples.

In the lowest possible transverse mode, TEM_{00}, there is no phase reversal across the beam (the beam is "uniphase"), the spatial coherence is the highest possible, and the beam can be focused to the smallest spot size and reach the highest power density. If, in addition, the laser also oscillates in the lowest possible axial mode, the cavity will select, and amplify, out of several resonant frequencies only one frequency. This results in the highest possible temporal coherence, that is, the light will be as monochromatic as light can be.

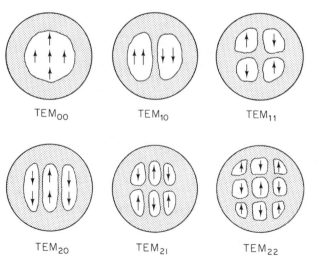

Figure 5.5-7   Schematic representation of different TEM modes.

Gain.   The gain of a laser depends on several factors. Foremost among them is the separation of the energy levels that provide laser transition. If the two levels are farther apart, the gain is higher because then the laser transition contains a larger fraction of the energy compared to the energy in the pump transition. If the two levels are closer, the gain is less.

In a way, gain is the opposite of *absorption*. Consider once more the exponential law,

$$\phi' = \phi_0 e^{-\alpha x} \tag{5.5-19}$$

and note that the exponent $(-\alpha x)$ is negative and, therefore, the absorptivity α positive. This is true for thermal equilibrium (normal Boltzmann distribution), where $N_2 < N_1$. Conversely, in population inversion, where $N_2 > N_1$, the exponent is positive and α negative. Indeed, laser emission could be considered *negative absorption* and Equation [5.5-19] be written

$$\phi' = \phi_0 e^{\beta x} \tag{5.5-20}$$

where β is the *gain coefficient*, the negative of the "absorption coefficient," $\beta = -\alpha$.

As the wave is reflected back and forth between the mirrors, it will lose some of its energy, mainly because of the limited reflectivity of one of the mirrors. (The limited reflectivity, of course, is needed to let radiation pass through the mirror to be utilized.) If the two mirrors have reflectivities r_1 and r_2, then with each round trip the initial power in the cavity, ϕ_0, decreases by a factor $r_1 r_2$ and the resultant power becomes

$$\phi' = \phi_0 r_1 r_2 \tag{5.5-21}$$

Eliminating ϕ'/ϕ_0 between Equations [5.5-19] and [5.5-21] and calling γ the loss per round trip, we find that

$$r_1 r_2 = e^{-\gamma} \tag{5.5-22}$$

Hence,

$$\gamma = -\ln(r_1 r_2) \tag{5.5-23}$$

where γ, like α in absorption, is positive.

For the system to lase, the gain must equal or exceed the sum of the losses. Thus we extend Equation [5.5-19] to read

$$\phi' = \phi_0 e^{(\beta - \alpha)x} \tag{5.5-24}$$

If $\beta > \alpha$, the radiation field inside the medium builds up and the system will lase. If $\beta < \alpha$, the oscillations will die. The equality $\beta = \alpha$, hence, is the *threshold condition* necessary to sustain laser emission.

TYPES OF LASERS

Solid-state lasers.   The classical example of a solid-state laser is the *ruby laser*.* Ruby is synthetic aluminum oxide, Al_2O_3, with 0.03 to 0.05% of chromium oxide, Cr_2O_3, added to it. The Cr^{3+} ions are the active ingredient; the aluminum and oxygen

---

*First described by T. H. Maiman, "Stimulated Optical Radiation in Ruby," *Nature (London)* **187** (1960), 493–94, and "Stimulated Optical Emission in Fluorescent Solids: I. Theoretical Considerations," *Phys. Rev.* **123** (1961), 1145–50.

atoms are inert. The ruby crystal is made into a cylindrical rod, several centimeters long and several millimeters in diameter, with the ends polished flat to act as cavity mirrors. Pumping is by light from a xenon flash tube.

As shown in Figure 5.5-8, chromium-doped ruby has three energy levels, E_1, E_2, and E_3. The uppermost level, E_3, is fairly wide (it will accept a wide range of wavelengths) and has a short lifetime; the excited Cr^{3+} ions rapidly relax and drop to the next lower state, E_2. This transition is nonradiative.

The E_2 state is *metastable* (that is, nearly stable); it has a lifetime of about 10^{-3} s, considerably longer than that of E_3, and the Cr^{3+} ions remain that much longer in E_2 before they drop to the ground state, E_1. The $E_2 \to E_1$ transition is radiative; it produces the spontaneous, incoherent red fluorescence typical of ruby, with a peak near 694 nm. But as the pumping energy is increased above a critical threshold, population inversion occurs in E_2 with respect to E_1 and the system lases, with a sharp peak at 694.3 nm.

The *neodymium:YAG laser* has four energy levels. As shown in Figure 5.5-9, the laser transition begins at the metastable state and ends at an additional level somewhat above the ground state.

The active ingredient is trivalent neodymium, Nd^{3+}, added to an yttrium aluminum garnet, YAG, $Y_3Al_5O_{12}$. As before, excitation raises the system from the ground state, E_1, to the highest of the four levels, E_4. From there the system returns to the metastable state, E_3. Since both E_4 and E_2 drain rapidly, laser emission will commence as soon as E_3 has reached population inversion with respect to E_2. This inversion can be maintained even with moderate pumping. That allows such lasers to operate at high repetition rates as well as in the *continuous wave* (c.w.) mode, in contrast to most three-level lasers, which emit only pulses. The Nd:YAG laser is fairly efficient; its emission is at 1.064 μm.

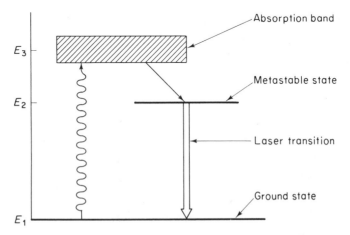

Figure 5.5-8   Three-level energy diagram typical of ruby. Pumping transition is shown by wavy line, nonradiative transition by simple arrow, laser transition by hollow arrow.

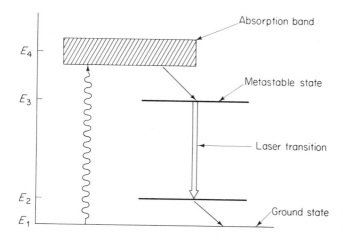

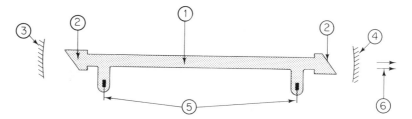

Figure 5.5-9   Four-level system typical of neodymium.

Gas lasers.   The widely used *helium-neon laser* is an example of an atomic gas laser.* Typically, it consists of a tube about 30 cm long and 2 mm in diameter, with two electrodes on the side and fused silica windows at both ends. The tube contains a mixture of 5 parts helium and 1 part neon, kept at a pressure of 133 Pa (approximately 1 mm Hg). The mirrors are placed outside the tube, as shown in Figure 5.5-10.

The end windows are set at Brewster's angle. This causes light, returning from the outside mirrors back to the cavity and oscillating *normal* to the plane of incidence to be reflected away from the cavity. Consequently, the component oscillating *parallel* to the plane of incidence becomes dominant and will sustain laser emission; the radiant energy that emerges from the laser, therefore, is linearly polarized.

Pumping is by electric current. Initially the excitation affects only the helium. But, two of the excited helium states have about the same energies as two of the higher neon states,

Figure 5.5-10   Helium–neon laser with (1) gas discharge tube, (2) Brewster windows, (3) fully reflective mirror, (4) partially transparent mirror, (5) electrodes connected to power supply, (6) output.

---

*First described by A. Javan, W. R. Bennett, Jr., and D. R. Herriott in "Population Inversion and Continuous Optical Maser Oscillation in a Gas Discharge Containing a He-Ne Mixture," *Phys. Rev. Lett.* **6** (1961), 106–10, who discovered emission at 1.15 μm. The familiar red radiation at 632.8 nm was obtained a year later by A. D. White and J. D. Rigden, "Continuous Gas Maser Operation in the Visible," *Proc. IRE* **50** (1962), 1697.

and thus the excited helium atoms will, by collision, transfer their energy to the neon, raising the neon while the helium returns to the ground state. The neon E_3 states are metastable and quickly reach population inversion relative to E_2. The neon returns to the lower states by various transitions, most notably by laser emission at 632.8 nm (Figure 5.5-11). The essence is that the helium atoms, which are fairly light, can easily be excited; the neon atoms, which are much heavier, could not be pumped up efficiently without them.

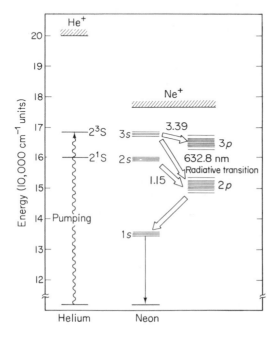

Figure 5.5-11   Energy level of helium–neon laser. Hollow arrows: radiative transitions.

Another type of a gas laser is the *argon laser*. It generates a strong turquoise-blue line at 488 nm and a green line at 514.5 nm, in either pulsed or c.w. operation. Still another type is the *helium–cadmium laser;* it emits a brilliant blue at 441.6 nm.

The *carbon dioxide laser* is virtually in a class by itself because of its high power. The first CO_2 lasers had a continuous output of a few milliwatts. Today we have powers of some 200 kW, more than enough to cut through steel plates several centimeters thick in a matter of seconds. In fact, CO_2 lasers may be the most practical lasers of all. Their efficiency in converting electrical energy into radiation is better (more than 10%) than that of any other laser. Their construction and operation are relatively simple. Excitation is by way of nitrogen, which takes the place of the helium in the He-Ne laser, and subsequent collision with CO_2. Often other gases, such as He, Xe, and H_2O, are added because they enhance the homogeneity of the output and the yield. Emission is at 10.6 μm.

A special type is the *TEA CO_2 laser*, the acronym standing for **t**ransverse **e**xcitation at **a**tmospheric pressure. At that pressure the gas has more molecules per unit volume and the electrodes can be placed closer together, which means a lower pumping voltage and still high efficiency.

Excimer lasers contain rare-gas halides such as XeCl, KrF, or others. These molecules are unstable in the ground state but bound in the excited state. That makes these lasers exceedingly powerful, with outputs as high as several GW. They emit in the ultraviolet.

Semiconductor lasers.   Still another class of lasers is derived from the light-emitting diode, LED. These diodes are small (less than 1 mm in diameter); they are semiconductors as we have discussed them before on page 364.

The first LEDs were made from gallium arsenide, GaAs. By adding a small amount of a dopant, additional (**negatively** charged) free electrons can be supplied to the conduction band; that results in an *n*-type material. Removing electrons from the valence band leaves (**positively** charged) "holes" in their place; that results in a *p*-type material. Both electrons and holes can be considered charge carriers, a hole moving from atom to atom the way a bubble moves through a liquid. With the *n*-side of a *p-n* junction connected to the negative terminal of a battery, and the *p*-side to the positive terminal, electrons will flow from *n* to *p*, and holes from *p* to *n*. As the electrons combine with the holes, excess energy is released in the form of light and, as even more current is applied, excited hole–electron pairs that have not had time to recombine spontaneously are forced together causing *stimulated emission.*

Much as in a conventional laser, the light produced is trapped and reflected back and forth between the end faces of the crystal; the light leaves the junction as shown in Figure 5.5-12. Emission was first observed in the near IR,* but depending on the material and the dopant, diode lasers can be made to emit almost anywhere in the spectrum, from the UV to the IR, and with an efficiency much higher than with optical pumping (around 40% versus 3%). The main application of diode lasers is in waveguides and integrated optics.

Tunable lasers.   The first tunable lasers were dye lasers. They often had a diffraction grating replacing one of the mirrors. Today there exist several types of tunable solid-state lasers. Of particular interest is the *parametric oscillator* which, in contrast to the dye laser, is more compact, less expensive, and easier to operate; it delivers coherent radiation over a tuning range much wider than that of a single dye. The first medium used was $LiNbO_3$; at present $AgGaS_2$ is preferred in the IR, urea at shorter wavelengths. *Color center*

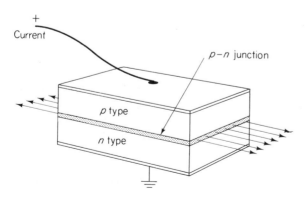

Figure 5.5-12   Schematic diagram of a *p-n* junction laser.

---

*R. N. Hall, G. E. Fenner, J. D. Kingsley, T. J. Soltys, and R. O. Carlson, "Coherent Light Emission from GaAs Junctions," *Phys. Rev. Lett.* **9** (962), 366–68.

lasers contain alkali halides such as KF, LiF, KCl, or NaCl. They can be tuned over wide bands in the UV, the visible, and the IR. Tunable lasers are most welcome to spectroscopists, who for long have wanted a laser that could be tuned over any set of resonances desired.

The *free-electron laser,* finally, derives its radiation from a stream of intense, highly energetic electrons passing through an oscillating electromagnetic field. These lasers have high powers (of the order of megawatts), they are very efficient, and can be tuned through a wide range of wavelengths.

Since Maiman first used ruby, numerous other materials have been found to lase, emitting radiation from the extreme ultraviolet (XUV), with wavelengths as short as a few nanometers, to millimeter waves. All gaseous elements and many molecules are known to lase, and the same is true of most metals. Any dye that fluoresces will lase if only enough power is supplied. Today an estimated 5000 laser transitions are known. The types of lasers most commonly used are listed in Table 5.5-1.

TABLE 5.5-1 SUMMARY OF TYPICAL LASERS

Ion	Matrix	Principal wavelength
Cadmium	Helium	441.6 nm
Argon		514.5 nm
Neon	Helium	632.8 nm
Chromium	Aluminum oxide (ruby)	694.3 nm
Gallium arsenide		Near IR
Neodymium	YAG	1.064 μm
Carbon dioxide	Nitrogen	10.6 μm

APPLICATIONS

Compared to radiation from other sources, laser radiation stands out in several ways. It is highly coherent, both spatially and temporally. It can be generated in the form of very short pulses, at high powers and, because of its high spatial coherence, at very high power densities. These are among the properties that make lasers the unique tool they are for a great many practical applications.

Beam shape.   There are two parameters characteristic of a laser beam as it leaves the cavity. First, the beam has a certain *profile,* that is, a certain energy distribution across its diameter. With the laser operating in the TEM_{00} mode, the energy has a *Gaussian distribution:* at a given distance r from the axis, the irradiance I falls off exponentially,

$$I(r) = I_0 e^{-(2r/w)^2} \qquad [5.5\text{-}25]$$

This distribution is completely described by the parameter w, the distance from the axis at which I has dropped to $1/e^2$ of I_0, the irradiance in the center (Figure 5.5-13).

A beam retains its profile both inside and outside the cavity; the profile merely contracts or expands. For example, with a confocal cavity (two concave mirrors of the same radius of curvature) the beam has its least diameter, or *beam waist,* halfway between the

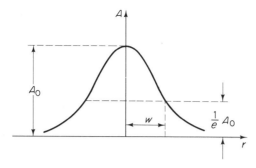

Figure 5.5-13 Amplitude distribution across laser beam oscillating in the TEM_{00} mode.

mirrors. With a hemispherical cavity (concave-plane) the beam waist lies on the plane mirror.

The beam's radius, w, changes as a function of distance, z, from the waist, according to

$$w(z) = w_0 \sqrt{1 + \left(\frac{\lambda z}{\pi w_0^2}\right)^2}$$ [5.5-26]

where λ is the wavelength and w_0 the radius at the waist.* For a confocal cavity, this simplifies to

$$w_0 = \sqrt{\frac{L\lambda}{2\pi}}$$ [5.5-27]

where L is the distance between the mirrors.

Twice the radius w_0 is the diameter of the beam at the waist. The size of that diameter accounts for (part of) the *divergence* of the beam, simply because the shape of the beam outside the cavity is an extension of its shape inside, a relationship illustrated in Figure 5.5-14.

Farther away from the laser, in the *far field* where the beam's parameters can be considered linear functions of the distance, the beam's contour subtends with the axis an angle θ,

$$\theta = \frac{\lambda}{\pi w_0}$$ [5.5-28]

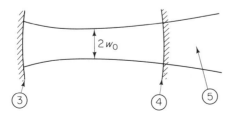

Figure 5.5-14 Confocal cavity causes beam to contract to minimum beam diameter $2w_0$. (3), Fully reflective mirror; (4), semitransparent mirror; (5), laser output.

---

*H. Kogelnik and T. Li, "Laser Beams and Resonators," *Appl. Opt.* **5** (1966), 1550–67.

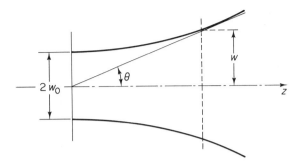

Figure 5.5-15   Relationship between beam waist and divergence of a laser beam.

This angle determines the "far-field half-angle divergence" illustrated in Figure 5.5-15. Twice that angle is the *full-angle divergence*,

$$2\theta = \frac{4\lambda}{\pi d_0} \qquad [5.5\text{-}29]$$

an expression where I have replaced the radius at the waist by the diameter, d_0.

Note that the divergence of the beam is inversely proportional to d_0. With a small waist, the divergence is large. Conversely, for a well collimated beam (of small divergence), the waist must be large—as we have already seen from Figure 5.5-14.

Now we place a converging lens in the path of the light, causing the beam to contract to a "focus." That focus, though, is merely another waist where the beam's wavefronts are essentially, if temporarily, plane. Beyond the waist, the beam expands again. Assume the beam has a radius r and the lens a focal length f. The half-angle vergence of the beam is then

$$\theta = \frac{r}{f} \qquad [5.5\text{-}30]$$

Taking twice that angle, combining it with Equation [5.5-29] by eliminating 2θ, and solving for $2r$ gives

$$2r = \frac{4f\lambda}{\pi d_0} \qquad [5.5\text{-}31]$$

where $2r$ is the beam's diameter at the focus.

Another part of the beam's divergence is due to diffraction. Its magnitude is found from Rayleigh's criterion,

$$\theta \approx 1.22 \frac{\lambda}{D} \qquad [5.5\text{-}32]$$

where D is the diameter of the laser's aperture, or of the lens if it limits the beam. Combining Equations [5.5-30] and [5.5-32] gives

$$r \approx 1.22 \frac{f\lambda}{D} \qquad [5.5\text{-}33]$$

With most gas lasers, the diffraction divergence is about twice as large as the beam-waist divergence.

Figure 5.5-16 Beam expander. Type shown is inverted Galilean telescope.

Equation [5.5-32] tells us that the diffraction divergence can be reduced if the beam is passed through a *beam expander* (which makes D larger). A beam expander is typically an inverted telescope, either of the astronomical or the Galilean type. The astronomical type is used when a spatial filter is needed (which then is placed in the focal plane common to the two lenses). The inverted Galilean type, shown in Figure 5.5-16, is preferred with high-power lasers because the radiation is not brought to a focus (which might cause ionization and breakdown of the air).

Power and power density.   A typical laser pulse contains about 10 J of energy. That does not seem to be very much, but if this energy is delivered within a pulse only 0.5 ms long, the output power is 20 kW. Moreover, this power, because of a laser's high spatial coherence, can be focused to produce exceedingly *high power densities.*

Some lasers generate 300 J in 5 ns, which translates into 50 GW (gigawatt, 10^9 W). But even pulses as short as 8 femtoseconds (8×10^{-15} s) have been reached, and powers as high as 2500 TW (terawatt, 10^{12} W). That is an incredible amount, about as much as the output of all power plants in the world combined. These plants, of course, generate power continually, whereas a laser pulse may last for perhaps only a fraction of a nanosecond.

Such high powers are generally produced by *Q switching,* which means compressing the energy into a very short period of time. One possibility of doing this is by using a rotating mirror at one end of the laser cavity (Figure 5.5-17). Then, after the system has been pumped, oscillation cannot occur *except* when one of the facets of the rotating mirror is exactly parallel to the stationary mirror; at that instant all of the accumulated energy is dumped into one single, very short "giant" pulse.

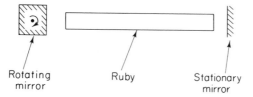

Rotating mirror   Ruby   Stationary mirror   **Figure 5.5-17** *Q*-switched laser.

Nonlinear effects.   Ordinarily, the refractive index and the absorptivity of a material are properties of the material, and independent of the intensity of the light that passes through. With very intense light, that can be different and either the index or the absorptivity or both may become *nonlinear* functions of the intensity. This causes a number of phenomena, long predicted by theory but observed only recently since the advent of high-powered lasers. Light from such lasers can in fact change the refractive index, even of completely transparent material. Heat and thermal expansion, as they occur with absorbing materials, are not involved.

Some lasers produce power densities so high that the radiation field strips electrons from atoms and molecules creating a *plasma* (a mixture of ions and free electrons rarely found in nature except in the atmosphere of the sun). The electric field strength can be as high as 10^9 V m^{-1}, more than enough to cause the breakdown of air (3×10^6 V m^{-1}). The Coulomb force between the nucleus and the electron in an hydrogen atom, from Problem 5.3-6, for example, is approximately

$$\mathbf{F} \approx 8 \times 10^{-8} \, \text{N} \qquad [5.5\text{-}34]$$

Converting this force into electric field strength, $\mathbf{E}$, gives

$$\mathbf{E} = \frac{\mathbf{F}}{q} = \frac{8 \times 10^{-8}}{1.6 \times 10^{-19}} = 5 \times 10^{11} \, \text{V m}^{-1} \qquad [5.5\text{-}35]$$

about the same as the electric field holding a crystal together.

Nonlinear effects worth noting are *self-focusing,* where a beam of light contracts into thin, short-lived, powerful threads of light that quickly shatter the material through which they pass. *Optical bistability* arises in saturable systems (whose absorptivity increases as the material becomes more excited). The resulting increase in absorption makes the material yet more excited, causing even more absorption, and so on.

Phase conjugation refers to a molecular reflection of light. Consider light that is passed through a randomly irregular medium. That causes the beam to break up into blotches of light. But imagine that the light comes from an high-intensity laser and after passing through the medium is sent down a tube containing methane under high pressure. The beam is scattered by the methane molecules and reflected, the gas acting like a mirror, though a mirror of rather unusual properties: The reflected wavefronts are now distorted opposite to those in the incident beam, an effect called phase conjugation or *wavefront reversal.* The distortions, in short, are canceled as the light retraces its path through the random medium once again.*

Frequency doubling is the generation of second harmonics. For example, when red light from a ruby laser (694.3 nm) is passed through a KDP crystal, part of the light is converted into 347.15-nm UV, exactly one-half the wavelength of the incident light. Similarly, the 1.064-μm emission from an Nd:YAG laser can by successive frequency doubling, first in a crystal of KDP and then in potassium dideuterium phosphate, be converted to its fourth harmonic, 266 nm. And even the fifth and the seventh harmonics have been reached, which means wavelengths *as short as 38 nm,* well within the X-ray range.†

Industrial applications.   The high powers available from lasers have led to a great many industrial and medical applications, from more mundane tasks such as the *machining* of materials to the exotic, such as containing a plasma and trying to release energy from nuclear fusion.

A well-known example of *cutting* with a laser beam appears in the James Bond movie *Goldfinger,* where the secret agent's nemesis cuts his way into the gold bullion depository at

*V. V. Shkunov and B. Ya. Zel'dovich, "Optical Phase Conjugation," *Sci. Am.* **253** (Dec. 1985), 54–59; D. M. Pepper, "Applications of Optical Phase Conjugation," *Sci. Am.* **254** (Jan. 1986), 74–84.

†J. Reintjes, S. Chiao-Yao, and R. C. Eckardt, "Generation of Coherent Radiation in the XUV by Fifth- and Seventh-Order Frequency Conversion in Rare Gases," *IEEE J. Quantum Electron.* **QE-14** (1978), 581–96.

Fort Knox, using a laser mounted atop a personnel carrier. In the mind of the public, this episode has come to represent the wondrous capabilities of a laser. Today, some industrial lasers may be equal to the task of cutting through an armored door, but their principal use lies in a great variety of other applications.

Drilling by laser is commonplace. Diamond, about the hardest material known, has been drilled before, but the process is tedious and time consuming. A laser does it quickly, producing traces of (black) graphite which facilitates absorption. Holes have been drilled also into teeth, paper clips, even into single human hairs.

Welding by laser meant earlier only welding on a microscopic scale, such as in integrated circuits. But in recent years, as lasers have become much more powerful, even heavy steel plates have been welded. For example, steel plates 5 cm thick can be fused using a 90-kW laser, at speeds higher than 2.5 m per minute. Lasers are often guided by a computer programmed to move across the workpiece and perform various operations in sequence.

Because of their high directionality and high frequency, laser beams looked at first promising for *communications.* However, turbulence in the air severely limits the amount of information that can be transmitted, unless the beam is confined to a fiber waveguide. *Optical radar,* known as "lidar," for **l**ight **d**etection **a**nd **r**anging, makes it possible to obtain echoes from clouds, haze, and atmospheric pollutants too small to be detected by conventional (microwave) radar.

Important among practical applications are *precision measurements,* including alignment and the measurement of distances, thicknesses, angles, and velocities. The distance to the moon, for example, has been determined to an accuracy *better than 15 cm.*

Medical applications.   Laser effects on biological tissue are either thermal or nonlinear. Thermal effects depend on absorption, mainly in pigments such as melanin (as it occurs in the skin and the iris and choroid of the eye) and hemoglobin (blood). Absorption causes a conversion of radiant energy into heat, increasing the temperature of the tissue and resulting in a denaturation of proteins called *coagulation.* With the process produced by light, it is called *photocoagulation.* *

Nonlinear effects are entirely different. The rapid expansion of a plasma that is characteristic of a nonlinear effect produces a shock wave which mechanically (rather than thermally) causes a disruption of the tissue called *photodisruption.* The difference between thermal and nonlinear effects can be shown by aiming a c.w. laser at the tip of a match and seeing it burst into flame. Firing a *Q*-switched laser (which because of its short pulse length has much higher power) at a match makes the tip crumble, without ignition.

Medical applications include the treatment of *retinal detachment,* where the focused beam causes small burns that on healing keep the retina back in place by the formation of scar tissue. Similar is the treatment of *diabetic retinopathy,* where local distentions of small

---

*Photocoagulation is of interest especially in ophthalmic surgery, much of it pioneered by Gerd Meyer-Schwickerath (1920–   ), German ophthalmologist, longtime chairman of the Department of Ophthalmology at the University of Essen. Meyer-Schwickerath's most dramatic result came when he was asked to treat a one-eyed ophthalmologist who had a completely updrawn iris in an aphakic eye, leaving him blind. Because of his darkly pigmented iris, laser irradiation caused a sudden, almost explosive reaction, producing a hole 2 mm in diameter and instantly opening the line of sight. The patient jumped up from the table, stared at Meyer-Schwickerath and said: "Now I can see you! Thank you very much! You look very young."

blood vessels *(microaneurysms)* and newly formed blood vessels *(neovascularization)* are obliterated by coagulation. A cloudy posterior capsule, often left in place after removal of a cataract, can be opened quickly and without discomfort by *posterior capsulotomy.* Lasers are used also in the treatment of *melanoblastoma,* a heavily pigmented tumor of the skin and the choroid of the eye that sometimes erupts into highly malignant growth.

Both photocoagulation and photodisruption occur where the light is *focused,* rather than where the beam enters the eye. Therefore, the light can be aimed, without making any incision at all, at the diseased part *inside* the eye, truly a noninvasive kind of surgery.

Laser speckle.   Laser light, as we have seen, is very coherent and very intense. So it comes as a disappointment when we notice that this light, when projected on a screen, is rather mottled and uneven. Laser light has a definitely grainy structure, a phenomenon called *speckle.*

The theory of laser speckle is very complex. It is a matter of the statistics of random processes. With some simplification we can say that any surface has a certain roughness, with a vast number of individual facets and scatterers. The light reflected from such a surface consists of contributions from all of these facets and, since the light has high spatial coherence, these contributions form loci of interference, distributed at random in three-dimensional space in front of the surface. For example, look at a screen diffusely illuminated by laser light. While looking, slowly move your head from side to side as if saying "no." The path lengths traversed by these contributions will change and the speckle pattern will move, an effect aptly called a *"red snowstorm."*

The same effect is seen when, instead of moving the head, the screen is moved slowly transverse to the line of sight. If the head, or the screen, is moved slowly in one direction and the speckle grains are seen to move in the *opposite* direction, the observer is myopic (nearsighted). If the grains move in the *same* direction, the observer is hyperopic (farsighted). This can be explained by the position of the far point of the eye, which, as shown on page 173, in myopia is located in front of the eye, and in hyperopia behind it. The apparent motion of the speckle pattern, therefore, can be used to test for visual refractive deficiencies.

LASER SAFETY

Any light, no matter where it comes from, can cause damage. But *laser radiation is particularly dangerous,* mainly because of its high spatial coherence (which means that it can be focused down to very high power densities). In addition, many lasers emit in a region of the spectrum (red) where the light just does not seem to be as powerful as if it were white.

Of all the parts of the human body, clearly the *eye* is the most vulnerable. Which part of the eye is subject to injury is a matter of wavelength. In the IR, above 10 μm, much of the energy is absorbed by water. Since water is the main constituent of most any biologic tissue, it is the *cornea* that is severely damaged first.

Light between 400 nm and 1.4 μm, on the other hand, will penetrate through the eye and be absorbed in the pigment epithelium next to the *retina.* The pigment epithelium then virtually explodes, ejecting black granules into the vitreous, forming vapor bubbles and causing hemorrhages all around the destroyed tissue.

Example

Compare the irradiances at the retina that result when looking:
(a) Directly at the sun.
(b) Into a 1-mW He-Ne laser.

Solution.   (a) The sun subtends an angle of $0.5° = 0.0087$ rad. Outside the Earth's atmosphere, the sun's irradiance is 1.4 kW m^{-2}. At the Earth's surface, it is approximately 1 kW m^{-2} or, easier to comprehend, 1 mW mm^{-2}. Assume that the pupil of the bright-adapted eye is 2 mm in diameter. Its area, therefore, is

$$A = \tfrac{1}{4}\pi D^2 = \tfrac{1}{4}(\pi)(2 \text{ mm})^2 \approx 3 \text{ mm}^2$$

With an irradiance of 1 mW mm^{-2}, a power

$$\phi = (1 \text{ mW mm}^{-2})(3 \text{ mm}^2) = 3 \text{ mW}$$

will pass through the pupil. The focal length of the eye, from Example 2, page 340, is 22.5 mm. The image of the sun on the retina, therefore, has a diameter

$$D = (22.5 \text{ mm})(0.0087 \text{ rad}) \approx 0.2 \text{ mm}$$

and an area

$$A = \tfrac{1}{4}(\pi)(0.2 \text{ mm})^2 \approx 0.03 \text{ mm}^2$$

The irradiance at the retina then is the power (3 mW) divided by the area of the sun's image,

$$E = \frac{\phi}{A} = \frac{3 \text{ mW}}{0.03 \text{ mm}^2} = \boxed{100 \text{ mW mm}^{-2}}$$

(b) The laser may have a beam 2 mm in diameter. Using Equation [5.5-31] and setting $\lambda = 633$ nm, $f = 22.5$ mm, and $d_0 = 2$ mm, the focal spot has a diameter

$$2r = \frac{4f\lambda}{\pi d_0} = \frac{(4)(22.5)(633 \times 10^{-6})}{(\pi)(2)} = 9 \text{ }\mu\text{m}$$

which corresponds to an area

$$A = \tfrac{1}{4}(\pi)(9 \times 10^{-3} \text{ mm})^2 = 64 \times 10^{-6} \text{ mm}^2$$

and an average irradiance

$$E = \frac{1 \text{ mW}}{64 \times 10^{-6} \text{ mm}^2} \approx 0.016 \times 10^6 \text{ mW mm}^{-2} = \boxed{16 \text{ W mm}^{-2}}$$

That is *160 times the irradiance brought about by the sun!* Clearly, **EXTREME CAUTION MUST BE USED IN ANY WORK INVOLVING ANY TYPE OF A LASER!**

SUGGESTIONS FOR FURTHER READING

M. BERTOLOTTI, *Masers and Lasers; An Historical Approach* (Bristol, England: Adam Hilger Ltd., 1983).

J. HECHT, *The Laser Guidebook* (New York: McGraw-Hill Book Company, 1986).

A. E. SIEGMAN, *Lasers* (Mill Valley, CA: University Science Books, 1986).

L. V. TARASOV, *Laser Physics* (Moscow: Mir Publishers, 1983).

W. J. WITTEMAN, *The CO₂ Laser* (New York: Springer-Verlag New York, Inc., 1987).

Y. R. SHEN, *The Principles of Nonlinear Optics* (New York: Wiley-Interscience, 1984).

J. T. LUXON and D. E. PARKER, *Industrial Lasers and Their Applications* (Englewood Cliffs, NJ: Prentice-Hall, Inc., 1985).

J. C. DAINTY, editor, *Laser Speckle and Related Phenomena,* 2nd edition (New York: Springer-Verlag New York, Inc., 1984).

D. C. WINBURN, *Practical Laser Safety* (New York: Marcel Dekker, Inc., 1985).

PROBLEMS

5.5-1. If laser action occurs by the transition from an excited state to the ground state, $E_1 = 0$, and if it produces light of 693 nm wavelength, what is the energy level of the excited state?

5.5-2. Transition occurs between a metastable state E_3 and an energy state E_2, just above the ground state. If emission is at 1.1 μm and if $E_2 = 0.4 \times 10^{-19}$ J, how much energy is contained in the E_3 state?

5.5-3. Certain lasers can be *optically* pumped by a chemical reaction, exposing the active material to a sudden flash of light. What are the advantages and disadvantages of doing that?

5.5-4. Rate, in order of decreasing efficiency, the filling provided by different cavity configurations.

5.5-5. Ruby has a refractive index of $n = 1.765$. If the crystal is 4 cm long, the wavelength 694.3 nm, and the laser operating in the TEM$_{00}$ mode, what is the least number of standing waves that fit into the cavity?

5.5-6. Determine the frequency difference that occurs with a laser that is 1.5 m long and contains a gas of $n = 1.0204$.

5.5-7. Emission from a semiconductor laser occurs in the narrow junction between the two types of material. If the wavelength is 905 nm and the "slit width" 5.2 μm, what is the full-angle divergence of the output?

5.5-8. What is the divergence of a beam that at first is 2 mm wide and, after a distance of 10 m, has spread to a diameter of 16 mm?

5.5-9. What is the full-angle divergence of a He-Ne laser, oscillating in the TEM$_{00}$ mode, whose beam is 1 mm in diameter?

5.5-10. If the beam produced by a CO₂ laser is 4 mm wide, what should be the focal length of a lens that will focus the beam into a spot 0.25 mm in diameter?

5.5-11. If light of 1 mrad divergence is passed through an inverted 8× telescope, how much has it spread when it hits a target 5 km away?

5.5-12. If the light from a He-Ne laser is 1.93 mm in diameter and limited only by diffraction, what is the divergence of the beam:
(a) As it leaves the aperture?
(b) After it has passed through a 5× telescope used as a beam expander?

5.5-13. If a laser delivers 1.48 mW in a beam 3.6 mm in diameter, what is the power density in a spot 1 μm in diameter, assuming a loss of 20% in the focusing system?

5.5-14. If a camera lens can focus collimated light to a spot 25 μm in diameter, what is the power density produced by a pulse from a Q-switched 100-MW laser?

5.5-15. Show, in the form of a simple schematic drawing, what the wavefront of a laser beam may look like:

(a) Before the beam enters a sheet of ground glass.

(b) After the beam emerges from the ground glass.

(c) After phase conjugation.

5.5-16. While looking at the speckle pattern produced on a diffuse screen, an uncorrected *myopic* observer turns his/her eyes slowly from left to right. That will cause the speckle grains to move.

(a) How do the loci of interference, generated in front of the screen, move relative to the screen?

(b) How do their conjugate images move on the retina?

(c) Therefore, what does the observer see?

5.5-17. A He-Ne laser of only 1 mW power projects a beam of 2 mm diameter and 6 mrad divergence into the eye. Using the data of the reduced eye on page 34, how large a focal spot will form on the retina?

5.5-18. Continue with Problem 5.5-19 and assume that the light suffers a loss of 35% before reaching the retina. What is the power density on the retina?

6.1

Relativistic Optics

WE HAVE ALMOST REACHED THE END of our introduction to classical and modern optics. And now comes a major constraint. Whenever the light source or the observer or any part of the system is moving, virtually all the laws of optics change. The example of relativistic optics cited most often is the Michelson–Morley experiment. Actually, the Michelson–Morley experiment, while of historic interest, has not materially contributed to the development of the theory of special relativity and has little bearing on our thinking today. Accordingly, we go directly to a discussion of the facts of relativity, both Galileo's and Einstein's, and then continue with a series of examples, from satellite navigation to laser gyroscopes, that illustrate how relativistic optics replaces the laws of conventional optics.

TRANSFORMATIONS

Galileo's transformation. Assume that somewhere in three-dimensional space there occurs a single physical *event,* such as a collision of two particles. To describe the event fully we need four coordinates, three space coordinates, x, y, z, and a time coordinate, t. It is understood that the coordinate system is that of the laboratory including the observer and that it is an *inertial system,* with a frame of reference in which Newton's first law, the law of inertia, holds true. Such a system may either be stationary or it may be in uniform motion, but it may not contain any element subject to acceleration: A space vehicle merely drifting along, without spinning and with its engines cut off, is a good example of an inertial system.

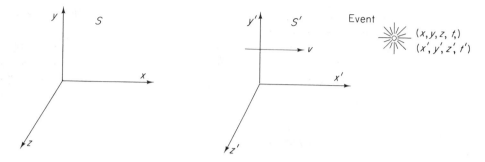

Figure 6.1-1 Two inertial frames: (*left*), S is stationary; (*right*), S' moves to the right at velocity v.

Now consider a given inertial frame, S, and another inertial frame, S', that is in uniform motion relative to S. For simplicity assume that the three sets of axes are parallel to each other and that the frame S' moves, at velocity v, along the x-x' axis, in the direction of $+x$ (Figure 6.1-1). The observer attached to S ascribes to the event the coordinates x, y, z, t. But to an observer moving along with S' the same event occurs at x', y', z', t'. Since the two frames did coincide at time $t = t' = 0$ and since $x = vt$,

$$
\boxed{
\begin{aligned}
x' &= x - vt \\
y' &= y \\
z' &= z
\end{aligned}
}
$$

[6.1-1]

which is *Galileo's space coordinate transformation*. The time at which the event is seen to occur, to both the observers in S and S', is the same; thus we add the statement that

$$
\boxed{t' = t}
$$

[6.1-2]

Now assume that instead of a point we have an extended (one-dimensional) object such as a meter stick and that we wish to determine its length. The endpoints of the stick are called A and B and these are at rest in the S frame. If the stick is parallel to the x axis, the observer in S assigns to these points the coordinates x_A and x_B and the observer in S' the coordinates x'_A and x'_B. Using Galileo's transformation with respect to x, we find that

$$x'_A = x_A - vt_A \qquad \text{and} \qquad x'_B = x_B - vt_B$$

Subtracting the first of these equations from the second gives

$$x'_B - x'_A = x_B - x_A - v(t_B - t_A)$$

Since the two endpoints, A and B, are measured at the same time,

$$t_A = t_B$$

the last term in the preceding equation drops out and

$$x'_B - x'_A = x_B - x_A$$

[6.1-3]

The meter stick, therefore, has the same length in both the S frame and the S' frame.

Similar arguments can be made for the velocity, the acceleration, momentum, angular momentum, kinetic energy—in short, for Newton's laws and all other laws of mechanics that follow from them. These laws are the same in all inertial frames. The laws of electrodynamics, however, are not. This dilemma could be resolved by *Einstein's postulates* (assumptions).

Einstein's postulates.   The two Einstein postulates lie at the heart of the special theory of relativity.* They state that:

1. The laws of physics, including electrodynamics, are the same in all uniformly moving coordinate systems (inertial frames). There is no preferred frame. It is not possible to detect any absolute motion of bodies in space but only relative motions of one body with respect to another. This is the *principle of special relativity.*
2. The velocity of light in free space is the same in all inertial frames and is independent of the motion of the source emitting the light. This is the *principle of the constancy of the speed of light.*† The unique aspect of Einstein's work is that it was not based on prior experimental evidence. Rather, it led to a *prediction* of experiments whose results, over the years, have conclusively shown that Einstein was right.

Lorentz' transformation.   If we measure the length of an object *at rest,* in its own frame of reference, S, we measure its *proper length* ("rest length"). But the object may be moving, in a frame S', relative to S. This requires the two observers, in S and S', to determine the coordinates of the endpoints of the object simultaneously. If the endpoints are not determined at the same time, then, because of the finite speed of light, one point may be seen to have moved more, or less, than the other point and the length measured may not be the proper length.

Simultaneity is the crucial word. But how can we tell that two events are simultaneous or, in a more general sense, how can we *synchronize* two clocks? Within the same frame this is easy. But we have two frames, moving relative to each other. If it were possible to transmit a signal with infinite speed, there would be no problem. But we do not know any signals that could reach all points of the universe in zero time. The fastest signals known are those trans-

---

*Albert Einstein (1879–1955), German-born physicist, mathematician, philosopher, and humanitarian. From 1902 to 1909, Einstein worked as a patent examiner in the Swiss Federal Patent Office in Bern, then became professor at the universities of Zürich and Prague, the Swiss Polytechnic Institute, and the Prussian Academy of Sciences in Berlin. In 1933, Einstein came to the United States and joined the then newly organized Institute for Advanced Studies in Princeton, New Jersey, where he remained for the rest of his life.

The principal paper that laid the foundation for the special theory of relativity is A. Einstein, "Zur Elektrodynamik bewegter Körper," *Ann. Phys.* (4) **17** (1905), 891–921, available in an English translation in A. Einstein et al., *The Principle of Relativity* (New York: Dover Publications, Inc., 1981). By his own account, Einstein began his work at the age of 16 and continued working on it, off and on, for 10 years.

†At first sight, this is hard to comprehend. If a bullet is fired from a moving truck, the velocity of the bullet is different depending on whether the gun is pointed forward or backward with respect to the motion of the truck. This is not so with light. Although experiments have been tried to prove the "ballistic theory," all of them failed, as shown, for example, by G. C. Babcock and T. G. Bergman, "Determination of the Constancy of the Speed of Light," *J. Opt. Soc. Am.* **54** (1964), 147–51.

mitted by light or other electromagnetic radiation, such as radio waves. No faster method of sending a signal has ever been found.

It is actually classical physics which makes the fictitious assumption of zero time (science fiction). Relativistic physics requires a finite, limiting speed. Nature itself shows that relativistic physics is *real* and not a philosophical concept.

Will the synchronized clocks of the observer in S also be in synchrony with the synchronized clocks of the observer in S'? Because of the finite speed of light, they will not. This means that simultaneity is not independent of the frame of reference. Time, and space, are both relative; to find a connection between the unprimed and the primed coordinates, we somehow have to tie the concept of time into the concept of space.

To do this we use a hypothetical *light clock*. This "clock" is illustrated in Figure 6.1-2 (and will come up again, somewhat modified, in the Michelson-Morley experiment, Figure 6.1-7). A source A emits a pulse of light that travels through distance L to a mirror, M, and back again to A. To an observer in the same frame as the clock, the time interval for the pulse to go from A to M and back is

$$\Delta t = \frac{2L}{c} \tag{6.1-4}$$

Now let the clock be moving, in frame S'. To the observer in S the path from A to the mirror and back to A' is *longer*. The new time interval is found from the triangle in Figure 6.1-2, right, and Pythagoras' theorem:

$$\left(c\,\frac{\Delta t}{2}\right)^2 = L^2 + \left(v\,\frac{\Delta t}{2}\right)^2 \tag{6.1-5}$$

$$c^2\Delta t^2 - v^2\Delta t^2 = 4L^2$$

$$\Delta t^2 = \frac{4L^2}{c^2 - v^2}$$

and thus

$$\Delta t = \frac{2L}{\sqrt{c^2 - v^2}} = \frac{2L}{c}\,\frac{1}{\sqrt{1 - (v/c)^2}} \tag{6.1-6}$$

Figure 6.1-2 Light clock at rest (*left*) and moving at velocity v (*center*). (*Right*) graphical construction.

Substituting $2L/c = \Delta t$ yields

$$\Delta t' = \Delta t_0 \sqrt{1 - (v/c)^2} \qquad [6.1\text{-}7]$$

which means that to the observer in S the (moving) clock in S' *runs slow,* a phenomenon called *time dilation.*

Consider again the two endpoints, A and B, of the object, and their coordinates, x_A and x_B. In the S frame, these coordinates determine the proper length, L_0, of the object:

$$L_0 = x_B - x_A \qquad [6.1\text{-}8]$$

To the observer in S, it takes time interval Δt for the object to move past a given point:

$$\Delta t = \frac{L_0}{v}$$

But in frame S', it takes

$$\Delta t' = \frac{L'}{v} \qquad [6.1\text{-}9]$$

We solve Equation [6.1-9] for L', substitute Equation [6.1-7], and set $v\Delta t_0 = L_0$. This gives

$$\boxed{L' = L_0 \sqrt{1 - \left(\frac{v}{c}\right)^2}} \qquad [6.1\text{-}10]$$

which is the *FitzGerald–Lorentz length contraction.*[*]

Evidently, an object has its greatest length when seen in its rest frame (where its velocity is zero). When seen from another frame, its length, measured in a direction parallel to the motion, becomes less by a factor $\sqrt{1 - (v/c)^2}$.

Does the object *really* become shorter? No, it retains its proper length for an observer flying along with it in the S' frame. But to an observer in S the object not only *appears* to be shorter, it really *is* shorter. The length contraction is not an illusion; it is real for an observer not moving along with the object.

If we apply the contraction factor to the x dimension only, and leave y and z unchanged, we obtain a new set of equations:

---

[*]Named after George Francis FitzGerald (1851–1901), Irish physicist and professor of natural philosophy at Trinity College in Dublin, and Hendrik Antoon Lorentz (1853–1928), Dutch physicist and professor of mathematical physics at the University of Leiden, winner of the Nobel Prize in physics in 1902. FitzGerald and Lorentz, working at the time of intense search for the aether, used their contraction to explain the Michelson-Morley experiment (page 433). H. A. Lorentz, "De relatieve beweging van de aarde en den aether," *Versl. Zitt. Wis. Natuurkundige Afdeeling Koninklijke Akad. Wetenschappen, Amsterdam* **1** (1892), 74–79.

$$x' = \frac{x - vt}{\sqrt{1 - (v/c)^2}}$$

$$y' = y$$

$$z' = z$$

[6.1-11]

the *Lorentz' space coordinate transformation equations.*

OPTICS AND THE SPECIAL THEORY OF RELATIVITY

Headlight effect.   Now we come to the specific applications of the theory of relativity to optics. When a firework explodes in the sky, it scatters sparks uniformly in all directions. But when the firework explodes while in rapid motion, it scatters most of its sparks in the forward direction.* It is much the same with a source of light.

Consider an isotropic point light source (which, by definition, emits light uniformly in all directions). Then imagine a screen with a hole in it that is placed over the source so that one half of the light goes out to the left of the screen, the other half to the right (Figure 6.1-3, left). With the source at rest, the screen is flat and the angle subtended by the screen and the +x axis, θ, is 90°.

Now let the source and the hypothetical screen travel to the right at velocity v. The screen then folds into a cone (with the angle θ becoming smaller than 90°), but it still divides the light into equal parts, one half emitted *outside* the cone and the other half *inside* it. A

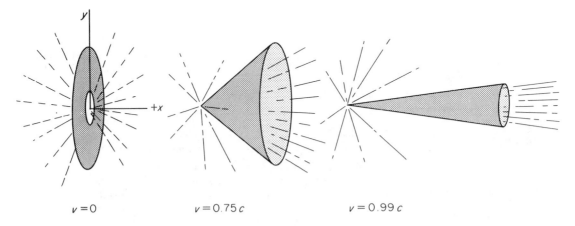

$v = 0$ $v = 0.75\,c$ $v = 0.99\,c$

Figure 6.1-3   Light emitted from moving source becomes concentrated in the forward direction.

---

*As quoted from W. Rindler, *Essential Relativity: Special, General, and Cosmological,* 2nd edition, p. 260 (New York: Springer-Verlag New York, Inc., 1977).

photon traveling along the cone's surface covers, per unit time interval Δt, a distance $c\Delta t$. The projection of $c\Delta t$ on the $+x$ axis is Δx. But Δx is also the distance the *source* travels, $\Delta x = v\Delta t$. In the first order, the cosine of the apex half-angle is the ratio of these two distances:

$$\cos\theta = \frac{v\Delta t}{c\Delta t} = \frac{v}{c} \qquad [6.1\text{-}12]$$

Hence, as the source moves faster (Figure 6.1-3, center and right), the cone becomes tighter and the light more concentrated in the forward direction, a result that gave the phenomenon its name, *headlight effect*.

Reflection of light on a moving mirror. When a plane mirror moves in a direction parallel to its surface, nothing much happens. But when the mirror moves normal (perpendicular) to its surface, several things change. First consider the *angle of reflection*. As the mirror moves *toward* the source, the angle becomes *less* (Figure 6.1-4). If, as before, I_1 is the angle of incidence, I_2 the angle of reflection, and v the velocity of motion of the mirror, then *

$$\frac{\tan\frac{1}{2}I_1}{\tan\frac{1}{2}I_2} = \frac{c+v}{c-v} \qquad [6.1\text{-}13]$$

If $v = 0$, $I_1 = I_2$. This is the conventional case, the first law of reflection in an inertial frame. As the mirror *recedes* from the source, the angle of reflection becomes *larger* than the angle of incidence and may even exceed $90°$.

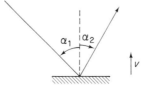

Figure 6.1-4 With the mirror moving toward the source, the angle of reflection becomes less, $\alpha_1 > \alpha_2$.

The same conclusions can be drawn from a construction of wavefronts. As a wavefront, inclined at a certain angle, approaches the mirror, one edge of the wavefront reaches the mirror earlier than the opposite edge does. But this makes the mirror *appear* (to the light) as if it were tilted with respect to the real mirror and reflection on the now "virtual" mirror yields the same ray trace as that obtained from Equation [6.1-13] and the real mirror.

Example

What velocity is needed to observe the effect, assuming the angle of incidence is $45°$?

---

*Following W. Rindler, *Introduction to Special Relativity*, p. 54 (New York: Oxford University Press, 1982).

Solution. The most sensitive goniometer, that is, angle-measuring instrument, is the Michelson stellar interferometer. It can resolve angles as small as 0.022 arc sec, or $0.022/3600 \approx$ 0.000 006°. Then, setting $I_2 = 45° - 0.000\ 006°$ and using Equation [6.1-13], we have

$$\frac{0.414\ 213\ 562}{0.414\ 213\ 501} = 1.000\ 000\ 147 = \frac{300\ 000\ 000 + v}{300\ 000\ 000 - v}$$

$$300\ 000\ 000 + v = 300\ 000\ 044 - 1.000\ 000\ 147\ v$$

so that

$$v \approx \frac{44}{2} = \boxed{22\ \text{m s}^{-1}}$$

a surprisingly low velocity.

There is another effect seen with a moving mirror: The light reflected from the mirror changes in *wavelength*. This is easy to understand. As the mirror is moving toward the source and *against* the flow of photons incident on it, the momentum of the mirror *raises* the energy of the reflected photons, increasing $h\nu$, thus increasing the frequency and reducing the wavelength (Figure 6.1-5, left). As the mirror moves away from the source and *with* the flow of the photons, the opposite happens and the wavelength increases (right). The two wavelengths, λ_1 for the incident light and λ_2 for the reflected light, are related, in the first order, as

$$\frac{\lambda_1}{\lambda_2} = \frac{1 + (v/c) \cos I_1}{1 - (v/c) \cos I_2} \qquad [6.1\text{-}14]$$

If $v = 0$, then $\lambda_1 = \lambda_2$, which, again, is the conventional case.

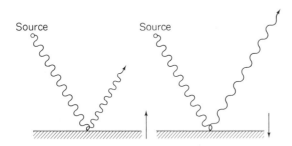

Figure 6.1-5 With the mirror moving toward the source, the wavelength of the reflected light becomes shorter (*left*); with the mirror moving away, it becomes longer (*right*).

Doppler effect. Light reflected from a mirror can be considered as coming from a virtual source located behind the mirror. Thus, as the mirror moves, the virtual source moves as well (at twice the velocity of the mirror) and the light is subject to the *Doppler effect.*

Let us call c the velocity of the wave emitted by the source, not necessarily the velocity of light, and x_0 the distance from the source, real or virtual, through which the wave travels during time interval Δt:

$$x_0 = c\Delta t$$

With the source at rest, the distance x_0 contains N waves of length λ_0,

$$x_0 = N\lambda_0$$

Thus

$$\lambda_0 = \frac{x_0}{N} = \frac{c\Delta t}{N} \qquad \text{[6.1-15]}$$

Now let the source be moving at velocity v,

$$v = \frac{\Delta x}{\Delta t}$$

The new distance x' contains the same number of waves but now of length λ':

$$x' = x_0 - \Delta x = N\lambda'$$

$$\lambda' = \frac{x_0 - \Delta x}{N} \qquad \text{[6.1-16]}$$

Dividing Equation [6.1-15] by [6.1-16] gives

$$\frac{\lambda_0}{\lambda'} = \frac{c\Delta t}{x_0 - \Delta x}$$

Substituting $x_0 = c\Delta t$ and $\Delta x = v\Delta t$, canceling Δt, and solving for λ' yields

$$\lambda' = \lambda_0\left(\frac{c - v}{c}\right) = \lambda_0\left(1 - \frac{v}{c}\right) \qquad \text{[6.1-17]}$$

and, setting $\lambda_0 - \lambda' = \Delta\lambda$,

$$\boxed{\Delta\lambda = \lambda_0\frac{v}{c}} \qquad \text{[6.1-18]}$$

which is the *classical, first-order Doppler effect.** With the source approaching, $\lambda' < \lambda_0$, which gives a "blue shift." Conversely, with the source receding, $\lambda' > \lambda_0$, which gives a "red shift."

Relativity has added a correction to the first-order Doppler effect, Equation [6.1-18]. The reason is that the source of any wave is by necessity an oscillator (a "clock") and, when moving, is subject to time dilation and *runs slow*. The distance $\Delta x = v\Delta t$ referred to before must then be modified and the wavelength of the light emitted by a receding source becomes

$$\lambda' = \lambda_0\frac{1 + v/c}{\sqrt{1 - (v/c)^2}} \qquad \text{[6.1-19]}$$

---

*Named after Johann Christian Doppler (1803–1853), Austrian high school mathematics teacher, later professor of experimental physics at the University of Vienna. In his publication, "Ueber das farbige Licht der Doppelsterne und einiger anderer Gestirne des Himmels," *Abh. Königl. böhm. Ges.* (Prag: Borrosch und André, 1842), p. 465, Doppler attributed the different colors of certain stars to their motion toward or away from Earth. He was wrong; the speed of light is so high that in order to change color, the stars would have to move at velocities too high even on an astronomical scale.

If the source moves purely radial, along the line of sight, this reduces to

$$\lambda' = \lambda_0 \sqrt{\frac{1 + v/c}{1 - v/c}}$$ [6.1-20]

which is the *longitudinal relativistic Doppler effect.*

If the direction of motion of the source subtends an angle θ with the direction of the light, a $\cos \theta$ factor is needed to modify the v/c term in the numerator of Equation [6.1-19]. This means that when the source moves at right angles of the line of sight, that is, when $\cos \theta = 0$, Equation [6.1-19] simplifies to

$$\lambda' = \lambda_0 \frac{1}{\sqrt{1 - (v/c)^2}}$$ [6.1-21]

which is the *transverse Doppler effect;* it accounts for a red shift even if there is no first-order Doppler effect.

Example

Certain double stars can be resolved by the Doppler effect even if they are too close together to be resolved otherwise. Spectra obtained from such *spectroscopic binaries* indicate their motion relative to Earth by characteristic blue and red shifts as illustrated in Figure 6.1-6.

One of the largest red shifts on record is shown by the galaxy 3C123. There, lines normally found in the UV are shifted into the red, the actual figures indicating that this galaxy, the farthest out known, is moving away at a velocity of 0.637c, more than one-half the speed of light.

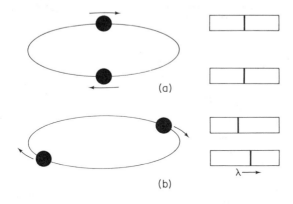

(a)

(b)

Figure 6.1-6 Schematic representation of the motion of a binary star system and of the Doppler shifts associated with it. (a) Sidewise motion, no first-order Doppler shift; (b) motion toward and away from observer produces blue and red shift.

The Michelson–Morley experiment.   Of all the interference experiments connected with relativistic optics the *Michelson–Morley experiment* is the most famous.* A Michelson interferometer, as described on page 213, is set up so that one arm is parallel to

---

*Albert A. Michelson, at the time of the experiment, was professor of physics at the Case School of Applied Science in Cleveland; Edward Williams Morley (1838–1923) was professor of chemistry at Western Reserve University, also in Cleveland. The pertinent publication is A. A. Michelson and E. W. Morley, "On the Relative Motion of the Earth and the Luminiferous Aether," *Philos. Mag.* (5) **24** (1887), 449–63. The Michelson-Morley experiment was later repeated with extraordinary care by Joos, with the same result. G. Joos, "Die Jenaer Wiederholung des Michelsonversuchs," *Ann. Phys.* (5) **7** (1930), 385–407.

the orbital motion of Earth and the other arm normal to it (Figure 6.1-7). The two arms are adjusted to equal length.

Then, as Earth turns, the beamsplitter and the two mirrors reach new positions, shown by the dashed lines. The time needed for the light to make a round trip in the upper (perpendicular) arm is that shown earlier with Equation [6.1-6],

$$t_\perp = \frac{2L}{\sqrt{c^2 - v^2}} = \frac{2L}{c} \frac{1}{\sqrt{1 - (v/c)^2}} \qquad [6.1\text{-}22]$$

In the parallel arm it is

$$t_\parallel = \frac{L}{c + v} + \frac{L}{c - v} = \frac{2cL}{c^2 - v^2} = \frac{2L}{c} \frac{1}{1 - (v/c)^2} \qquad [6.1\text{-}23]$$

If we expand both Equations [6.1-22] and [6.1-23] by the binomial theorem and drop terms higher than $(v/c)^2$,

$$t_\perp = \frac{2L}{c} \left[1 + \frac{1}{2} \left(\frac{v}{c} \right)^2 \right] \qquad [6.1\text{-}24]$$

and

$$t_\parallel = \frac{2L}{c} \left[1 + \left(\frac{v}{c} \right)^2 \right] \qquad [6.1\text{-}25]$$

The interferometer is now turned through 90°, which means the time difference between the two positions is

$$\Delta t = t_\parallel - t_\perp \approx \frac{2L}{c} \left(\frac{v}{c} \right)^2 \qquad [6.1\text{-}26]$$

The quantity of interest is the optical path difference, Γ:

$$\Gamma = c\Delta t = 2L \left(\frac{v}{c} \right)^2 \qquad [6.1\text{-}27]$$

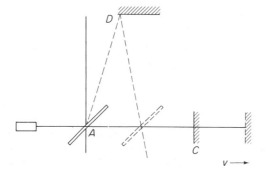

Figure 6.1-7   Michelson–Morley experiment. Note similarity to the "light clock" shown in Figure 6.1-2.

and since the number of fringes, m, is related to Γ as $m = \Gamma/\lambda$,

$$m = \frac{2L}{\lambda}\left(\frac{v}{c}\right)^2 \qquad \text{[6.1-28]}$$

In the actual experiment, L was 11 m, $\lambda = 590$ nm, and $v = 30$ km s^{-1}. This should have given a fringe shift of

$$m = \frac{(2)(11)}{590 \times 10^{-9}}\left(\frac{3 \times 10^4}{3 \times 10^8}\right)^2 = 0.37 \text{ fringe}$$

which would have been easy to see.

But *there was no fringe shift*. This result of the Michelson–Morley experiment, at that time, came as a complete surprise. Today we know that, according to the second postulate, the velocity of light is the same in all inertial frames and independent of the motion of the source.

Velocity of light in moving matter.   In 1818, Fresnel predicted that light would be "dragged along" by a moving medium and that its new phase velocity should depend both on the velocity of the medium and on its refractive index.* If c is the speed of light in free space, w the velocity of the medium, and n the index, then, Fresnel predicted, the new phase velocity, v', should be

$$v' = \frac{c}{n} + w\left(1 - \frac{1}{n^2}\right) \qquad \text{[6.1-29]}$$

The phase velocity of light is defined as c/n. Thus, as we see from Equation [6.1-29], Fresnel assumed that the phase velocity does not change by the full velocity of the medium, w, but only by a fraction thereof, $1 - 1/n^2$, called the *Fresnel drag coefficient*. The higher the refractive index, furthermore, the larger the drag coefficient. The experiment was later performed by Fizeau,† who confirmed Fresnel's prediction.

In Fizeau's experiment, illustrated in Figure 6.1-8, light from a source, top, is divided by a beamsplitter B. One beam is traveling clockwise, the other counterclockwise. The beams recombine at B and form interference fringes that are observed through a telescope, bottom. When the water is flowing, one of the beams is traveling *with* the direction of flow and the other *against* it. If then the flow of the water is reversed, the fringes are seen to shift.

Fizeau's experiment can be interpreted in terms of r:radiation (page 297). With the medium at rest, the light encounters a certain number of molecules. But if the medium is moving toward the source, the light encounters *more* molecules per unit time; hence, its refractive index appears to be higher.

---

*A. Fresnel, "Sur l'influence du mouvement terrestre dans quelques phénomènes d'optique," *Ann. Chim. Phys.* (2) **9** (1818), 57–66.

†H. Fizeau, "Sur les hypothèses relatives à l'éther lumineux, et sur une expérience qui parait démontrer que le mouvement des corps change la vitesse avec laquelle la lumière se propage dans leur intérieur," *Compt. rendu* **33** (1851), 349–55.

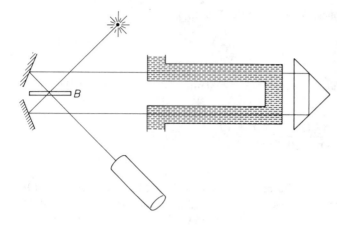

Figure 6.1-8 Fizeau's experiment (shown as modified by Michelson and Morley) for measuring the velocity of light in running water.

If each of the tubes has a length L, and the drag coefficient is called D, the time for one beam to go through the path is

$$t_1 = \frac{2L}{(c/n) - wD}$$

and for the other beam

$$t_2 = \frac{2L}{(c/n) + wD}$$

The difference between the two is

$$\Delta t = t_1 - t_2 = \frac{4LwD}{(c/n)^2 - (wD)^2}$$

Since $(wD)^2$ is much smaller than $(c/n)^2$,

$$\Delta t \left(\frac{c}{n}\right)^2 \approx 4LwD$$

and

$$\Delta t \approx \frac{4Ln^2w(1 - 1/n^2)}{c^2} \qquad [6.1\text{-}30]$$

Within a time interval Δt a wave of velocity c proceeds through a path difference Γ,

$$\Gamma = c\Delta t$$

Since from Equation [3.2-1]

$$\Gamma = m\lambda$$

a fringe shift m and the time difference Δt are related as

$$m = \frac{c\Delta t}{\lambda} \qquad [6.1\text{-}31]$$

Inserting Equation [6.1-30] in [6.1-31] then leads to

$$m \approx \frac{4Ln^2w(1 - 1/n^2)}{c\lambda} \qquad [6.1\text{-}32]$$

which is the fringe shift seen when light passes through moving matter.

Rotating systems. A good example of a rotating system is *Sagnac's interferometer*. As in Fizeau's experiment, one of the beams travels clockwise and the other counterclockwise (Figure 6.1-9). The whole apparatus is mounted on a rigid support that can be rotated about a vertical axis. The rotation causes one beam to travel farther than the other so that its frequency falls. The frequency of the other beam correspondingly rises. The result, again, is a fringe shift.

For simplicity assume the path to be circular. We call v the tangential velocity and L the length of a section of the path. Then the time it takes the light to travel through that section in direction of rotation is $t_1 = L/(c - v)$; in the opposite direction it is $t_2 = L/(c + v)$. The difference between the two is

$$\Delta t = t_1 - t_2 = \frac{L}{c - v} - \frac{L}{c + v} = \frac{2Lv}{c^2 - v^2} \approx \frac{2Lv}{c^2} \qquad [6.1\text{-}33]$$

If R is the radius of the loop and ω the angular velocity,

$$v = R\omega$$

For a total length of path, $L = 2\pi R$, thus, from Equation [6.1-33],

$$\Delta t \approx \frac{4\pi Rv}{c^2} = \frac{4\pi R^2\omega}{c^2} = \frac{4A\omega}{c^2} \qquad [6.1\text{-}34]$$

where A is the area enclosed by the path. Inserting Equation [6.1-34] in [6.1-31] gives

$$m \approx \frac{4A\omega}{c\lambda} \qquad [6.1\text{-}35]$$

which is *Sagnac's formula*. It describes the fringe shift seen, no matter what the shape of the loop. Sagnac's formula has become the basis for today's *fiber gyroscopes*.*

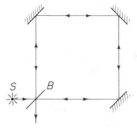

Figure 6.1-9 Sagnac's interferometer.

*Georges Marc Marie Sagnac (1869–1928), French physicist, professor of physics at the University of Lille, later at Paris. In one of his experiments, Sagnac used a square path about 1 m on each side. It took a velocity of 120 rpm to see the fringes shift but Sagnac, who retained a lifelong dislike for relativity, attributed the result to the "aether

Miscellaneous other effects. A special case of light emitted by a moving source is that of *Vavilov–Cerenkov radiation*. The light is emitted by particles that move rapidly through a medium, at a velocity higher than the phase velocity c/n in that medium, $v > c/n$. The result is a faint, eerie blue light, easily seen in swimming-pool nuclear reactors.

Snell's law changes in a way similar to the reflection of light on a moving mirror. When a rare-to-dense boundary moves toward the source, the angle of refraction becomes smaller than predicted by the conventional form of Snell's law, and when the boundary moves away from the source, the angle becomes larger.

The equation

$$E = mc^2 \qquad [6.1\text{-}36]$$

is Einstein's much quoted *mass-energy relation*. If we combine this expression with Planck's

$$E = h\nu \qquad [6.1\text{-}37]$$

we obtain

$$h\nu = mc^2$$

Replacing ν by v/λ and c by v leads to

$$h\,\frac{v}{\lambda} = mv^2$$

and thus

$$\lambda = \frac{h}{mv} \qquad [6.1\text{-}38]$$

This is an important concept. It assigns a wavelength to particles (moving with a momentum mv), an idea that originated with Louis de Broglie, who thought that if light is acting at some times as waves and at others as particles, perhaps "real" particles would show some wave behavior too. The waves themselves are called *de Broglie waves* and the wavelength *de Broglie wavelength.**

De Broglie's hypothesis has been extensively tested and completely verified. It applies to all particles, of any size, even to a 10-ton truck, although *its* diffraction pattern can probably never be observed. But electrons, neutrons, and similarly small particles, when incident on a crystal of the proper interplanar spacing, are diffracted much the same as X rays are diffracted—which shows that such "real" particles have wave properties as well.

---

wind." G. Sagnac, "L'éther lumineux démontré par l'effet du vent relatif d'éther dans un interféromètre en rotation uniforme," *Compt. rendu* **157** (1913), 708-10.

Some time later, Michelson and Gale, in order to detect the lesser angular velocity of Earth, used a longer path, enclosing an area 339 × 613 m², near the present O'Hare airport in Chicago. A. A. Michelson and H. G. Gale, "The Effect of the Earth's Rotation on the Velocity of Light," *Astrophys. J.* **61** (1925), 137–45. The heart of a modern fiber gyroscope is an optical fiber some 1000 m long, coiled for compactness. See, for example, D. Z. Anderson, "Optical Gyroscopes," *Sci. Am.* **254** (Apr. 1986), 94–99.

*Louis Victor Pierre Raymond Duc de Broglie (1892–   ), French physicist. After graduating from the Sorbonne with a degree in medieval history, de Broglie, pronounced "d' Bro'lie," turned to science. He wrote a thesis on quantum theory, became professor at the Sorbonne, and in 1929 won the Nobel Prize in physics. L. de Broglie, "A Tentative Theory of Light Quanta," *Philos. Mag.* (6) **47** (1924), 46–58.

The principle of complementarity. The "dualistic nature of light" is a term that has intrigued many writers and implies some kind of dichotomy. I take a more *unitary* view and think of light as something *unique*. Light is neither a true wave (its oscillations come in finite wavetrains, like no other wave) nor is light made up of conventional particles (quanta can only move at the speed of light, while "real" particles can have any speed less than *c*). Light's wave nature and quantum nature are not mutually exclusive; they are, as Bohr put it, *complementary*.

Light is both, quanta and waves. The quanta contain the energy. The waves guide them. At shorter wavelengths, and in the emission and absorption of light, the quantum aspect becomes predominant. At longer wavelengths, and in diffraction and interference, the wave aspect becomes predominant. But wavelength is not the divisive factor: Light of the same wavelength can act as waves in one experiment and as quanta in another.

Terrell rotation. Think of a cube, made out of transparent material and moving to the right as shown in Figure 6.1-10. The vertical edges of the cube are numbered 1, 2, 3, 4 and the front, side, and rear faces carry the letters F, S, and R, respectively. The width (length) of the cube is L_0 and its velocity v. We are looking at the cube from the side, at a right angle to its direction of motion, viewing the letter S.

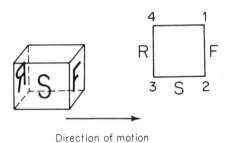

Direction of motion

Figure 6.1-10 Transparent cube with lettering R-S-F seen in perspective view (*left*) and from the top.

If the cube is at rest, face S appears as an exact square (Figure 6.1-11, left). The rear face, R, between edges 3 and 4, cannot be seen. But then assume that the cube moves at high speed and that we look at it from the same direction as before or (which is easier to see) that we take a photograph of it using a camera with a high-speed shutter. This requires the photons that build up the image to arrive on the retina (or on the photographic film) *simultaneously*. But simultaneous *arrival* precludes simultaneous *emission* because edge 4 is farther away than edge 3. Thus light from edge 4 must have left edge 4 when it was still at 4′, to the left of 4. This makes the cube appear as if it had turned, an effect which I will call *Terrell rotation*.[*]

In addition, the side face, but not the front or rear face, is subject to Lorentz contraction (bottom right). Thus the total apparent width of the cube moving at velocity v is

$$\Sigma L' = L_0 \frac{v}{c} + L_0 \sqrt{1 - \left(\frac{v}{c}\right)^2} \qquad [6.1\text{-}39]$$

---

[*]J. Terrell, "Invisibility of the Lorentz Contraction," *Phys. Rev.* **116** (1959), 1041–45. V. F. Weisskopf, "The Visual Appearance of Rapidly Moving Objects," *Phys. Today* **13** (Sept. 1960), 24–27.

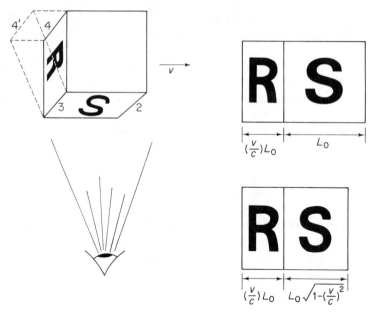

Figure 6.1-11   Cube moving to the right with velocity v and seen by observer looking at side face S.

At rest, $v = 0$, and at the velocity of light, $v = c$, the total apparent width is L_0. At intermediate velocities, the total width will be *greater* than L_0. In order to find the velocity at which maximum elongation occurs, we write Equation [6.1-39] in the form of

$$y = \left(\frac{L_0}{c}\right)v + \frac{L_0}{c}\sqrt{c^2 - v^2}$$

and differentiate with respect to v:

$$\frac{dy}{dv} = \frac{L_0}{c} + \frac{L_0(-2v)}{2c\sqrt{c^2 - v^2}} \qquad [6.1\text{-}40]$$

If we then set the right-hand term equal to zero, we find that

$$v = \sqrt{c^2 - v^2} = \frac{c}{\sqrt{2}} \approx 0.71c \qquad [6.1\text{-}41]$$

The second derivative,

$$\frac{d^2y}{dv^2} = -\frac{L_0}{c}\left[\frac{c^2}{(c^2 - v^2)^{3/2}}\right] \qquad [6.1\text{-}42]$$

is negative at $v = c/\sqrt{2}$, which shows that at that velocity there is indeed a maximum.

If the cube is seen from a direction other than from the side, the results are even more startling. At certain angles θ, subtended by the forward direction and the line of sight, and above certain threshold velocities, the front face becomes reverted and an F appears as ꟼ. Below the threshold the front face is seen as F, and the R is reverted, Я. At velocities higher than $v = c \cos \theta$, the front face is reverted, ⅂, and the R is seen correctly. In short, Terrell rotation and other distortions make an object look different than expected from Lorentz contraction alone.

OPTICS AND THE GENERAL THEORY OF RELATIVITY

Newton's law of gravitation tells us that the force F causing acceleration between two masses m_1 and m_2 is given by

$$F = G \frac{m_1 m_2}{R^2}$$ [6.1-43]

where G is the universal constant of gravitation, 6.672×10^{-11} N m^2 kg^{-2}, and R is the distance between the two masses. But does a force between masses act instantaneously? Probably not, because that would violate one of the basic tenets of relativity that no signal can go faster than light.

According to the *general theory of relativity,* the effects of acceleration cannot be distinguished from the effects of a gravitational field. This holds for light as well as for conventional masses.* Just as a projectile moves through a gravitational field in the form of a parabola, so does light, in an accelerated frame, follow a *curved path* rather than a straight line (Figure 6.1-12). For light passing by near the edge of the sun, Einstein predicted a deflection of 1.745 arc sec. Indeed, stars close to the solar disk were found to be displaced 1.70 ± 0.10 arc sec away from the sun, a brilliant confirmation of Einstein's theory.

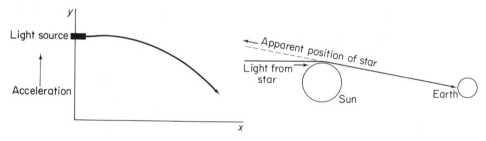

Figure 6.1-12   Bending of light under acceleration (*left*) and in a gravitational field.

Sir Arthur Eddington, British astronomer, was once asked: "Is it true, Sir Arthur, that you are one of three men in the world who understands Einstein's theory of relativity?" The astronomer appeared reluctant to answer. "Forgive me," said the questioner, "I should have realized that a man of your modesty would find such a question embarrassing." "Not at all," said Eddington, "I was just trying to think who the third could be."

---

*A. Einstein, "Über den Einfluss der Schwerkraft auf die Ausbreitung des Lichtes," *Ann. Phys.* (4) **35** (1911), 898–908.

SUGGESTIONS FOR FURTHER READING

A. EINSTEIN, *Relativity* (New York: Crown Publishers, Inc., 1961).

H. A. LORENTZ, A. EINSTEIN, H. MINKOWSKI and H. WEYL, *The Principle of Relativity* (New York: Dover Publications, Inc., 1981).

A. P. FRENCH, *Special Relativity* (New York: W. W. Norton & Company, Inc., 1968).

W. RINDLER, *Introduction to Special Relativity* (New York: Oxford University Press, Inc., 1982).

W. RINDLER, *Essential Relativity: Special, General, and Cosmological*, 2nd edition (New York: Springer-Verlag New York, Inc., 1977).

V. A. UGAROV, *Special Theory of Relativity* (Moscow: Mir Publishers, 1979).

J. VAN BLADEL, *Relativity and Engineering* (New York: Springer-Verlag New York, Inc., 1984).

PROBLEMS

6.1-1. A passenger walks forward in the aisle of a train at a brisk pace, 1.5 m s^{-1}. If the train moves along a straight track at 80 km/h, how fast is the passenger moving with respect to ground?

6.1-2. A gun with a muzzle velocity of 20 m/s, mounted on a vehicle, subtends an angle of 45° with the forward direction. If the vehicle moves at 40 km/h, what is the velocity of the projectile as it leaves the barrel?

6.1-3. If a 15.3-m-long space vehicle passes a planet at a velocity of 6×10^7 m s^{-1}, how long does the vehicle appear to an observer on that planet?

6.1-4. How fast must a meter stick travel for an observer to conclude that it is only one-half as long?

6.1-5. A linear object, subtending an angle of 45° with the direction of motion, travels at $0.8c$. What is the angle in the S frame?

6.1-6. A good example of an inertial system is a space vehicle in orbit, drifting along with its engines cut off. But the vehicle must not be spinning. Explain why not.

6.1-7. What is the (total) apex angle of a hypothetical cone equally dividing the light emitted by a point source traveling at 96% of the speed of light?

6.1-8. What velocity is needed to concentrate one-half of the light into a cone no wider than 5°?

6.1-9. Light is incident at an angle of 40° on a plane mirror moving toward the source at a velocity of $0.15c$. Find the angle of reflection.

6.1-10. A plane mirror is receding in a direction normal to its surface at a velocity of one-half the speed of light. Show how the light is reflected if it is incident on the mirror at an angle of:
(a) 30°.
(b) 60°.

6.1-11. If in Problem 6.1-9 the incident light has a wavelength of 500 nm, what is the wavelength of the reflected light?

6.1-12. If microwaves (radar) are reflected off an automobile traveling at 108 km/h, what is the relative change of frequency, in terms of (v/c), of the return signal?

6.1-13. A certain star shows a first-order Doppler shift of 1.4 Å involving the red (656 nm) hydrogen line. How fast does the star move relative to Earth?

6.1-14. If a distant star is receding from Earth at a velocity of 2.4×10^7 m s^{-1}, by how much would the 656 nm hydrogen line shift if we consider:

(a) The first-order Doppler effect?

(b) The relativistic Doppler effect?

6.1-15. The spectrum of a distant nebula shows the 434-nm hydrogen line to have shifted to 752 nm. How fast is the nebula moving away from Earth?

6.1-16. How fast do you have to go through a red traffic light ($\lambda = 600$ nm) so that it appears to you green (500 nm)? Consider the first-order effect only.

6.1-17. A simple way of introducing the Michelson–Morley experiment is by the *rowboat analogy*. A man is rowing through a distance of 100 m, at a velocity of $c = 0.2$ m s^{-1}, across a river flowing at $v = 0.1$ m s^{-1} (Figure 6.1-13).

(a) What is the effective velocity of the boat?

(b) What is the time needed to complete a round trip?

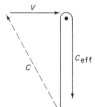

Figure 6.1-13

6.1-18. Continue with Problem 6.1-17 and now let the man row *parallel* to the river, downstream and upstream, again covering each way a distance of 100 m. How much longer does that take him?

6.1-19. Fizeau in his 1851 experiment used tubes each 1.5 m long, water flowing at a rate of 7 m/s, and light of wavelength 530 nm. How much of a fringe shift would you expect?

6.1-20. Actually, Fizeau saw a fringe shift of 0.23.

(a) How large, therefore, was the drag coefficient he found?

(b) By what percentage does that differ from Fresnel's prediction?

6.1-21. Sagnac performed his experiment with light of 530 nm wavelength. How much of a fringe shift did he see?

6.1-22. A laser gyroscope contains a coil, 50 mm in diameter, of 100 turns of optical fiber of index 1.49. If the coil spins at 3600 rpm and $\lambda = 633$ nm, by how much will the fringes shift?

6.1-23. Cerenkov radiation can be compared to the shock wave set up by an aircraft flying at supersonic speed. Assume that the particles which produce the radiation travel at a velocity of $v = 0.86c$ and that the medium is water ($n = \frac{4}{3}$). What is the apex half-angle of the shock wave?

6.1-24. In which direction must a dense-to-rare boundary move so that obliquely incident light is not refracted at all?

6.1-25. What is the de Broglie wavelength of an electron ($m = 9.1 \times 10^{-31}$ kg) traveling at one-tenth the speed of light?

6.1-26. What velocity is necessary for a proton ($m = 1.67 \times 10^{-27}$ kg) to attain a de Broglie wavelength of 0.99 Å?

6.1-27. A cube, carrying the letter F on its front face, is seen from an angle of 45° with the forward direction. If the cube moves at $v = 0.9c$, what does the F look like to a stationary observer?

6.1-28. Imagine looking at a cube of length (width) $L_0 = 1$ moving along at high speed. Plot:

(a) The width of the rear face of the cube as a function of its velocity, increasing from 0 to c.

(b) The width of the side face, for the same velocities.

(c) The total width of the cube.

Well, dear reader, we have come to the end of the third edition of my *Introduction to Classical and Modern Optics.* How did you like it? Do you have any comments or criticisms? If you do, I would appreciate it if you would write them down and send them either to the publisher or to me. Many thanks!

Jurgen R. Meyer-Arendt

Answers
to Odd-Numbered
Problems

Chapter 1.1

1.1-1. 6.25 Hz   **1.1-3.** 1697; 5×10^{14} Hz   **1.1-5.** 4 m   **1.1-7.** 6.25 m
1.1-9. 25 cm   **1.1-11.** 15 mm

Chapter 1.2

1.2-1. 1.645   **1.2-3.** 1.31   **1.2-5.** 61°   **1.2-7.** 22°   **1.2-9.** 3°
1.2-11. 8°   **1.2-13.** 1.1°

Chapter 1.3

1.3-1. -80 m^{-1}; -5 m^{-1}; -0.625 m^{-1} **1.3-3.** infinity   **1.3-5.** $+2.20$ m^{-1}
1.3-7. 1.6   **1.3-9.** -8 diopters   **1.3-11.** $+50$ mm   **1.3-13.** $+4$ cm
1.3-15. -125 m^{-1} **1.3-17.** 7.5°   **1.3-19.** -300 cm   **1.3-23.** $+20$ cm
1.3-25. $+5$ diopters   **1.3-27.** 120 cm; 12 mm

Chapter 1.4

1.4-5. -4 cm   **1.4-7** 18 mm   **1.4-9.** 10 cm   **1.4-11.** 60 cm
1.4-13. $+12$ diopters   **1.4-15.** $+9.26$ m^{-1}; $+7.04$ m^{-1}; 3.4 cm left of second lens
1.4-17. 12 diopters   **1.4-19.** $+5\times$ **1.4-21.** at center of curvature
1.4-23. 4.675 diopters   **1.4-25.** 1.578

Chapter 1.5

1.5-1. -6 cm   **1.5-3.** 8 cm   **1.5-5.** -45 cm   **1.5-7.** -20 cm; -60 cm
1.5-9. 4 m   **1.5-11.** 7°   **1.5-15.** -2.75 diopters   **1.5-17.** 50 diopters
1.5-19. 91 cm   **1.5-23.** 22.5°   **1.5-25.** spherical mirror's is shorter   **1.5-27.** -20 diopters

Chapter 1.6

1.6-1. $f/12$ **1.6-3.** 0.4%, the same for both   **1.6-5.** 53.13°   **1.6-7.** 20°
1.6-11. $+3$ cm; 30 mm

Chapter 2.1

2.1-1. 32 mm   **2.1-3.** 11.8 mm   **2.1-5.** 15°   **2.1-7.** 19.8°
2.1-9. 0.0402; 0.0803; 0.9960

Chapter 2.2

2.2-1. 1 mm   **2.2-3.** $+3$; 0   **2.2-5.** -1; $+1.053$; $+3.5$ cm; $+136$ cm
2.2-9. $+1.50$ diopters   **2.2-11.** 17.3 mm   **2.2-13.** 12 mm
2.2-15. $+2.5$ diopters   **2.2-17.** $+10.8$ diopters
2.2-19. blue focus, red halo; red focus, blue halo   **2.2-21.** $+11.8$ diopters; -5.8 diopters
2.2-23. 4 cm; 6 cm

Chapter 2.3

2.3-1. 2.6 mm   **2.3-3.** 40 cm   **2.3-5.** 2.8; immaterial
2.3-7. ratio of lines inside/outside of lens' image   **2.3-9.** 0.248; 14.37°
2.3-11. concentric circle of radius $3f$

Chapter 2.4

2.4-5. $-52.5°$ **2.4-7.** 24 mm   **2.4-9.** 106.3; 105.3; 103.7; 101.2; 97.8; 93.2

Chapter 2.5

2.5-1. $+1.5$ diopters; $+7.5$ diopters   **2.5-3.** 6.25 mm; $+20$ diopters
2.5-5. 1.5 cm away from objective   **2.5-7.** $-10\times$ **2.5-9.** 6 cm
2.5-11. $+1.6\times$; 18.75 mm   **2.5-13.** 2 mm   **2.5-15.** -2.4 cm; $62.5\times$
2.5-17. 50 mm; $30\times$ **2.5-19.** second Purkinje image forward and smaller
2.5-21. -2.5 diopters   **2.5-23.** 300 mm; $+60$ mm   **2.5-25.** 5 cm to right of primary

Chapter 2.6

2.6-1. 0.0625 mm^{-1} **2.6-3.** 0.2°   **2.6-5.** 20%   **2.6-7.** 25%; 40%
2.6-9. 25%; 25%; no   **2.6-11.** 40%   **2.6-13.** not much; severe degradation; not much

Chapter 3.1

3.1-1. 15 m; 2.2 cm   **3.1-3.** 0.13 s   **3.1-5.** 1.16 cm
3.1-7. 50 Hz; 8 mm; $0.02 \sin[100\pi(x/0.4 - t)]$ **3.1-9.** 3.46 cm; 20π cm s^{-1}

Chapter 3.2

3.2-1. 720 nm; red   **3.2-3.** 10.8 mm; 9 mm   **3.2-5.** 2.75 μm
3.2-7. same as double-slit pattern, but more intense   **3.2-9.** 1.414 A
3.2-11. $\mathbf{y} = \sin x(1 + \cos x)$ **3.2-13.** 776 638   **3.2-15.** 700 nm
3.2-17. 25   **3.2-19.** 1.0004

Chapter 3.3

3.3-1. (a) 1; (b) 608 nm   **3.3-3.** 36°; 1; 570 nm   **3.3-5.** 0.02 mm   **3.3-7.** 1.6
3.3-9. 2.85 mm   **3.3-11.** narrow and dark on bright background   **3.3-15.** 55 cm
3.3-17. 4.9 mm   **3.3-21.** 6.25%   **3.3-23.** 9.7%; 0

Chapter 3.4

3.4-1. 0.8 mm   **3.4-3.** 13.2 μm; 4.4×10^{-14} s   **3.4-5.** 5 units   **3.4-7.** 2 m
3.4-9. 88 m

Chapter 3.5

3.5-1. 2λ **3.5-3.** blue inside, red outside   **3.5-5.** triangle 6; pentagon 10; hexagon 6
3.5-7. 30 μm   **3.5-9.** 8 km   **3.5-11.** 10 cm   **3.5-13.** 0.067; 1.280
3.5-15. 4 mm   **3.5-17.** 3.6 mm; any   **3.5-19.** ratio obstacle : moon = 1 : 10

Chapter 3.6

3.6-1. 625   **3.6-3.** 500 lines/mm   **3.6-5.** prism: blue spread out; grating: linear
3.6-7. 1.2 mm   **3.6-9.** 3   **3.6-11.** 450 nm; 0.15 mm   **3.6-13.** 0.2 nm
3.6-15. 1.735 Å   **3.6-17.** 29°

Chapter 4.1

4.1-1. -7.2×10^{-4} N; $+9 \times 10^{-5}$ N   **4.1-3.** 8.854 187 818 $\times 10^{-12}$ C^2 N^{-1} m^{-2}
4.1-5. 1 : 4.16; 1 : 17.3   **4.1-7.** 40°   **4.1-9.** 7%   **4.1-11.** 1–2; 1–2, 2–3; 2–3; none

Chapter 4.2

4.2-1. = circle   **4.2-3.** 2.8×   **4.2-5.** 14.51   **4.2-7.** 10%   **4.2-9.** no benefit

Chapter 4.3

4.3-1. 0° unpolarized; 45° and 135° partial linear   **4.3-3.** 50°; 40°   **4.3-5.** 36°
4.3-7. horizontal   **4.3-9.** 22.5   **4.3-11.** 45°; 135°
4.3-15. 0.15 mm and multiplies thereof   **4.3-17.** 4.15 mm   **4.3-19.** 45°

Chapter 4.4

4.4-1. 2 nm   **4.4-3.** 0.27 Å   **4.4-5.** 12.6 kHz; yes   **4.4-7.** $y = 2A \sin(\omega t) + A \sin(3\omega t)$

Chapter 4.5

4.5-1. not at all; clockwise; spread out vertically   **4.5-3.** (b) not resolved
4.5-7. only the circles' vertical segments are visible   **4.5-11.** 0.9 mm

Chapter 4.6

4.6-1. 1.9 μm   **4.6-3.** 1 : 1.49
4.6-7. with 800 nm more, with 600 nm less magnification   **4.6-9.** 2×10^{10}

Chapter 5.1

5.1-1. positive carbon hotter = point light source   **5.1-3.** W m^2; m K
5.1-5. 3.75 μm; 2.28 μm   **5.1-7.** 9.3 μm   **5.1-9.** yes; higher
5.1-11. none; no current; diverted to other maxima   **5.1-13.** 500 nm
5.1-15. 0.62 eV; 4.96 eV   **5.1-17.** 400 nm

Chapter 5.2

5.2-1. 1.38 kW   **5.2-3.** 8 m W sr^{-1} **5.2-5.** 0.1 sr   **5.2-7.** 4 W
5.2-9. 76 lm/W   **5.2-11.** no change; no change   **5.2-13.** 6.4 lx   **5.2-15.** 25 $\times$
5.2-17. 4.4 lx   **5.2-19.** 5 cd   **5.2-21.** 1.54 lx

Chapter 5.3

5.3-1. 92 nm   **5.3-3.** 10 nm   **5.3-5.** 2.18 $\times$ 10^6 m s^{-1} **5.3-7.** 0.85 eV

Chapter 5.4

5.4-1. 0.42   **5.4-3.** 25%   **5.4-5.** 36%   **5.4-7.** 45%   **5.4-9.** 20%
5.4-11. 21%   **5.4-13.** 3.87 m^{-1} **5.4-15.** bright, on a dark background

Chapter 5.5

5.5-1. 2.87 $\times$ 10^{-19} J   **5.5-3.** light weight; destruction of laser   **5.5-5.** 2 $\times$ 10^5
5.5-7. 20°   **5.5-9.** 0.8 mrad   **5.5-11.** 62.5 cm
5.5-13. 1.5 kW mm^{-2} **5.5-17.** 9 μm

Chapter 6.1

6.1-1. 85.4 km/h   **6.1-3.** 15 m   **6.1-5.** 59°   **6.1-7.** 32.52°   **6.1-9.** 30°
6.1-11. 390 nm   **6.1-13.** 64 km/s   **6.1-15.** $v = \frac{1}{2}c$ **6.1-17.** 0.1732 m/s; 19.245 min
6.1-19. 0.2   **6.1-21.** $\frac{1}{20}$ of a fringe   **6.1-23.** 60.7°   **6.1-25.** 0.24 Å   **6.1-27.** ⊐

Index

Abbe's cube, 325
Abbe's method, 61
Abbe's number, 22
Abbe's sine condition, 116
Abbe's test plate, 343
Abbe's theory, 332
Aberrations, 108
Absorbance, 392
Absorption spectra, 390
Absorptivity, 393
Achromat, 123
Acoustic holography, 352
Adaptive optics, 183
Airy, G. B., 253
Amplitude, 5
Ångström (unit), 5
Angular magnification, 56
Antireflection coatings, 232
Aperture stop, 81, 87
Aspherical mirrors, 74
Astigmatism, 117
Atanasoff, J. V., 144
Atmospheric refraction, 129
Atomic transitions, 386
Autocollimation, 61

Babinet's principle, 266
Back vertex power, 53
Band-pass filter, 339
Bartholin, E., 309
BASIC, 146
Basov, N. G., 405
Beam expander, 417
Beer's law, 392
Bell, A. G., 139
Bessel's method, 60
Best form (of lens), 113
Biot–Savart's law, 287
Birefringence, 311
Blackbody radiator, 357
Blazed gratings, 278
Bohr's model (atom), 385
Boltzmann's constant, 458
Boltzmann's distribution, 400
Bouguer's law, 392
Bragg, W. L., 280
Bragg's law, 280
Brewster's angle, 307
Broglie, L. de, 438
Bunsen, R., 381
Bunsen photometer, 378

Camera lenses, 174
Candela (unit), 370
Cartesian sign convention, 30
Cassegrain's telescope, 180
Cataract, vision through, 244
Cavity (laser), 406
Centrad (unit), 20
Cerenkov radiation, 438
Change of vergence, 53
Chief ray, 40
Chromatic aberration, 122
Circle of least confusion, 119
Clerk Maxwell, J., 285
Coddington factors, 113
Coherence, 237
Collimation, 61
Color, 395
Color schlieren methods, 136
Color temperature, 359
Coma, 115
Complementarity, 439
Computer-aided lens design, 144
Conjugate points, 29
Contrast, 186
Convergence, 28

Cornu spiral, 258
Correlation, 341
Coulomb's law, 286
CO_2 laser, 411
Critical angle, 16
Curvature of field, 120
Cylinder lenses, 117
Czerny-Turner mount, 278

Depth of focus, 81
Descartes, R., 30
Detectors, 361
Determining focal length, 60
Diffraction, 247
    Fraunhofer, 249
    Fresnel, 255
    grating, 270
Dipole radiation, 296
Direction cosines, 99, 103
Dispersion, 21
Distance of best vision, 56
Distortion, 121
Divergence, 28
Doppler effect, 431
Double refraction, 309
Double-slit interference, 205
Drag coefficient, 435
Duffieux, P.-M., 191

Eddington, A., 441
Edison, T. A., 357
Einstein, A., 426
Einstein's coefficients, 402
Einstein's mass relation, 438
Einstein's photoelectric equation, 363
Einstein's postulates, 426
Einstein's relations (laser), 403
Electrooptics, 317
Ellipsoidal mirror, 75
Entrance pupil, 83
Equivalent power, 49
Euler's formula, 200
Excimer laser, 412
Exitance, 368
Exit pupil, 83
Exponential law, 392
Extinction, 390
Eye, 172
Eyepieces, 161

Fabry-Perot interferometer, 223

Faraday effect, 318
Fermat's principle, 14
Fiber optics, 137
Field of view, 82
Field stop, 87
Filters:
    color, 230
    interference, 230
    spatial, 335
FitzGerald, G. F., 428
Fluorescent lamps, 357
Focal points:
    lens, 35
    single surface, 28
"Footcandle" (obsolete unit), 377
Foucault grid (MTF), 192
Fourier spectroscopy, 321
Fraunhofer diffraction, 249
Fraunhofer lines, 391
Frequency, 6
Fresnel-Arago laws, 315
Fresnel diffraction, 255
    equations, 289
    integrals, 258
Front vertex power, 53
f-stop number, 82

Gabor, D., 346
Gain (laser), 409
Galileo's telescope, 163
Galileo's transformation, 424
Gas lasers, 411
Gauss, K. F., 32
General relativity, 441
Geodesic lens, 140
Glass, 22
Gradient-index lenses, 132
Graphical ray tracing:
    by computer, 148, 156
    lenses, 39
    mirrors, 71
Grating, 270
Green flash, 302
Grimaldi, F. M., 247
GRIN optics, 129
Gullstrand, A., 51
Gyroscope, 437

Half width, 230
Halo, 19
Harmonic wave motion, 197
Headlight effect, 429
Helmholtz, H. von, 42

He–Ne laser, 411
Herapath, W. B., 309
Hertz, H., 286
Holography, 346
Hurter–Driffield curve, 397
Huygens' principle, 247
Hydrogen series, 383
Hyperfocal distance, 82
Hyperopia, 174

Illuminance, 373
Image, 28
Integrated optics, 140
Intensity, 368
Intensity interferometer, 244
Interference, 205
Interference filters, 230
Interference fringes, 211
Internal reflection, 16
Interval of Sturm, 119
Inverse-square law, 373
Irradiance, 373

Joule (unit), 368

Kaleidoscope, 74
Kemeny, J. G., 146
Kepler's telescope, 158
Keratometer, 71
Kerr cell, 318
Kirchhoff, G., 381
Kirchhoff's law (emission), 360
Kurtz, T. E., 146

Lagrange's theorem, 42
Lambert's cosine law, 372
Lambert's exponential law, 392
Land, E. H., 309
Lasers, 400
    applications, 418
    components, 405
    safety, 420
    speckle, 420
    theory, 401
Laue diagram, 279
Leeuwenhoek, A. van, 167
Lens clock, 63
    combinations, 47
    design, computer-aided, 144

thick, 47
thin, 35
Lens-makers formula, 36
Lensometer, 63
Light-emitting diode, 413
Light scattering, 295
    sources, 356
Line emission spectra, 381
Lorentz contraction, 428
Lorentz transformation, 426
Lumen (unit), 368
Luminance, 371
Luminous efficiency, 370
Luneburg lens, 140

Mach–Zehnder interferometer,
    217
Magnification:
    angular, 56
    axial, 43
    microscope, 168
    telescope, 161
    transverse, 41
Maksutov telescope, 182
Malus' law, 312
Mangin mirror, 182
Maxwell's equations, 285
Meyer-Schwickerath, G., 419
Michelson interferometer, 213
Michelson–Morley experiment,
    433
Michelson stellar interferome-
    ter, 242
Microscope, 167
Mie scattering, 300
Mirage, 129
Mirrors:
    aspherical, 74
    plane, 73
    relativistic, 430
    spherical, 67
    vergence, 69
Modes (laser), 407
Modulation transfer, 188
Moiré fringes, 280
Monoyer, F., 33
Moor-Hall, C., 123
Multilayer coatings, 234
Myopia, 173

Newton's lens equation, 37
Newton's rings, 228
Nicol prism, 311
Nodal slide, 60

Nonlinear effects, 417
Numerical aperture, 171

Obliquity factor, 267
Optical activity, 316
Optical data processing, 332
Optical density (absorbance),
    392
Optical path length, 10
Optical transfer function, 191
Optical tube length, 167

Paraboloidal mirror, 74
Paraxial rays, 31
Pasteur, L., 317
Patents, 194
Path difference, 217
Path length, 9
Pattern recognition, 353
Penumbra, 7
Period, 6
Petzval condition, 121
Phase contrast, 340
Phase transfer, 190
Photochromic glass, 398
Photocoagulation, 419
Photodetectors, 364
Photoelectric effect, 362
Photography, 396
Photometry, 367, 378
Planck's radiation law, 358
Pockels effect, 318
Poisson's spot, 266
Polarized light, 304
Polaroid, 309
Population inversion, 403
Power, 367, 417
Poynting vector, 288
Principal planes, 48
Prisms, 17–20
Prokhorov, A. M., 405
Propagation of light, 5
Pupils, 79, 82

Q switch (laser), 417
Quantum detectors, 361
Quarter-wave plate, 313

Radiance, 371
Radiometry, 367, 378
Rayleigh criterion, 254
Rayleigh interferometer, 220

Rayleigh ratio, 299
Rayleigh scattering, 295
Ray tracing:
    computer-aided, 149
    skew rays, 99
    trigonometric, 92
Reading machines, 354
Reconstruction, 349
Reduced distance, 10
Reduced eye, 34
Reflecting microscope, 180
Reflection, 14
Reflection coefficient, 225
Reflectivity, 232, 291
Refraction, 15
Refractive index, 17
Relativistic optics, 424
Reradiation, 297
Resolvance:
    Fabry–Perot, 227
    Fourier, 323
    grating, 276
Ronchi grids, 135, 277
Ruby laser, 409
Rudolph, P., 175
Rutherford's model (atom),
    384
Rydberg equation, 382

Sagitta, 63
Sagnac's interferometer, 437
Scattering:
    Mie, 300
    Rayleigh, 295
Schematic eye, 172
Schmidt corrector, 182
Seidel aberrations, 109
Selective absorption, 395
Semiconductor, 364
Semiconductor lasers, 413
Shadows, 7
Sign convention, 30
Simple harmonic motion, 197
Sine condition, 116
Single refracting surface, 27
Single-slit diffraction, 249
Skew rays, 99
Slide projector, 79
Smakula, A., 232
Smith–Helmholtz relationship,
    42
Snell's law, 16
Snell's vector form, 102
Spatial filtering, 335
Speckle, 420

Spectrophotometry, 396
Spectroscopy, 24
Spherical aberration, 109
Spherocylinder, 119
Spread function, 189
Stefan–Boltzmann's law, 359
Stellar interferometer, 242
Steradian (unit), 369
Stimulated emission, 400
Stops, 79
Sturm's interval, 119
Submarines, 234
Superposition, 207
Surface power equation, 32

Taylor–Cooke triplet, 175
Telephoto lens, 178
Telescopes:
    astronomical, 158
    reflecting, 179–80
    terrestrial, 163
Terrell rotation, 439
Tessar camera lens, 176
Theory of image formation,
    332
Thermal detectors, 361
Theta modulation, 337

Thick lenses, 47
Thin films, 221
Thin-lens equation, 38
Thin lenses, 35
Thin prisms, 20
Throughput, 377
Töpler's method, 134
Toric surface, 120
Total internal reflection, 16
Townes, C. H., 405
Transfer functions, 188
Transformations:
    Fourier, 327
    Galileo's, 424
    Lorentz', 426
Transmittance, 392
Transverse magnification, 41
Trigonometric ray tracing, 92
Tunable laser, 413
Turbidity, 297
Tyndall effect, 295

Vander Lugt filter, 354
Velocity of light, 6
    in moving matter, 435
Vergence, 27
    change of, 53

Vertex, 29
Vertex depth, 63
Vignetting, 79
Virtual image, 28
V number, 22

Watt (unit), 368
Wave equation, 202
Wavelength, 5
Wave motion, 198
Wide-angle lens, 176
Wien's displacement law, 358
Wire-grid polarizer, 308
Wood, R. W., 133
Work function, 363

X-ray diffraction, 278

Young's double slit, 205

Zernike, F., 340
Zone plate, 263
Zoom lens, 176